Cases on Successful E-Learning Practices in the Developed and Developing World: Methods for the Global Information Economy

Bolanle Olaniran
Texas Tech University, USA

Information Science REFERENCE

INFORMATION SCIENCE REFERENCE
Hershey · New York

Director of Editorial Content:	Kristin Klinger
Senior Managing Editor:	Jamie Snavely
Assistant Managing Editor:	Michael Brehm
Publishing Assistant:	Sean Woznicki
Typesetter:	Kurt Smith
Cover Design:	Lisa Tosheff
Printed at:	Yurchak Printing Inc.

Published in the United States of America by
Information Science Reference (an imprint of IGI Global)
701 E. Chocolate Avenue
Hershey PA 17033
Tel: 717-533-8845
Fax: 717-533-8661
E-mail: cust@igi-global.com
Web site: http://www.igi-global.com/reference

Library of Congress Cataloging-in-Publication Data

Cases on successful e-learning practices in the developed and developing world
: methods for the global information economy / Bolanle Olaniran, Editor.
p. cm.
Includes bibliographical references and index.
Summary: "This book provides eclectic accounts of case studies in different contexts of e-learning"--Provided by publisher.
ISBN 978-1-60566-942-7 (hardcover) -- ISBN 978-1-60566-943-4 (ebook) 1. Education--Computer network resources--Case studies. 2. Internet in education--Case studies. 3. Education--Developing countries--Computer network resources--Case studies. 4. Internet in education--Developing countries--Case studies. I. Olaniran, Bolanle, 1964-
LB1044.87.C37 2010
371.33'44678--dc22

2009043363

British Cataloguing in Publication Data
A Cataloguing in Publication record for this book is available from the British Library.

Editorial Advisory Board

List of Reviewers

Table of Contents

Section 2
Addressing Various Divides in E-Learning

Section 3
User Centered Focus in E-Learning

Detailed Table of Contents

This case outlines the development of a pre-internet education initiative in New Zealand that linked eight rural schools, each with declining enrollments, to collaborate through audio technology in sharing specialist high school teachers. The collaborative structure that was formed enabled senior high school students in the intranet to access courses not available on-site, thereby expanding their range of curriculum options. Replication of the New Zealand model in rural Atlantic Canada, enhanced by the Internet, enabled senior students in an intranet to access four Advanced Placement (AP) science subjects, each taught from a participating site. Within the New Zealand and Canadian intranets collaborative teaching and learning has developed. The creation of virtual educational structures that support and enhance traditional classes has expanded the capacity of participating rural schools and reduced the significance of their physical locations. The New Zealand and Canadian initiatives highlight the possibilities of inter-school collaboration to sustain education in small rural communities.

This case describes the development of a re-purposable learning object for higher education. There is evidence of an increasingly diverse student population in UK higher education, where the sector is currently faced with re-positioning itself in order to meet the challenges of higher education in the 21st century. This has resulted in a new emphasis in education on supporting the learner, in collaboration with peers and tutors, through a lifetime of education, both within and outside the classroom. These factors, together with personal experience in teaching students data analysis have been instrumental in the formation, by the authors, of the conception of the Analyse This!!! learning object described in this case study. In June 2008 Analyse This!!! was successfully launched, and it is hoped that it will prove to be a useful resource for students and staff alike, across many different subject disciplines and across different institutions.

The chapter reports on a UK project which was designed to explore innovative ways of getting teachers to develop their use of new technology in subject teaching. The outcomes of this project suggest that in the area of developing teachers' use of ICT in subject teaching, simply providing support for teachers, in the form of time to explore the potential of ICT, to meet together to discuss ICT in subject groupings, and freedom to focus on their preferred ICT agendas, may be a more effective way forward than prescribing lists of required competences and providing generic 'training' type courses. This goes against the grain in an era characterised by 'top-down', centrally directed national strategies, high levels of accountability and auditing of teachers, and 'coverage' models of competence (Ball, 2003), but given the disappointingly sluggish and modest outcomes of such programmes, in the UK and elsewhere, such approaches may be worth exploring more extensively.

This chapter presents and contrasts descriptions of two cases of online affective support provided to support students engaged in higher level learning tasks. The cases are set in different cultures, centre upon different intended learning outcomes, and follow different tutorial styles. One (Eastern) tutor acted as a "shepherd leader" in response to needs arising in the Confucian Heritage Culture as the teacher promoted critical thinking, according to the Western model. The other (Western) tutor provided Rogerian facilitation of reflective learning journals, kept by students seeking to develop personal and professional capabilities. In both styles, affective support features strongly. The cultural and pedagogical comparisons between the cases have proved useful to the writers. These distinctions together with the similarities between the two online styles emerge in the comparisons.

In an age seemingly defined by near-ubiquitous access to Internet-based communication, there is little wonder that adult educators are turning to online, distance education as a means to reach their participants. In the traditional academy, post-secondary institutions increasingly include online courses and programs as elements, or comprising the entirety, of both undergraduate and graduate degrees (Allen & Seaman, 2006). Even in the realm of non-formal adult education, "hacktivism[1]" has become one of the most effective mechanisms through which engagement for social change – especially on a global scale – occurs (Day, 2004; Ganesh, Zoller & Cheney, 2005). Ironically, rather than truly integrating the philosophy of emancipatory and transformative adult education, cyber-education environments as typically implemented throughout the academy, overwhelmingly – if unwittingly – reproduce and reinforce the hegemony of traditional teacher-pupil power relations. By examining the mechanism of hegemony, and its pervasive presence in contemporary pedagogical technologies, this case will demonstrate how organized power is maintained through these mechanisms. In contrast, a case will be offered that demonstrates how engaged intellectuals can reconstruct the cyber-education environment in order to challenge the pretensions of entrenched academic power, and manifest adult education principles. In particular, the case will explore how the many years of research on how adults learn can be applied with the use of technology, so that the cyber learning milieu is as dynamic, personal and collaborative as the physical presence classroom context can be in the hands of a skilled adult educator.

This chapter presents a case study of developing and teaching an intercultural communication (IC) course online, within the context of a department in a large research University in the U.S. In so doing, we discuss a broadened and recursive model of cultural access and divides in E-learning. Expanding on Van Dijk's (2005) framework, the authors present several ways in which their IC course attempts to address multiple pathways of E-learning access, including motivational, material, skills and usage access. They describe both the successes and challenges of meeting the goals of e-learning access with specific examples of the content, activities, assignments, pedagogical strategies, and student assessment in this online course. Finally, they identify challenges of this e-learning at the micro and macro level context—in the course, university writ large and in the communication discipline.

these interactions to provide ways that online instructors can structure courses to optimize genuine and engaging online discourse. Additionally, it suggests that students and instructors who assume a Connected Stance show a depth of learning within the computer-mediated framework. Finally, it provides a unique format for analyzing online discussion boards.

Section 4
Special Considerations in E-Learning and Development

This case shows how a long-term, campus-wide effort balanced technological, pedagogical, financial, and political considerations to develop and implement a system for online course quality assessment at a medium-sized public university in the Midwest. The case shows how the need for an assessment system came to be recognized, and how the committee charged with creating the system arrived at a solution which took into account both course design and instructor performance. Thus, the institution now has, for the first time, a tool for improving the quality of its online courses. Moreover, it is hoped that administrators, faculty, and faculty developers will see that the quality assessment system joins a course management software suite development effort and a series of faculty training workshops in a wide-ranging list of tools for enhancing faculty competence as users of technology for teaching and learning.

In contrast to the formal school setting where learning is often linear, structured and controlled (be it online or face-to-face), for the 'net generation,' (Google, MySpace, MSN, YouTube and Yahoo) learning is often incidental and a sense of 'fun' is frequently of great importance. Such students' learning is often non-linear, unstructured and explained well by the tenets of Anderson's theory of online learning. This research discusses the benefits of fostering non-linearity in an online learning environment. A case study of an online business communication course at a university in Hong Kong is used to illustrate the importance of non-linear online learning by demonstrating how participants in this course adopted learning approaches that are consistent with, and a reflection of, the theory of online learning. Qualitative data from complete sets of online communication (including focus group interviews) collected over a one-semester, tertiary level course conducted at a university in Hong Kong are analyzed. The findings show that Chinese-speaking learners' online interactions, categorized into three broad areas (cognitive, affective and social), demonstrate that interactivity is a key feature of an online learning environment. Its nature is exposed and discussed, not least the finding that for the participants in this study, learning was incidental and a sense of 'fun' was important. The study suggests ways in which online theory can contribute to, as well as help in, understanding this phenomenon and makes recommendations for future research.

Spain, with students from many countries. As common to most math-related knowledge areas, teaching and learning Statistics and OR present difficult challenges in traditional higher education. These issues are exacerbated in online environments where face-to-face interactions between students and instructors as well as among students themselves are limited or non-existent. Despite these difficulties, as evidenced in the global growth of online course offerings, Web-based instruction offers comparative benefits to traditional face-to-face instruction. While there exists a plethora of literature covering experiences and best practices in traditional face-to-face instruction in mathematics, there is a lack of research describing long-term successful experiences in Statistics and OR online courses. Based on the authors' experiences during the last decade, this chapter aims to share some insights on how to design and develop successful online courses in these knowledge areas.

Preface

INTRODUCTION

The integration of communication technologies and Internet continues to create opportunity for organizations, institutions, and others to find novel use for specific technologies. Thus individuals and organizations or groups continue to integrate these technology into their activities and corporate practices. One of the key benefits of such integration includes increased use of information communication technology in learning and curriculum which is otherwise refer to as electronic leaning and more succinctly referred to as e-learning.

E-Learning, which involves knowledge dissemination and acquisition with the aid of communication technology or electronic medium, is fast becoming the norm in training and education in the global and information economy. One of the reason is to foster and create competent global work force. We all either work in an institution of higher learning or organizations instituting e-learning to disseminate learning contents and modules for training students and learners. From a related standpoint is the need and requirement for continued education and training for employees. This training requirement often circumvents traditional college training that requires participants to be in a specific location in order to access and participate in learning. Olaniran (2007a) contends that we do not have to look very far to see increase growth in Online Universities (e.g., AIU Online, Capella University, Devry University, Kaplan University, University of Phoenix, Walden University, and Westwood College Online) all of who are thriving and attractive to corporate travelers, non-traditional students, and expatriates.

Furthermore, the major draw of e-learning and important advantage is cost savings. However for e-learning to produce concrete results, there needs to be a way to account for effectiveness of the learning process. **Nevertheless, effectiveness of e-learning cannot be adequately assessed without regards to the contextual environments where it is taken place.** Therefore, in a global information economy with e-learning, we are either contributors, consumers, or both. Within this process, effective communication and content delivery is essential to the success of e-learning. However, one area that is crucial to such effectiveness is attention to the context (e.g., culture, learning styles, and preferences). Some teachers or consumers of e-learning found out the hard way that inattention to this issue creates pitfalls, while others simply ignore the idea altogether. On the other hand, attention to specific user contexts can make the difference toward successful e-learning experiences for all concern. It is in this regards that case studies appear to be appropriate and beneficial for informing publics, learners, teachers, and practitioners on how to avoid common pitfalls of e-learning while enhancing the possibility of success in deployment of e-learning and the technologies that powers them.

It is important to look at case study and its appropriateness as pedagogical method. A case study involves an examination of a single group, incidents, and or community (Green, 2003). A case study approach is popular in the social sciences because contrary to experiment or survey method, case study approach does not follow a rigid set of protocol but rather offers a way to observe data or information in their natural environment. For instance, case study approach allows individuals to take an in-depth look at events in a way that allows researcher to have a better understanding of the event and also to identify a new course of direction for future research. For practitioners however, a case study approach offers a general direction for putting in place safeguard mechanism to ensure successful practices or to learn from others mistakes. Methodologically, case study approach offers a way to generate new theories and even test specific hypothesis (Flyvbjerg, 2006; Glaser & Strauss, 1967). While case study as a methodological approach can be restrictive to data collection in a particular context, it can also be broad in the sense that it can serve as way of exploring events and their occurrence over a period of time. In other words Case study can be both introspective and retrospective *sense-making* (see, Weick, 1995). As such, the approach taken in this collection is a broad one that cast its nets around both. For instance the collections include specific instance of cases that looks at particular e-learning context and then evaluate outcomes for both learners and instructors. On the other hand, there are those collections that cast a wider net and offer a broad look at aftermath of incorporating ICT for e-learning. In either case, all contributions provide readers with useful implications on moving forward while avoiding major pitfalls for individuals and readers as they plan on deploying or embracing e-learning

The main goal of the book is to offer assessment of e-learning with the hope of offering ideas in terms of practical guide and points of good practices, while addressing potential pitfalls to avoid. Therefore, organizations, practitioners, and individuals alike should be aware of what constitutes good and effective e-learning practices and how to design them for specific contexts and audiences in the global information economy. At the same time, the collections within the book strive to address the issues of different divides that affects e-learning in the economically developed and the less economically developed world. As such, the book calls for maintaining open access to e-learning architecture platforms in less economically developed countries where computer access is sporadic at best, but mobile device penetration is in the uptrend. However, the way forward in e-learning involves the idea of open access while maintaining or not sacrificing privacy and security is a must. While few understand the challenges with e-learning and the related communication factors that affect it, such understanding is a necessity for individuals and organizations to better invoke or implement successful e-learning in the global information age and economy. For this reason, collections of case studies in e-learning whether successful or not offer a way that can reveal communication practices that can benefit instructors, learners, and vendors in the current information and global economy.

The collections in this series are therefore divided into four major categories including: Innovative uses of e-learning, Addressing various divides in e-learning, user centered focus in e-learning, special considerations in e-learning and development. These four categories appear to be good taxonomy scheme for making sense out of the idiosyncratic collections represented in the book. As such, they offer readers and practitioners alike to make sense and efficiently follow along. The next section provides a brief description or synopsis of each case and identifies their unique contribution to the book and e-learning as a whole.

CHAPTER SYNOPSIS

The first section **Innovative uses of e-learning** offers readers insights on creative approaches for e-learning and a way for identifying how best to apply or introduce e-learning in knowledge awareness and dissemination. Individual will walk away with idea of good points of practice to address in the process and deployment of e-learning. This particular section opens with "Virtual Structures and Collaborative Processes to Enhance Teaching and Learning Across Dispersed Sites" by Ken Stevens. Ken addresses the challenges facing students in different countries and how the problem is exacerbated by lack of adequate resources to offer educational opportunities for students in small schools especially those located in rural communities. In his work he points to the fact that sometimes government or social structure cannot justify commitment of prohibitive resources to provide teachers in the schools because there few students in the schools to start with. As a result, he explores the use of virtual structures in particular e-learning in both Canada and New Zealand schools. He argues that this method allow senior students to access similar curriculum other than those in core area of studies as those students that reside in metropolitan areas. As a result, senior students from rural schools do not have to leave their communities to attend larger schools in urban centres for them to complete their high school educations in order to have opportunity to participate in post-secondary education. According to the chapter, the author argue that the e-learning initiatives in both New Zealand and Atlantic Canada allowed creation of virtual structures that support collaborative teaching and learning that challenge the educational significance of school sizes and their locations.

The next chapter is "Development and Evaluation of a Generic Re-Purposable E-Learning Object on Data Analysis" by Griffiths and Craven. The case study identified the contextual drivers for the re-purposable e-learning objects. The authors discuss the approach leading to the development of the "Analyse This!!!" which was created to foster student diversity and use of pedagogical theories and e-learning. In the case, the authors describe how they identified the need for a generic learning object following interaction with stakeholders (staff at Manchester Metropolitan University and students). They discuss the importance of using feedback from stakeholder groups to refine and improve the learning object, and how important decisions came about in the final delivery - as a generic and re-purposable e-learning object with the possibility of delivery on mobile devices.

The next chapter is "Getting Teachers to Use New Technology byJust Giving Them Time" by Haydn and Barton. The case reports on a UK project which was designed to explore innovative ways of getting teachers to develop their use new technology in teaching. It focuses on the need to provide adequate support for teachers in terms of time to study and explore ICT potential and a forum to discuss their respective concerns. The authors suggest that these criteria may be more effective than prescribing to do list or required competencies and offering generic training.

Next is the chapter "Dealing with Affective Needs in E-Learning: Contrasting Two Cases, in Two Cultures" by Cowan and Chiu. The authors contrast two cases, in two different cultures. Both authors facilitate e-learning using eastern and western cultural philosophies. One area of similarity that they found is that in both cultural contexts meeting affective needs of learners in online environment is important and thus, offers implications from this standpoint.

The concluding chapter in this section is "A Cyber-Apple for the Teacher: A Case Study for Anti-Hegemonic Adult Education Practices" by Federman and Laiken. First, the author examined the idea of hegemony its mechanism in contemporary pedagogical technologies. Second, they discuss how hegemony is maintained or sustained. Finally, they offer a case that demonstrate how intellectuals can reconstruct the cyber-education environment in order to challenge the entrenched power in academic environment, and foster effective adult education principles. Specifically, the case explore how years of research on how adults learn can be applied to the use of technology and e-learning in a way that fosters dynamic, personal, and collaborative learning in adult education as the one offered in traditional classroom context with the physical presence of instructors.

The second section - **Addressing Various Divides in E-Learning**, acknowledges the different divides that faces e-learning and its use across contexts. Typically, when speaking of divides one looks at the haves and the have-nots and accessibility to technologies and e-learning. Here in this section, one finds out that the divide includes, culture, access, competency, among others, that is crucial to implementing successful e-learning. The section opens with "Cultural Implications of E-Learning Access (& Divides): Teaching an Intercultureal Communication Course Online" by Cheong and Martin. The authors present a case study of developing and teaching an intercultural communication (IC) course online. They discuss a broadened and recursive model of cultural access and divides in E-learning. They present how the IC course attempts to address multiple pathways for E-learning access, including motivation, material, skills and usage access. They then describe the successes and challenges of meeting the goals of e-learning access with specific examples of the content, activities, assignments, pedagogical strategies, and student assessment in the e-learning course.

The next chapter in this section is the "Application of VoiceXML in E-Learning Systems" by Azeta, Ayo, Atayero and Ikhu-Omoregbe. They look at the application of Voice based e-learning systems (VoiceXML) to address the needs of blind students in a less economically developed country and an environment where the blinds are often left out either because the government could not afford or fail to commit resources to their learning needs. The authors as a case study offer a prototype voice-based e-Learning application for course registration and examination for the blind. They then present the outcome of the usability evaluation. They conclude that the voice-based e-Learning technology will improve accessibility to education, including distance learning for learners who are visually impaired in the school for the blind.

The next case "Technophobe to Technophile: Entering the Internet Culture" is by Anderson-Mejias. The case explores how 16 fearful learners became confident and successful through the instructor's taking time for preparing and establishing sense of achievement using the technology, and fostering interconnections with peers to demonstrate the usefulness of the virtual class over the face-to-face class. The author hopes that the principles found can help educators to prepare their traditional students for the culture of virtual learning environments, thus, expanding options while addressing university administrative concerns about student retention.

Next case is "E-Training Support Program for Regional and Local Development" by Syrris and Tsobanopoulou. In this case study, they analyze the technological, cultural, and social issues involved in an online distance training program implemented to address the needs of agricultural, unemployed, and low-salary workers. Their distance consulting focuses on development of entrepreneurial skills and personal training using mobile satellite videoconferencing for training. The essence of their case is intervention for employment purposes and bridging geographical and technological distances.

The last contribution in this section is " The E-Learning Puzzle in Turkey: Déjà Vu?" by Ozdemir. As indicated earlier, this is not a traditional case study per se, but it offers a closer look at how government in this case Turkey implement technology to support e-learning effort. It points to the common pitfalls that face e-learning especially when resources are scarce and the deployment decision is made as a quick fix. This contribution offers a way to avoid similar challenges for those who care to heed the advice.

The third section addresses the notion of **User Centered Focus in E-learning**. In this section, the emphasis is on how to meet users and learners needs in e-learning environments. The section opens with a case on "Users' Satisfaction with E-Learning: A Case Study of the University of Botswana" by Tella. The case study draws from 415 undergraduate students' experience with e-learning at university of Botswana. Satisfaction was measured from the standpoint of outcome variables such as: Perceived usefulness, perceived ease of use, system quality, content quality teaching, and learning effectiveness. However, the author points to Challenges emanating from technology as well as recommendations for dealing with them.

The chapter is "A Case Study Analysis of the Use of Online vs. Proctored Final Exams in Online Classes" by Gold. The author focuses on the issue of security and integrity of online exams when compared to traditional exams. The effort includes analysis of 100 online courses and 1800 students. The author suggests measures for processes and procedures to allow online exams to be comparable to instructor proctored examination.

The next contribution in this section is Perez's "Sharing Insights: Teachers' problems and accomplishment…". The case study focuses on groups of teachers' postings and interactions about their online teaching. She points to problems relating to misappropriation of technologies by learners, course management and administrative issues. Through her analysis, she addresses technological, organizational, and cultural issues that accompany online teaching and e-learning, and offer recommendations for how institution and the individual teachers can address them.

The next chapter is by Hewett and Graham and addresses "The Effects of E-Learning on African-American Males: Three Cases Studies". The authors focus on e-learning use for meeting certain student ethnic group in a rural secondary school. The students experiences prior to, during, and post e-learning were showcased. The case also offers lesson learned from the implementation of the e-learning program.

The last case in this section by Wegmann looked at "Cross Talk Online". Author explored the complex nature of asynchronous e-learning platform in e-learning looking at a single student. It analyzed the student's interactions with peers, and uses these interactions to provide ways that online instructors can structure courses to create engaging online discourse. Additionally, it suggests that students and instructors who assume a connected stance show a depth of learning within the computer-mediated learning environment.

The last section deals with **Special Considerations in E-Learning and Development**. In this section, attempt is made to address concerns in e-learning application or use. The section opens with contribution from Rodgers "Building Quality Assessment". The case shows how a long-term, campus-wide effort balanced technological, pedagogical, financial, and political considerations to develop and implement a system for online course and develop quality assessment. The case details how the committee charged with creating the system arrived at a solution that account for course design and instructor performance. As a result of the effort, the institution now has in place a tool for improving the quality of its online courses. Recommendations are offered accordingly.

The next case by Jakubowicz is "Case Study of the CUForum @ CUHK". The case focuses on the benefits of an online learning environment at a university in Hong Kong. The findings reveal important learning style preferences. Specifically, Chinese-speaking learners' online interactions were found to include *cognitive, affective and social, and hence claimed that* interactivity is key an online learning environment.

The next chapter by Shambaugh explores "Using Activity Theory to Guide E-Learning Initiatives". The case showcases how activity theory can be used as a tool to help educators understand the issues in deploying online learning programs. The central tenet of the case was a guide on how activity theory can be used to represent overlapping goals of faculty, students, and administrators, and to understand the contextual issues of different roles for putting academic programs online.

Next is a case by Gannon Cook and Crawford on "Addressing Online Student Learning Environments and Socialization Through Developmental Research". The authors in this case study focused on developing expansive e-learning environment for adult learners and their socio-cultural environments as well as a call for targeted content materials. The study identified the strengths and potential weaknesses of the online course employed in the study.

The last chapter in this section is by Juan, Faulin, Fonseca, Steegmann, Pla, Rodríguez, and Trenholm. They explored teaching of statistics and operations research online and in a purely online University. In spite of its specific subject use, they offered that online courses offer comparable benefits to traditional face-to-face instruction. Based on their experiences with these courses, they shared their insights for fostering success and on how to design and develop successful online courses in these knowledge areas.

CONCLUSION

With the eclectic collections of cases and the title of the book as cases on successful E-Learning Practices. One thing that is certain is that while all the cases offer useful implementation strategies e-learning is not void of challenges and problems. However, with careful consideration, the experience can be worthwhile and offer learners and practitioners unique opportunity to think outside the general norm of a way to deliver and acquire knowledge via computer and other technology-mediated e-learning environments. There is more to do in this specific learning domain. It is hoped that this book is a step in the right direction for those already using or contemplating the use of e-learning.

REFERENCES

Flyvbjerg, B. (2006). Five Misunderstandings about Case Study Research. *Qualitative Inquiry, 12*(2), 219-245.

Glaser & Strauss (1967). *The Discovery of Grounded Theory*. New York: Aldine de Gruyter.

Greene, R. W. (2003). *Sociology and You*. Ohio: Glencoe McGraw-Hill.

Olaniran, B. (2007a). Challenges to implementing e-learning in lesser-developed countries. In A. Edmundson (Ed.), *Globalized e-learning cultural challenges* (pp. 18-34). Hershey, PA: Idea Group, Inc.

Weick, K. (1995). *Sensemaking in organizations*. Thousand Oaks, CA: Sage.

Section 1
Innovative Uses of E-Learning

Chapter 1
Virtual Structures and Collaborative Processes to Enhance Teaching and Learning Across Dispersed Sites:
Some Implications for Rural Societies

Ken Stevens
Memorial University of Newfoundland, Canada; Victoria University of Wellington, New Zealand

EXECUTIVE SUMMARY

This case outlines the development of a pre-internet education initiative in New Zealand that linked eight rural schools, each with declining enrollments, to collaborate through audio technology in sharing specialist high school teachers. The collaborative structure that was formed enabled senior high school students in the intranet to access courses not available on-site, thereby expanding their range of curriculum options. Replication of the New Zealand model in rural Atlantic Canada, enhanced by the Internet, enabled senior students in an intranet to access four Advanced Placement (AP) science subjects, each taught from a participating site. Within the New Zealand and Canadian intranets collaborative teaching and learning has developed. The creation of virtual educational structures that support and enhance traditional classes has expanded the capacity of participating rural schools and reduced the significance of their physical locations. The New Zealand and Canadian initiatives highlight the possibilities of inter-school collaboration to sustain education in small rural communities.

BACKGROUND

Over the last two decades the introduction of e-learning to small schools has enhanced their capacity to provide extended curriculum opportunities for senior students in rural communities in several parts of the developed world (Asher, 2005; Dell, 2005, Dorniden, 2005). E-learning has changed the nature of small schools by enhancing their teaching and learning capacities (Hawkes & Halverson, 2002)

DOI: 10.4018/978-1-60566-942-7.ch001

and, thereby, their futures, as viable educational institutions at a time when many are under threat of closure because of declining enrolments.

Schools that are small in terms of the number of students who attend, in person, on a daily basis, can, through the introduction of e-learning, become large educational institutions through the expanded range of teaching and learning opportunities they can provide. The enhancement of small schools in rural communities has implications for the sustainability of regional economies. If schools in rural communities are perceived to be as viable in terms of teaching, learning and the range of curriculum options they can make available as their counterparts in urban centres, it becomes increasingly possible to attract skilled workers and professional people to the local economy. The attraction and retention of skilled workers and professionals is important in countries like Canada, New Zealand and Australia where much of the national wealth is in natural resources, often located in remote areas.

Setting the Stage: Rural Education in New Zealand and Atlantic Canada

Rural New Zealand and rural Atlantic Canada have many small schools located in remote communities. In both countries a prominent social issue has been the sustainability of small, local schools that serve these places. Because of the importance of agriculture and forestry in New Zealand and fishing and mining in Atlantic Canada, small schools are central to both the national and regional economies. The development of virtual classes between them, accompanied by collaborative teaching and learning, was initiated in New Zealand and subsequently partially replicated in Atlantic Canada. It is beyond the scope of this chapter to undertake a comparative analysis of rural education in each country, but the following brief case studies outline recent changes in each place that have been shaped by the introduction of information technologies in small schools. In the New Zealand case study, small schools in rural communities in the Canterbury region formed a common, collaborative structure that led to the creation of enhanced learning experiences for students (Stevens & Moffatt, 1996). Publication of developments in rural New Zealand (New Zealand Ministry of Education, 2002; Renwick, 1993; Stevens, 1995b) led directly to the implementation of a collaborative structure in rural Atlantic Canada. The Canadian initiative therefore forms a second, directly-related, case study to the developments that preceded it in rural New Zealand.

The changes that took place in rural New Zealand schools preceded the introduction of the internet and were intended to challenge an environment in which traditional, autonomous schools competed with one another for students, in favour of a more collaborative approach between institutions. By the time the Canadian initiative commenced the Internet was being examined for its teaching and learning potential in schools. The linking of rural schools in pre-internet days in New Zealand provided a useful model to be considered in Canada.

Case Description: A New Zealand Model for Rural School Collaboration

Agriculture, horticulture, fishing and forestry have always been important aspects of the New Zealand economy and the provision of education in its many rural communities is, accordingly, an important economic as well as social consideration. During the latter half of the twentieth century New Zealand society became increasingly urbanized and increasing numbers of rural families migrated to cities to take advantage of educational and vocational opportunities. Student enrolment in many small schools declined and some were closed. The provision of quality education in small communities that were distant from major centres of population became increasingly difficult. Rural educators, parents and students were motivated to explore new ways of accessing educational opportunities based on collaborative structures and processes.

Before the introduction of the internet in the New Zealand education system a collaborative structure was developed that enabled 12 small schools located in different communities within the South Island of that country to explore the possibility of both teaching and learning collaboration (Stevens, 2003b). The twelve small rural schools were located in the Canterbury region that is bounded by mountains to the west and the Pacific Ocean to the east. Each participating school was provided with a dedicated phone line by Telecom New Zealand for the purpose of organizing audio-graphic networking. Students in each of the schools that became, in effect, sites in the network, could hear one another and participate in learning using a graphic tablet that could be seen by students on other sites during a lesson. At the time, many of the participating schools faced closure because of falling enrolments, ensuring support for the initiative from students, teachers, principals and parents. With this pre-internet technology, students on one site within the ten-school network were able to join a class that was delivered from another place. Teachers taught beyond their own classrooms and students met peers through the audio-graphic network from other schools in the Canterbury region of New Zealand. In spite of what, by current standards, was simple and rudimentary technology, this initiative heralded a major conceptual change in the nature of classrooms, schools and the work of both teachers and students (Tate, 1993). Small rural schools in this region of New Zealand academically and administratively opened to one another for that part of the school day during which classes were either delivered or received. With the advent of computers in schools and the realization of the internet's potential for teaching and learning, this early audio-graphic network and the pioneering virtual classes within it were enhanced. Teachers were introduced to the idea of teaching beyond their classrooms to students in real time and students were able to access subjects that were not locally available.

The development of the inter-school teaching and learning network was known as "Cantatech" (The Canterbury Area Schools Technology Project). The project brought twelve rural schools together to explore ways in which teaching and learning resources could be shared. Each school was able to provide its students with traditional face to face teaching in the core subjects on-site: English, Mathematics and Science, as well as instruction in one or two specialized subjects such as French, Japanese, Economics and Agriculture. By collaborating in the teaching of specialist subjects between the participating schools, senior students who wished to receive instruction in courses not locally available were able to access them from another site in the Cantatech network. By mutual consent, designated schools in the Cantatech network accepted responsibility for a particular area of the curriculum in which they had a qualified teacher. In return, schools that provided specialized expertise to other sites in a designated area of the curriculum could expect to receive from within the network other subjects for the benefit of their own students that they would not otherwise be able to provide. Subsequently, the Cantatech network developed extranets to polytechnics and other educational organizations to further extend educational opportunities for rural students throughout the Canterbury region.

By collaborating in the appointment of specialist teachers to each Cantatech site it was possible to avoid duplication of human resources and encourage the development of an expanded range of shared appointments. Each school in the network was, increasingly, able to provide its senior students with access to an extended range of learning opportunities. It became a condition of appointment for teachers in the Cantatech schools that courses were to be provided in both traditional face-to-face mode, on site, as well as to other parts of the network, using emerging technologies. The development of the Cantatech network brought about many changes in the administrative and academic life of each school. Each school in the network had to academically and administratively interface with each of the other schools and work collaboratively in the interests of the broader, regional educational community, not just its originally-designated local one. Because of the necessity of finding new solutions to the delivery of education to geographically isolated senior students, each school in the Cantatech network had to consider the potential role of new and emerging information technologies for delivery of the curriculum. For teachers, students, administrators as well as people in each of the participating communities, information technology became a means to enlarge local educational, and indirectly, vocational, opportunities for young people.

Following the success of the Cantatech project and several similar initiatives in other rural areas of New Zealand, e-learning in schools became a national educational priority In 1998 the *National Information and Communications Technology Strategy* was published, containing four objectives. The first objective was to improve learning outcomes through the use of information technologies in schools for teaching and learning. The second was to increase the effectiveness and efficiency of teachers and schools by helping them to use these technologies. The third objective was to improve the quality of teaching and leadership in schools by helping teachers and principals to identify their technology needs and to develop the skills necessary to meet them. The final objective was to increase opportunities for schools, businesses and government to work together in developing an information technology-literate workforce.

The national strategy developed several major initiatives. A resource centre to provide teachers and schools with a mechanism for the delivery of multi-media resources, including curriculum and administration resources, using the Internet, was created. The centre provided links to curriculum experts, bilingual discussion forums, databases, multimedia and a variety of other sites. Schools were encouraged to create their own websites and to have these placed in the on-line resource centre. A "Computer Recycling Scheme" was initiated to enable more schools to obtain computers for student and teacher use, at low cost, when they were replaced in other organization in the public and private sectors. The scheme was described as the "recycling and up-grading (of computers) in schools" (Ministry of Education, 1998, p12).

Professional development for implementation and planning was introduced during the first year of the national strategy. Principals and senior administrators in schools throughout New Zealand were provided with professional development opportunities in their local areas. These meetings focused on the use of information technologies for both teaching and learning. Finally, "Professional Development Schools" were initiated. Twenty-three schools, strategically selected from throughout New Zealand, were designated information technology professional development lead schools from late 1998. Their purpose was to assist the development of information technology in other schools within their local areas. Each lead school was chosen primarily for the technological expertise of its staff.

The development of virtual classes in rural New Zealand can be summed up as having origins in small schools in rural communities, an emphasis on collaborative teaching and learning and, from 1998, direct government involvement in e-learning. A range of initiatives to encourage on-line learning to complement existing educational structures and processes were promoted ((Ministry of Education, 2006); Stevens, 1995a).

Case Description: Partial Replication of the New Zealand Model in Atlantic Canada

The New Zealand initiative preceded the following developments in Atlantic Canada and directly influenced them. The concept of collaborative structures, or intranets, enabling teaching to be provided from one school to another was of particular interest to Canadian rural educators. Canada is a considerably larger country than New Zealand; it has a different history and many education systems as each province is responsible for its own schools. Like New Zealand, Canada has many small schools in rural communities that were experiencing out-migration. As in New Zealand, many Canadian rural communities faced problems maintaining viable local schools as enrolments in them declined.

The Canadian province of Newfoundland and Labrador has a predominantly rural social structure, a distinctive history and a unique culture. Newfoundland and Labrador is the eastern-most province of Canada and is situated on the North Atlantic Ocean. The province consists of a large, roughly triangular island (108,600 square kilometers), an irregularly-shaped mainland territory (294,300 square kilometers), and over seven thousand smaller islands (3,598 square kilometers). Declining enrolments in rural schools presents particular administrative issues in the province and has twice led to the re-organization of school districts from 27 in 1997 to 11 in 1998 to 5 in 2004. As with e-learning in New Zealand, several stages in the development of virtual classes can be identified in this part of Canada.

Initially, web-based courses were introduced in selected institutions in Newfoundland and Labrador in the belief that these would become "lead schools" and assist other schools in using new technologies in teaching and learning. After one year of lead schools the initiative was countered by teachers and principals who were attracted to the collaborative New Zealand model of clustering schools within internet-linked networks, or intranets. In moving from the development of web-based courses in selected schools to a more collaborative model, a decision was made in Newfoundland and Labrador to begin with advanced courses on-line to challenge senior rural students in the province.

The development of Advanced Placement (AP) Web-based courses in Biology, Chemistry, Mathematics and Physics took place within a development team in each subject area (see Note 1). In developing AP courses on-line, a lead science teacher in each discipline was paired with a recent graduate in the disciplines of Biology, Chemistry, Mathematics and Physics who possessed advanced computer skills including web page design, Java and HTML. The lead teacher and the graduate assistant were advised from time to time by Faculty of Education specialists at Memorial University of Newfoundland in each curriculum area and, where possible, scientists from the Faculty of Science. The extent to which each web-based course was developed by a team of four people varied; most development took place through interaction between lead teachers and the recent graduates in these disciplines. This focussed approach enabled the four courses to be developed over a sixteen-week summer recess period in time for the 1998-1999 school year.

Minimum specifications were adopted for computer hardware and network connectivity. All schools involved in the project had DirecPC satellite dishes installed to provide a high speed down-link. In most rural communities in this part of Canada, digital telecommunications infrastructures do not enable schools to have a high speed up-link to the internet. Appropriate software had to be identified and evaluated for both the development of the resources and the delivery of instruction within the Intranet. Front Page 98 was selected as the software package. Additional software was used for the development of images, animated gifs and other dimensions of course development.. The province of Newfoundland and Labrador has a high rate of use of satellite dishes per capita and there are many schools in this province with Local Area Networks (LANs). As a province, Newfoundland and Labrador provided excellent opportunities for the development of these technologies.

Many software packages were evaluated and finally WebCT was selected. This package enabled the instructor to track student progress, it contained online testing and evaluation, private E-mail, a calendar feature, public bulletin board for use by both instructor and student, a link to lessons and chat rooms for communication between teacher and student. For real time instruction, Meeting Point and Microsoft NetMeeting were selected. This combination of software enabled a teacher to present real-time interactive instruction to multiple sites. An orientation session was provided for students in June 1998, prior to the implementation of this project in September. Students had to learn how to communicate with each other and with their instructor using these new technologies before classes could begin.

In eight schools within the rural Vista school district of Newfoundland and Labrador (see Note 2), 55 students were enrolled in AP Biology, Chemistry, Mathematics and Physics courses. While AP courses are a well-established feature of senior secondary education in the United States and Canada, it is unusual for students to be able to enroll for instruction at this level in small schools in remote communities. It is rare to find high school students in small and remote communities anywhere in the world who are provided with instruction in university-level studies. In Iceland (Stefansdottir, 1993), New Zealand (Stevens, 1995a, 1995b) and Finland (Tella, 1995) there had previously been attempts to provide alternative models for the delivery of education to rural students and these were used to guide the creation of the Vista project in Newfoundland and Labrador. In addition, there were two visits to the Newfoundland and Labrador intranet by New Zealanders who were familiar with school networking and who were able to provide practical advice.

The Vista school district initiative challenged the notion that senior students in small schools had to leave home to complete their education at larger schools in urban areas. By participating in open classes in real (synchronous) time, combined with a measure of independent (asynchronous) learning, senior students were able to interact with one another through audio, video and electronic whiteboards. From time to time they met for social occasions and to spend some time with their science teachers in person. The creation of the Vista School District Intranet was an attempt to use information and communication technologies to provide geographically-isolated students with extended educational and, indirectly, vocational opportunities. This was part of a broader pan-Canadian initiative to prepare people in Canada for the "Information Age" (Information Highway Advisory Council, 1995;1997). The electronic linking of eight sites within the Vista School district to collaborate in the teaching of AP Biology, Chemistry, Mathematics and Physics created a series of open classes in rural Newfoundland. The development of the intranet within a single school district involved the introduction of an open teaching and learning structure to a closed one. Accordingly, adjustments had to be made in each participating site so that administratively and academically, AP classes could be taught.

It was recognized early in the 1998-1999 school year that a common schedule had to be adopted throughout the school district to allow students to interact with their instructors in the new intranet. Unfortunately, this was not fully realized until after classes commenced, with the result some instructors had to repeat classes for small numbers of students. The initial plan was to allow for five on-line sessions and five off-line sessions. This schedule was not followed in all schools. On-line sessions were scheduled in the morning when network traffic was at its lowest point. Off -line sessions were scheduled in the afternoon.

Students in the Vista School District Intranet were frequently subject to scrutiny by their peers as they responded through chat-rooms, audio, video and with their AP on-line teachers. The intranetprovided students with access to multiple sites simultaneously, as well as the opportunity to work independently of a teacher for part of the day. The need to prepare for classes before going on-line became increasingly apparent to both teachers and students if the open, synchronous, science classes were to succeed. The advent of the intranet had implications for students who had to interact with teachers and their peers in a variety of new ways. The teaching of each of the four AP Science disciplines in the Vista School District Intranet took place within classes that were open between participating sites. Many students experienced difficulty expressing themselves and, in particular, asking questions in open electronic classes when they did not know their peers from other small communities. The organization of social occasions for students learning science in open classes in the Intranet helped overcome these problems. As students became more comfortable with one another, inhibitions such as asking questions on-line were overcome.

The major change for the students in the first intranet in Newfoundland and Labrador was the opportunity to study advanced science subjects, as members of open classes, from their small, remote communities. An intranet has many implications for the management of education, based on the need to ensure all sites collaborate both academically and administratively. The most important administrative issue in the first year of the Vista School District Intranet was the co-ordination of timetables across participating sites. Research into the organization of senior students in a networked environment in New Zealand (Stevens, 1994) preceded the formation of the Vista School District Intranet in Canada. Senior students who teachers considered to be "independent learners" in New Zealand were found to learn effectively and were able to obtain satisfactory results in national examinations within an electronic network of small rural schools. In the New Zealand situation though, students usually had at least one teacher on site to assist with questions of an academic nature. In the Canadian Intranet, this was not always possible. A question facing teachers and researchers in the initial stage of the Vista School District Intranet was whether students who were not used to being unsupervised could cope with new freedom and accept increased responsibility for their learning. Students were unanimous at the conclusion of the Canadian school year, that to be successful in an AP on-line course, it was necessary to be able to learn independently, cope with a high volume of work and be willing to ask teachers and other learners questions as they arose (Stevens, 1999). The need for increased technical support for this new, open structure became increasingly urgent for teachers and students who were using information and communication technologies to teaching across dispersed sites. A particular problem was difficulty in securing and maintaining instructional design expertise in the preparation and upgrading of courses delivered through the intranet.

At the end of the first year, Canadian students were asked to reflect on their on-line experience. In spite of technological and administrative problems in establishing and maintaining the delivery of the AP on-line program, students were largely positive about the experience, although there were some suggestions for improvement for future delivery. Positive student comments included:

"I have been introduced to one of the best teachers I have ever had."

Online learning opened small rural classrooms to teachers at a distance, albeit within a single school district. Introducing students to new teachers, online, in addition to their traditional on site instructors was generally considered in positive terms.

"If you are planning on doing post-secondary education, do one of these courses"

The value of receiving Advanced Placement instruction was recognized by final year rural students aiming to enter a university. Success in Advanced Placement courses at school proved that a student was capable of meeting the demands of post-secondary education.

There were some critical comments at the end of the year by senior students, including:

"I think these courses are valuable, but there was much confusion early in the year"

There was administrative as well as technological confusion in the introduction of the initial Canadian intranet. One area of confusion was the integration of participating school timetables so some on-line classes had to be repeated during the first year. Subsequently this problem was solved by the school board, ensuring on-site and on-line instruction was co-ordinated between participating sites.

"The intranet is unreliable when communicating between numerous sites. There are some slow connections"

There were technological issues to be remedied and this put a strain on school district finances and personnel. Nevertheless, the initial year of Advanced Placement teaching on-line enabled students to complete their courses in spite of technological and administrative difficulties.

"More use of video would help so that we can see the teacher."

Teachers of the four science classes discovered that it was not necessary to be able to see their students on the screen and it was not considered important for students to be able to see one another across participating sites in the intranet. What was considered important was being able to hear one another and participate on interactive whiteboards or screens to collectively work on equations, formulas and the collaborative solving of science problems. The fragility of connections between sites and the extra band-width required led to minimal use of video links.

In the initial and subsequent years of the Canadian intranet, reservations were expressed by on-site teachers about the growing use of technology and its perceived future demands on teaching. There was awareness and, in some cases, disquiet, that traditional pedagogy was being challenged (Levin & Wadmany, 2005). A few teachers observed that they were not technicians and that online teaching depended on robust technology as well as technologists to administer and maintain school networks (Sandholtz & Reilly, 2004). The introduction of Internet technology in rural Newfoundland and Labrador had a different reception from students who were of the digital generation (Green & Hannon, 2007). In spite of initial technical problems in the network, Advanced Placement students were comfortable with their electronic classrooms. In several participating schools in the Vista network, students, as members of an increasingly 'wired' generation (Levin et.al., 2005) were able to assist teachers in the transition to the new online environment.

There were several observations from Principals regarding the Intranet as a new educational structure at the conclusion of the first year. Principal's comments included:

"The intranet opened my eyes to other possibilities for teaching and learning at this school. We could do more with the IT resources we now have than just AP subjects"

This view was widely shared at the end of the initial year of the intranet. The introduction of the AP curriculum, online, in small rural schools, provided learning opportunities that had previously not been possible. The potential of extending learning opportunities to other areas of the curriculum and to other levels of the school was open for examination.

"I don't see myself as the director of a virtual school. There needs to be someone at the Board Office level who is the director of the intranet"

The first year of the intranet engaged a lot of administrative time by board officers and technical staff in the absence of defined roles. Administrators and technologists responded to requests as required, sometimes at short notice, to keep the initial eight-site intranet functioning. The following year responsibilities for the intranet were assigned by the board office. The introduction of the Canadian intranet challenged traditional ways of providing education in face to face classes and invited reflection by educators on changes that had taken place in their schools (Hannay & Ross, 2001). Following a ministerial inquiry into what became known as "distance education in classrooms" (Government of Newfoundland and Labrador, 2000) in rural Newfoundland and Labrador schools, a decision was made by the provincial Department of Education to accelerate and extend the development of school district intranets. Teacher and student reflections on the Vista intranet were invited by the ministerial panel together with the views of principals and parents. The outcome was the creation within the Department of Education of Newfoundland and Labrador, of the Centre for Distance Learning and Innovation to promote on-line learning within and between new electronic educational structures throughout the province.

Principals acknowledged that the intranet has been a positive educational development for their schools and were able to identify advantages that it had brought to their students. It enabled participating schools to extend course offerings; it provided a new challenge for senior students in by bringing AP possibilities to them; it gave teachers, students and parents a psychological boost by providing them with evidence that their community school was progressive and it demonstrated that teaching expertise could be shared. While the first year of the intranet concentrated exclusively on the teaching of Advanced Placement subjects, providing unprecedented challenges for 'academic' students, it also brought new resources to the school, including the internet and, importantly, it demonstrated the long-term viability of small rural schools. Parents in each of the eight participating communities were provided with what some called a 'technologically-literate' school and an example of 'cutting edge' education.

The New Zealand and Canadian Initiatives

The significance of the developments of rural school e-learning in New Zealand and Canada has been primarily for the institutions involved and for the communities that they serve. Nevertheless, in seeking to extend educational opportunities for senior students in small schools in rural communities new teaching and learning structures have emerged in both countries in the form of school district intranets within which virtual classes have developed. In each society there is now a challenge to develop pedagogy that is appropriate for the integration of on-site and on-line teaching and learning (Ertl & Plante, 2004). New educational professionals are emerging in Canada: e-teachers (who teach through the Internet), m-teachers (who mediate on-site between e-teachers and local students) and instructional designers (Furey & Murphy, 2005). In New Zealand specialist teachers are being developed through government initiatives including online Maori-medium subject specialists for the indigenous population. As well as extending the educational opportunities of senior students in small schools, the pioneering nature of inter-school teaching and learning in Canada and New Zealand has implications for the teaching profession, for the delivery of courses, for the organisation of classrooms and schools and for the funding of education (Brown, et.al., 2001; Education Review Office, 2005; Lai, 2005).

The development of virtual environments between rural schools supported the introduction of e-learning, facilitating collaboration by students and mutual knowledge building as they participated in classes across multiple sites (Scardamalia & Bereiter, 2006). Virtual educational spaces between schools encourage collaboration between teachers and between students (Furey, 2008; Galway, 2004). There are many implications of e-learning in schools for the professional education of future teachers (Kozma, 2003; Stevens & Starkey, 2006). Teachers have traditionally been taught to teach in schools but the New Zealand and Canadian initiatives suggest that the space between them can extend traditional classes. Teachers in future will possibly teach not just in classrooms but also in the space between schools. This space may not necessarily be within a defined geographical area as in the New Zealand and Canadian cases and may, in fact, cross time zones. Teachers can teach both synchronously and asynchronously between schools that can be in close physical proximity or continents apart, challenging traditional pedagogy (Brown, 2004; Cuban, 2001; Van Manen, 2002). This has led to students in rural Atlantic Canada using the space between their school and a school in Iceland so that a shared lesson could take place in real time during the Canadian morning and the Icelandic afternoon. The space between the same rural Canadian school and a school in South Australia was further expanded overnight. While the Canadian students slept their Australian peers collaborated with them in a joint project, delivering their contributions to waiting computers in the Canadian classroom overnight. The international project was advanced as Canadian students worked on the Australian contributions while their peers in Australia slept in their distant time zone.

E-learning developed in rural New Zealand and Canada when communities were faced with the possible closure of their schools unless organizational changes were made. By creating collaborative educational structures, resources could be shared and learning opportunities enhanced.

Current Problems in Teaching and Learning Across Dispersed Rural Sites

In New Zealand and Canada teachers are prepared for teaching in traditional classrooms rather than in the collaborative and interactive teaching and learning environments facilitated by intranets. The concept of classes as sites in networked environments is not a familiar one to many teachers who work in non-rural communities. This can be a problem when urban teachers take up positions in rural schools that are part of intranets.

The professional education of pre-service teachers in Canada and New Zealand usually includes experience as interns in classrooms. Pre-service teaching experience in virtual classrooms between networked rural schools is much more difficult to obtain.

An enduring problem for some members of the teaching profession and their students in Canada and New Zealand in the development of teaching and learning across dispersed rural sites is reliance on computers and information technologies. This can be seen in the above comments by Canadian students about the initial year of learning in a networked environment. A high level of technical support is needed to ensure that teachers can teach and students can learn in intranets by using information technologies with confidence.

CONCLUSION

The development of rural school networks in New Zealand and, subsequently, Atlantic Canada, has been guided by the desire to extend educational opportunities for young people regardless of the location of their homes or the size of their communities (Healey & Stevens, 2002). The development of virtual structures and collaborative processes in New Zealand and Canada enabled traditional rural schools to be sustained when many of them faced closure. By linking small rural schools within intranets, economies of scale in terms of access to specialist teaching expertise has been addressed (Stevens, 2003a). A challenge facing educators now is evaluating the potential of intranets for all schools, regardless of their size and location. If the virtual structures and collaborative processes initiated in rural New Zealand and Canada are extended to city schools so that teaching and learning networks proliferate across and even between education systems, the concept of 'rural' education may become redundant.

REFERENCES

Asher, G. (2005). Inadequate Infrastructure and the Infusion of Technology into K-12 Education. In C. Howard, J.V. Boettcher, L. Justice, K. Schenk, P. L. Rogers, & G.A. Berg, (eds.), *Encyclopedia of Distance Learning,*(pp. 1061-1063). Hershey, PA: Idea Group Reference.

Brown, J., Sheppard, B., & Stevens, K. (2001). *Final Report: Effective schooling in a telelearning environment*. St. John's, Canada: The Centre for TeleLearning and Rural Education.

Brown, M. (2004). *The Study of Wired Schools: A Study of Internet-Using Teachers*. Palmerston North, New Zealand: Massey University.

Cuban, L. (2001). *Oversold and Underused: Computers in the Classroom*. Cambridge, MA: Harvard University Press

Dell, L. A. B. (2005). Connecting K-12 Schools in Higher Education. In C. Howard, J.V. Boettcher, L. Justice, K. Schenk, P. L. Rogers, & G.A. Berg, (eds.), *Encyclopedia of Distance Learning* (pp. 374-378). Hershey, PA: Idea Group Reference

Dorniden, A. (2005). K-12 Schools and Online Learning. In C. Howard, J.V. Boettcher, L. Justice, K. Schenk, P. L. Rogers, & G.A. Berg, (eds.), *Encyclopedia of Distance Learning,* (pp. 1182-11880. Hershey, PA: Idea Group Reference.

Education Review Office. (2005). *E-learning in Secondary Schools*. Wellington, New Zealand: Education Review Office.

Ertl, H., & Plante, J. (2004). *Connectivity and Learning in Canada's Schools*. Ottawa, Canada: Statistics Canada, Government of Canada.

Furey, D. (2008). From radio broadcasts to virtual reality: A case study of distance education in Hermitage Bay Schools. Paper presented May 2, 2008 at Symposium 2008, *From the Rhetoric to the Reality, Fifty Years of Educational Change in Newfoundland and Labrador*. St. John's, Canada: Memorial University of Newfoundland.

Furey, D., & Murphy, E. (2005). An introduction to K-12 virtual schools and schooling: Factors enabling growth and impacts on traditional systems, schools and roles. *The Morning Watch: Educational and Social Analysis*, *32*, 3–4.

Galway, G. (2004). E-Learning as a means of promoting educational equity in rural and remote communities: A Canadian case study. Paper presented at the *2003 OECD Workshop: Promoting Equity Through ICT in Education*, Budapest (pp. 70-77).

Government of Newfoundland and Labrador. (2000). *Supporting Learning: Report on the Ministerial Panel on Educational Delivery in the Classroom*. St John's. New Zealand: NL, Department of Education.

Green, H., & Hannon, C. (2007). *Their Space: Education for a Digital Generation*. London: Demos.

Hannay, L. M., & Ross, J. A. (2001). Internalizing Change Capacity in Secondary Schools Through Organizational Change. *The Alberta Journal of Educational Research*, *47*(4), 325–340.

Hawkes, M., & Halverson, P. (2002). Technology Facilitation in the Rural School: An Analysis of Options. *Journal of Research in Rural Education*, *17*(3), 162–170.

Healey, D., & Stevens, K. (2002). Student Access to Information Technology and Perceptions of Future Opportunities in Two Small Labrador Communities. *Canadian Journal of Learning and Technology / La Revue Canadienne de l'Apprentissage et de la Technologie, 28*(1,Winter), 7-18.

Information Highway Advisory Council. (1995). *The Challenge of the Information Highway*. Ottawa, Canada: Industry Canada.

Information Highway Advisory Council. (1997). *Preparing Canada for a Digital World*. Ottawa, Canada: Industry Canada.

Kozma, R. B. (2003). Technology and Classroom Practices: An International Study. *Journal of Research on Technology in Education, 36*(91), 1–14.

Lai, K.-W. (2005). *e-Learning Communities: Teaching and Learning With the Web*. Dunedin, FL: University of Otago Press

Lenhart, A., Hitlin, P., & Madden, M. (2005). *Teens and Technology: Youth Are Leading the Transition to a Fully Wired and Mobile Nation*. Washington, DC: Pew Internet and American Life Project.

Levin, T., & Wadmany, R. (2005). Changes in Educational Beliefs and Classroom Practices of Teachers and Students in Rich Technology-Based Classrooms. *Technology, Pedagogy and Education*, *14*(93), 281–307. doi:10.1080/14759390500200208

Ministry of Education. (1998). *Interactive Education – An Information and Communication Technologies Strategy for Schools*. Wellington, New Zealand.

New Zealand Ministry of Education. (2002). *ICT Draft Strategy for Schools - 2002-2004*.

Renwick, W. (1993). The Future of Face-to-Face and Distance Teaching in Post-Secondary Education. *The Impact of Information and Communication Technologies on Post-Secondary Education*. Paris: CERI/OECD.

Sandholtz, J. H., & Reilly, B. (2004). Teachers, Not Technicians: Rethinking Technical Expectations for Teachers. *Teachers' . College Record, 106*(3), 487–512. doi:10.1111/j.1467-9620.2004.00348.x

Scardamalia, M., & Bereiter, C. (2006). Knowledge-building: Theory, Pedagogy and Technology. In R. K. Sawyer (Ed.), *The Cambridge Handbook of the Learning Sciences*. New York: Cambridge University Press.

Stefansdottir, L. (1993). The Icelandic Educational Network – Ismennt. In G. Davies, & B. Samways, (Eds.), *Teleteaching – Proceedings of the IFIP TC3 Third Teleteaching Conference* (pp. 829-835). Amsterdam: Elsevier Science Publishers.

Stevens, K. (1999). Telecommunications Technologies, Telelearning and the Development of Virtual Classes for Rural New Zealanders. *Open Praxis, 1*, 12–14.

Stevens, K., & Moffatt, C. (1996). *TeleLearning and New Zealand Secondary Schools* – Some Pedagogical and Management Issues in the Development of Courses for Flexible Delivery. *Technology for Flexible Delivery Conference*. Palmerston North, New Zealand: Massey University.

Stevens, K. J. (1994). Some Applications of Distance Education Technologies and Pedagogies in Rural Schools in New Zealand. *Distance Education, 15*(4), 318–326. doi:10.1080/0158791940150210

Stevens, K. J. (1995a). Geographic Isolation and Technological Change: A New Vision of Teaching and Learning in Rural Schools in New Zealand. *The Journal of Distance Learning, 1*(1), 3–38.

Stevens, K. J. (1995b). The Technological Challenge to the Notion of Rurality in New Zealand Education Repositioning the Small School. In I. Livingstone (ed.) *New Zealand Annual Review of Education, 5*, 93 –102.

Stevens, K. J. (2003a). E-Learning and the Development of Open Classes for Rural Students in Atlantic Canada. In J. Bradley (ed.), *The Open Classroom - Distance Learning In and Out of Schools* (pp. 149-157). London: Kogan Page.

Stevens, K. J. (2003b). From Distance Education to TeleLearning – The Organization of Open Classes at Local, Regional and National Levels. In J.Bradley (Ed.) *The Open Classroom – Distance Learning In and Out of Schools* (pp. 171-180). London: Kogan Page.

Stevens, K.J. & Starkey, L. (2006). Three Stages in the Digital Integration of New Zealand Schools. *New Zealand Annual Review of Education - Te Arotake a Tau o te Ao o te Matauranga i Aotearoa, 16*, 105-117.

Tate, O. (1993). *A Vision of Distance Education in New Zealand*. Inaugural conference of Distance Education and Information and Communications Technology Providers, National Library of New Zealand, Wellington, New Zealand.

Tella, S. (1995). *Virtual School in a Networking Learning Environment*. Helsinki, Finland: University of Helsinki, Department of Teacher Education.

Van Manen, M. (2002). The Pedagogical Task of Teaching. *Teaching and Teacher Education, 18*(2), 135–138. doi:10.1016/S0742-051X(01)00058-0

Wellington, N. Z. Ministry of Education (2006). *Enabling the 21st Century Learner; An e-Learning Action Planfor Schools 2006-2010*. Wellington, New Zealand: Learning Media.

ENDNOTES

1. AP courses enable senior students to begin undergraduate degrees with part of their program completed from high school if courses are passed at grade levels specified by the university of their choice. Until this initiative, AP courses had not been taught on-line or in rural schools in either Canada or the United States.
2. The Vista School District (District 8) contained 18 schools ranging in student enrolment from 40 to 650. The region in which the Vista School District is located extends from Bonavista in the north, (the place where John Cabot landed in North America in 1497) to the Burin Peninsula in the South. It is a large geographic area covering about 7000 square kilometres. The region has a population of about 35,000 people and an economy supported by a diverse infrastructure including fishing, forestry, farming, mining, aquaculture and tourism. The Vista School District was formed in 1996 and became a legal entity in January 1997. There were 5165 students enrolled in the 18 schools in the district who were taught by 366 teachers. The Vista School District is approximately two hours by road from the capital city, St Johns, which is the location of Memorial University of Newfoundland. Eight schools within the Vista School district, together with the TeleLearning and Rural Education Centre of Memorial University of Newfoundland, formed a digital Intranet within which senior science courses were taught in open classes.

Chapter 2
Development and Evaluation of a Generic Re-Purposable E-Learning Object on Data Analysis

Jillian R. Griffiths
Manchester Metropolitan University, UK

Jenny Craven
Manchester Metropolitan University, UK

EXECUTIVE SUMMARY

This case describes the development of a re-purposable learning object for higher education. There is evidence of an increasingly diverse student population in UK higher education, where the sector is currently faced with re-positioning itself in order to meet the challenges of higher education in the 21st century. This has resulted in a new emphasis in education on supporting the learner, in collaboration with peers and tutors, through a lifetime of education, both within and outside the classroom. These factors, together with personal experience in teaching students data analysis have been instrumental in the formation, by the authors, of the conception of the Analyse This!!! learning object described in this case study. In June 2008 Analyse This!!! was successfully launched, and it is hoped that it will prove to be a useful resource for students and staff alike, across many different subject disciplines and across different institutions.

INTRODUCTION

E-learning requires continual evaluation and updating in a way that emphasizes 1) the goals of the organizations and 2) the goals of the users. Case studies are an effective method for the assessment of e-learning in terms of practical guide and points of good practices, whilst also addressing potential pitfalls to avoid. As such, the case study proposed here provides a practical guide to the development of a re-purposable e-learning object, 'Analyse This!!!' and is pertinent to the areas of e-learning and mobile learning.

Analyse This!!! is a free online tutorial, created by the Centre for Research in Library and Informa-

DOI: 10.4018/978-1-60566-942-7.ch002

tion Management (CERLIM) at Manchester Metropolitan University, and funded by LearnHigher (http://www.learnhigher.ac.uk/analysethis/), to help students develop data analysis skills for coursework studies, projects, and dissertations. It includes quantitative and qualitative data analysis, with some practical examples and advice on effectively analysing data. The project team identified that research methods and data analysis is included in a wide range of subject disciplines taught by many staff across Manchester Metropolitan University and could relate to a specific piece of assessed course work, supervision of project work, support given for dissertation writing and for students undertaking post-graduate degrees by research. It was apparent that whilst the theoretical and practical aspects of data analysis are more generic in nature and can therefore be transferred across subject areas, subject context was seen as important in underpinning the students' understanding and application for their particular area of study. By developing a re-purposable learning object, staff are able to adapt it according to different subject disciplines, student requirements, and level and type of study.

The motivation for developing the learning object was driven by the increasingly diverse student population in UK higher education, where the sector is currently faced with re-positioning itself in order to meet the challenges of higher education in the 21st century. In addition, the team was influenced by the new emphasis in education on supporting the learner, in collaboration with peers and tutors, through a lifetime of education, both within and outside the classroom.

The authors view this as the beginning of their work in this area. Consultations with students and staff have raised a number of issues which they would like to explore further, such as enhancements to the learning object in terms of increased interactivity through the use of screen shots, audio files and video clips. They see this as an ongoing evaluation process, which can be updated and repurposed accordingly in a way that emphasizes the goals of the organizations and users.

The case study described in this chapter will firstly discuss the contextual drivers which brought about the development of Analyse This!!! (student diversity, pedagogical theories and e-learning) before providing a practical approach to developing successful e-learning materials. It will describe how, having identified a need for a generic learning object following liaison with stakeholders (staff at MMU and LearnHigher), a prototype was developed and evaluated with further stakeholder groups (students and staff). Accessibility and usability issues will be discussed and the case study will show how feedback from stakeholder groups was used to refine and improve the learning object, and how important decisions were reached as to the final delivery - as a generic and re-purposable e-learning object, with the possibility of delivery via mobile devices.

BACKGROUND

Student Diversity

The UK higher education sector is currently faced with re-positioning itself in order to meet the challenges of higher education in the 21st century. In addition to the widening participation agenda, changing social contexts are contributing to a greater diversity of students entering higher education. Drivers of this change include legislative change (for example, the UK Disability Discrimination Act) and new forms of pre-entry qualification (especially the introduction of Curriculum 2000, the shifting emphasis towards parity of esteem for vocational qualifications and, more recently, the 14-19 diplomas) are also impacting on the nature of the student body. The dramatic growth in student numbers as a result of these changes has led to an increase in students from non-traditional and previously underrepresented groups (Callender, 2003; Laing, Chao, & Robinson, 2005). The widening participation agenda of the movement from an elitist to 'massification'

(Scheutze, 2000) of higher education system underpins issues of student diversity.

Most research into student diversity thus far has focused on getting students into higher education, with relatively few studies focusing on the student experience, needs and support requirements once studying on a course. Coupled with this is the relative infancy of research on pedagogical issues arising from the changing nature of the student population – do diverse students need a different approach in terms of content delivery and presentation? Haggis and Pouget (2002) noted that this area has not as yet been widely investigated, this is certainly changing but work needs to continue if we are to better understand the pedagogic needs of a diverse student population.

Students needs have changed and therefore it would not be unreasonable to explore what is taught, how it is taught and how learning is to be assessed. In a review of widening participation literature Gorard, Smith, & May (2006) found little evidence to suggest that teaching approaches are being adapted for a diverse student population. Further, that the dominant pedagogy in higher education is still the lecture (Lammers and Murphy, 2002). Some subject disciplines favour a strong framing, teacher-led pedagogic approach (Bernstein, 1996), but it is now argued that whilst this teacher-focused approach is efficient in presenting a lot of information to a lot of students in a short time it does not allow for exploration of the experiences that diverse students bring with them. Hockings, Cooke, & Bowl posit that "there is now considerable evidence that teacher-focused, transmission approaches are ineffective for all but the most 'academically motivated' students" (Hockings, Cooke, & Bowl, 2007).

The pedagogical approach that Haggis and Pouget (2002) suggests is one of collective inquiry into the nature of specific disciplines – a focus on collective forms of exploration in relation to different aspects of disciplinary practice, so that if the teacher is able to see how the students are thinking/talking/approaching particular aspects of a subject *and* if the students are able to see how other students are approaching it *and* if they can see how the teacher is approaching it then this may open up possibilities for new types of understanding.

Pedagogical Theories[1]

There are a huge number of theoretical approaches to learning, and it is possible to group these in different ways. One immediate caveat is that we need to recognise that different people learn in different ways, so that blending different approaches will be needed to achieve the best results for the whole group. A particularly useful theoretical approach is that provided by Pask in his Conversation Theory (Pask, 1975) which suggests that learning occurs through 'conversations' about a subject between learner and another agent – which may be human or machine, but which are also internal to the learner who then effectively converses with himself. Through these 'conversations' the learner 'comes to know', or to construct an interpretation or shared understanding of the world based upon a 'conversational framework'. It suggests that the teacher's role may be seen in the encouragement of conversational reflection on situations.

Pask's theory has been widely applied to e-learning environments, where the technology provides the social environment in which conversations may take place between a teacher and learner(s), or between learners. In an e-environment, there can also be considerable emphasis on collaborative tools, or tasks which can only be carried out successfully through collaboration between peers and/or others (teachers, experts, members of the community).

In a Situated Learning approach, learning takes place by participation in communities and the undertaking of developed social practices. It suggests that learning needs to be designed in the context of the learner's social engagement. Ideas such as Wenger's 'Communities of Practice' (Wenger, 1998) encapsulate this viewpoint. There are also strong philosophical underpinnings from philosophical approaches such as Wittgenstein's concept of the 'language game' – we learn to participate by observing and otherwise acquiring understanding of the meaning of what is taking place. These approaches recognise that we cannot separate out the learners from their context. This is particularly relevant to students undertaking individual research or dissertation projects, the context of their situation will be reflected in their choice of research topic, which will impact greatly on the research design, strategy and methods of data collection – which in turn will dictate the analysis they will need to undertake. Anecdotal evidence from one lecturer of Research Methods at MMU showed that students will often appreciate the full relevance of the Research Methods unit *only* when they begin their individual research work when they see it in the context of their choice of research.

E-Learning[2]

It has been noted that there is an emphasis in education on supporting the learner, in collaboration with peers and tutors, through a lifetime of education, both within and outside the classroom. All e-learning projects are built on certain philosophical assumptions, for example about the nature of knowledge and competence, the purposes of learning, how learning occurs, how students should and should not be treated, etc (for example, see Afaneh, Basile, & Bennett, 2007). In particular, learning should be:

- learner-centred
- fun
- personalised (for example, linked to students' everyday experiences)
- participative and collaborative, emphasising learning as a social activity requiring the development of social skills
- active
- creative
- reflective
- not necessarily classroom based
- facilitated by technology
- encouraging of self-awareness of the learner's own and others' cultures and of their place within these cultures
- inclusive
- constructively aligned with assessment

Analyse This!!! broadly adheres to the 'anytime, anywhere learning' approach – that is, to enable learning in places and at times of the learner's choice. The UK JISC *Innovative Practice with elearning* (http://www.jisc.ac.uk/eli_practice.html#downloads) initiative has suggested four key advantages of this approach:

- **Spontaneity**: Learning activities take place when the learner feels ready, or can be used to fill 'dead time'.
- **Immediacy:** Learning becomes possible at the point of need, regardless of location.
- **Increased access**: Learning resources can be accessed from the home or workplace and in the field, while travelling, and during classes and lectures.
- **Portability**: Communication with peers and tutors, and the capture, storage and retrieval of information in multimedia formats, are possible from one device in any location (http://www.elearning.ac.uk/innoprac/learner/anytime.html)

Learning Objects

There has been much interest in learning objects since their original development, with differences in opinion as to definition, concept and scope. However, as Boyle et al (2003) posited, a coherent view has emerged of learning objects as basic standalone units of learning organized around one educational objective or goal (for example, Dalziel, 2002; Polsani, 2003; Buzzetto-More and Pinhey, 2006). Such that we may define learning objects as "web-based interactive chunks of e-learning designed to explain a stand-alone learning objective" (http://www.rlo-cetl.ac.uk/joomla/press/FAQs.pdf) and "learning objects are self contained units of instructional content in e-learning they may be found as units of study located on a course website that a used either in a web-assisted or a hybrid course" (Buzzetto-More and Pinhey, 2006, p.96). It is this view of learning objects which has informed the development of Analyse This!!! – thus, the learning object may be used as part of a taught course, or accessed independently by students at their time and choosing.

Further to this, it is generally accepted that learning objects should be re-usable or re-purposable. That is, they should be "constructed to support and enhance re-usability" (Boyle et al, 2003). Wiley (2003) states that "This is the fundamental idea behind learning objects: instructional designers can build small (relative to the size of an entire course) instructional components that can be reused a number of times in different learning contexts". In light of this Analyse This!!! has been designed to be re-purposable, in terms of both the subject context and student level. The theoretical and practical aspects of data analysis are generic in nature and can therefore be transferred across subject areas, but it is acknowledged that subject context is important in underpinning the students' understanding and application for their particular area of study. By allowing re-purposing of the learning object, lecturers may adapt it according to different subject disciplines, student requirements, and level and type of study.

Analyse This!!! has also been re-purposed as a mobile learning object in a project which sought to assess the use and viability of learning objects delivered via mobile technologies[3]. Initial results of user evaluation conducted with a group of Manchester Metropolitan University students found that whilst uptake on use of mobile resources for studying was lower than expected, students responded well to the mobile version of the learning object.

Further work is planned to explore use of Analyse This!!! in different subject disciplines and to identify how the learning object has been re-purposed by different academics.

SETTING THE STAGE: DEVELOPMENT OF THE LEARNING OBJECT

Liaison with Staff

The project team identified that research methods and data analysis is included in units taught by many staff across Manchester Metropolitan University. This could relate to a specific piece of assessed course work, supervision of project work, support given for dissertation writing and for students undertaking post-graduate degrees by research. The theoretical and practical aspects of data analysis are of a more generic nature and can therefore be transferred across subject areas. By developing a re-usable learning object, staff can adapt it according to different student requirements and levels, therefore cascade benefits down to students.

Staff at Manchester Metropolitan University were informed about the project via the Staff Digest in early 2008. Interest was generated from a number of staff and three meetings were held to discuss the project and share ideas. The following learning objects in the form of online courses, and tutorials were identified as being useful to inform the development of the learning object: 1) Internet Detective; 2) Citing proficiency; 3) Info Skills; 4) Plagiarism; 5) Maths skills; 6) Writing skills.

Elements of research methods, and specifically data analysis methods, is currently taught within MMU. Examples include:

- Replacing an *IT Skills* unit and *Introduction to Quantitative Methods in Business* with *Information discovery, analysis and interpretation* (Year One).

- *Introduction to business research methods* (Year Two) culminating in the writing of a research proposal for the final year project.
- SPSS based programmes that allows for the easy generation of illustrative specified data analysis designs with user specified data parameters (e.g. number of cases, variables, type of distribution, means, standard deviations and correlations).
- *Research methods* unit on the MA Academic Practice, includes some data analysis.
- Research studies for undergraduate nurses and post registration health care students and take the lead on quantitative approaches.
- Data analysis to first year and final year students, in both cases based on quantitative data produced from lab experiments.

As a result it was felt that there is a need for a data analysis learning object and that setting it at undergraduate level (but with references to further in-depth reading) would ensure participation from a wide group of students (and staff).

Funding was granted from LearnHigher to develop a web-based course on data analysis in the form of a 'learning object' which provides different delivery methods to take into account different styles of learning. This would avoid duplication of effort by providing staff with a generic re-usable learning object on the topic of data analysis, allow staff to adapt the course according to the requirements levels of their students (undergraduate, postgraduate etc) and subject areas and provide opportunities for third stream income by offering a short course on data analysis to external bodies.

The following learning outcomes have been addressed in the development of the learning object:

- A deeper knowledge of theoretical basis of data analysis.
- Awareness of different methods of data analysis to help inform decisions in the selection of the most appropriate method for a specific study.
- Understanding of the advantages and disadvantages of the different methods to inform the selection of the most appropriate method for a specific study.
- Knowledge of data analysis tools available, ranging from simple tables in word/ five bar gates, through use of databases and spreadsheets, to powerful tools such as SPSS and Atlas-ti.
- Practical experience of the use of a range of tools through a series of exercises.
- Reinforcement of learning experience through revision tutorials.

The learning object also addresses a number of different learning areas including, doing research, critical thinking and reflection, independent learning/self directed learning and, report writing. It has also been developed to enable use of a range of different delivery methods, including traditional methods of presentations, lecture notes, reading and resource lists, practical experience through interactive exercises and examples and, reinforcement of learning through revision tutorials.

A decision was made to find a name for the learning object which would be both relevant and memorable. The Internet Detective (www.vts.intute.ac.uk/detective/) for example, is a catchy title which is easily remembered and provides a clue to the nature of its content. With this in mind the learning object developed for the Data Analysis for All project was named: **Analyse This!!!**

Analyse This!!! may be used by a tutor within a unit or as a standalone resource which students are able to access independently. The learning object has been developed on a web-based platform (hosted by LearnHigher) and can be used by anyone interested in learning more about data analysis. Through initial promotional activities the main target audience of the learning object is

MMU staff and students. This is to be extended to any of the LearnHigher partner institutions and beyond.

Feedback from Students on Content

Feedback on content of the learning object was obtained from post graduate students who took part in qualitative and quantitative lectures and hands-on exercises prior to transferring and re-purposing content into the web-based format. Comments included:

Introduction to the sessions:

"I have never used qualitative data before, so it was good to get some detailed information on it- especially with a view to including some in my dissertation."

"These [quantitative data methods] were useful as it is good to have a wide knowledge of a subject, even if you don't end up using all of it."

"... this was not a subject I was familiar with, so it was useful"

Different methods:

"If you are going to learn about the analysis of it, it is vital to have this part of the lecture as well, otherwise the analysis part would be pointless. It was good reinforcement to earlier lectures."

"It was useful to have some insight into the various databases to be used, but sometimes it is difficult to fully comprehend what can achieved when you don't have a full working knowledge/experience of the database."

Hands-on sessions:

"It helps the understanding and learning process when you can actually get involved with the software."

"I found Atlas-ti quite confusing, maybe more time could have been given? or an additional (optional) session was needed?"

" It was a brief, but useful exercise."

"..... I may have considered it for a higher level of study, or for an assignment when I intended to have vast quantities of data to collate."

Overall impression of the sessions:

"There is a lot of detail to take in and concentration is needed - what let the session down was the time of day that it was delivered; by five o'clock people were waning, but that wasn't something that could be prevented."

"... the disadvantage was time of day [late afternoon]. I did find it increasingly hard to concentrate. It would have been nice to hear about other quantitative data analysis tools. Excel was mentioned, but not really explained what it would have been useful for".

"A lot of ground was covered, from basics to the level we would need for an MA- with ample opportunities for questions".

Preferred methods of delivery included:

"Hands on following a short lecture - often the best way to learn is to get it wrong!"

"Hands-on"

"Hands-on combined with lectures"

"I probably like lectures best"

"Lecture, online-tutorials".

Conclusions from this feedback were that the amount of content was about right, the level of the sessions was about right, students like a mix of traditional lectures, online learning and hands-on and, students became tired by the end of each session – which reinforces the advantages of incorporating a learning object into the session, thus allowing self-directed learning at a time more convenient to them.

CASE DESCRIPTION

The choice of content to be included in the Analyse This!!! learning object has been influenced by a number of resources (for example, Higgison, 2002; Smith, 2004). Smith (2004) recommends that content is meaningful and will support the learning goal. Smith identifies nine points for consideration, thus: 1) it is important to recognize and address common preconceptions learners may have about the content being taught; 2) where possible, examples should be drawn from real-world data and case studies; 3) provision of complex scenarios that invite a range of opinion and provoke thought; 4) connect with the content and the learner's own life or situation (or guide the learner in making such a connection); 5) demonstrate new knowledge to the learner in a memorable way; 6) provide a conceptual framework for facts and ideas; 7) choose content and examples that are concrete rather than abstract; 8) build on learner's existing knowledge and, 9) keep content focused on how the world works and demonstrate how the learner may use the new knowledge in his or her own way.

Taking the above points into consideration, Analyse This!!! provides a range of activities within the object (see Figures 1, 2 and 3 below). This includes:

- Text explaining a concept of data analysis.
- Following links to explain areas or concepts in more depth.
- Following links to external resources.
- Examples of how theoretical concepts can relate to real world situations (e.g. analysis of comments provided in a focus group can be used to illustrate a particular topic or situation such as to demonstrate how a particular service is perceived by a typical group of people).

Figure 1. Analyse This!!! Qualitative Data

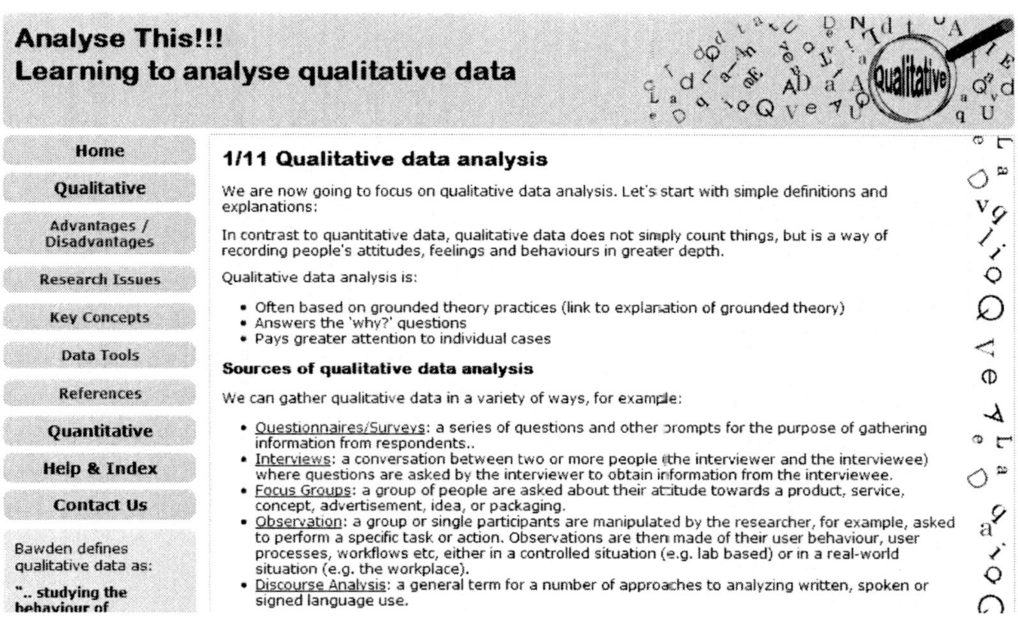

- Opportunities for learners to make choices or manipulate elements (NB: choices must be meaningful and relate to the learning goals).

Analyse This!!! offers the learner a choice of paths rather than an inflexible path taking through the learning object by simply following a 'next' arrow or 'proceed to next page'. Smith (2004) advises that an inflexible path does not give the learner 'ownership' of the learning process. Analyse This!!! avoids this by providing a menu to the left of the page allowing the learner to dip in and out of the topic areas provided and also to provide them with an overview of the learning object as a whole. Each page also allows users to 'proceed to the next page'. In this way, the learner can choose whether to follow a linear path through Analyse This!!! or to pick and choose the order of their learning experience. A site map provided by Analyse This!!! shows the relationships between ideas, content or sections.

Analyse This!!! provides feedback on revision and quiz exercises in a constructive way. For example: the learner is asked to make some choices where there is a right and wrong answer; the learner selects the wrong choice; feedback doesn't simply say 'wrong, try again' as this may encourage the learner to simply guess until they get it correct; instead the feedback will not only tell them the choice is wrong, but also WHY it is wrong so that they can learn from the mistake.

User Interface

The design of the user interface takes into account the following design guidelines:

- Page or screen will be visually balanced.
- Important items are larger and nearer the top or left edges of the page.
- Elements are organized in a logical way according to related content.
- White space is provided between groups of content.
- The same font and typeface is used throughout but allows for user control to adjust these.

Figure 2. Analyse This!!! Quantitative data

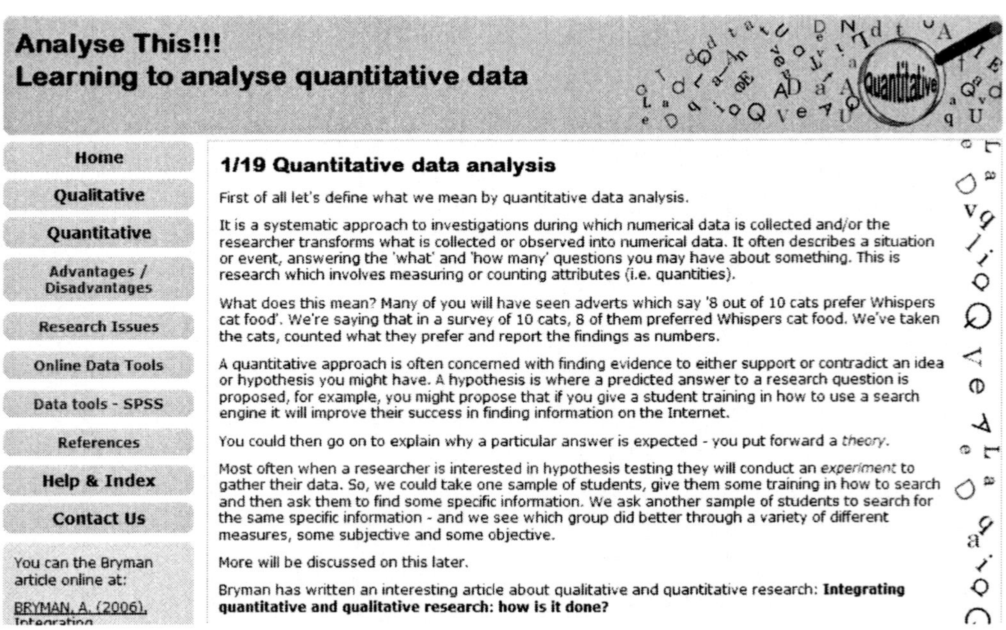

Figure 3. Analyse This!!! Quiz

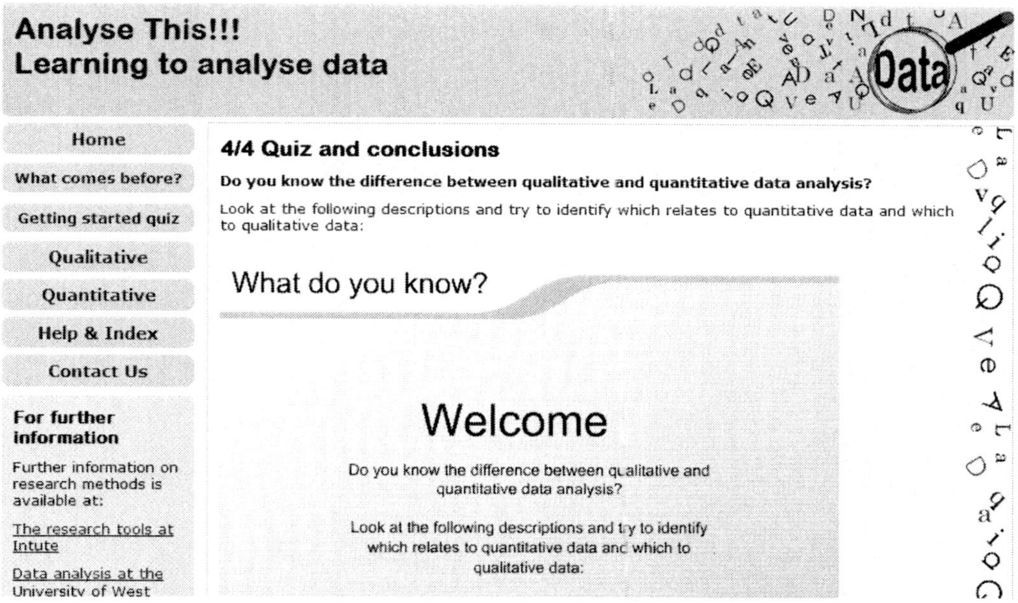

- Contrasting colours are used throughout and allows for user control to adjust these.

Although when designing for usability it is important to maintain consistency, the look and feel of the learning object can differ according to the type of activity it is displaying. For example, the Home page, Qualitative section and Quantitative section are in different colours as an aid to the learner (and teacher) to see which section they are visiting (text for each page Title also provides this information to aid navigation for the user and facilitator without having to rely on colours); text explaining concepts can differ to quiz pages, but fundamentals such as navigational elements and language are kept the same to maintain consistency. Good practice in design for accessibility and usability of e-learning objects has been followed (Smith, 2004; Kelly, Phipps, & Swift, 2004), and are illustrated in Table 1.

Re-Using and Re-Purposing Analyse This!!!

The aim of the Analyse This!!! project is to create a learning object that is reusable by different people. In this way it can be used by: 1) the same learner working on different tasks, course work, projects etc.; 2) learners at various levels of knowledge who may want to brush up on specific elements provided in the learning object; 3) learners in different disciplines where the generic nature of Analyse This!!! will apply to all; 4) developers or teachers who want to reuse particular elements of the learning object, perhaps to add to a lecture or to run a seminar session and, 5) developers or teachers who want to repurpose particular elements of the learning object, perhaps to use the core structure of Analyse This!!! to develop a more high-level learning object as required.

To enable Analyse This!!! to be reused the following points have been taken into consideration:

Table 1. Good practice in design for accessibility and usability

Navigation	Design	Format
• Provide each page with a meaningful title (this is helpful for screen readers in particular). • Allow learners to control their interactions, e.g. a choice of paths to navigate through Analyse This!!! • Make sure the navigation is consistent and allow learners to undo or quit actions wherever possible. • Place main navigational elements either horizontally along the top or vertically down the left side of the screen. • Place controls for quitting at the upper right or as the last item on the left hand side menu.	• Limit the number of icons. • Do not disable link underlining. • Make sure link colours are contrasting with text. • Minimize the amount of text on screen. • Write clearly using simple and natural language. • Leave white space around headers and between text sections. • Only use graphics where they add meaning. • Use colour with discretion. • Make the design visually pleasing. • Design for device independence – for example, so that learners can choose whether to use the mouse or keyboard.	• Provide alternative formats for visual and auditory content – for example, alternative text description for graphics or text captions or transcripts for audio • Allow learners to control content – including moving content. • Consider different formats of learning for different learners (for Analyse This!!! the quiz elements may need careful consideration). • Where a focus on the learner's needs rather than the digital resources themselves to address the accessibility of the learning experiences, rather than the accessibility of the e-learning resources.

- Each element of Analyse This!!! will be stand-alone, so elements can be reused without losing their meaning.
- Links to external resources can be used to acquire further knowledge rather than an integral part of the learning object; therefore if the external resource becomes unavailable it will not affect the core learning goal of Analyse This!!!
- Analyse This!!! is web-based so can be reused or repurposed accordingly.
- Analyse This!!! has been properly attributed so that the ownership and copyright licence (e.g. Creative Commons) always goes with it.
- Appropriate metadata has been applied to Analyse This!!! for the purposes of searching, retrieval, selection, tracking ownership, reuse, and repurposing.

Evaluating Analyse This!!!

Evaluation is an important part of the development of Analyse This!!!. In the early stages of the project feedback from students and liaison with MMU staff took place to help ensure the content was appropriate and to identify any useful resources for inclusion. Evaluation of the final draft version of Analyse This!!! was undertaken at the beginning of June. Participants were mainly students from MMU and who may be interested in using a resource like Analyse This!!!

The participants were asked to work through sections of the learning object and then provide feedback on the Introductory sections (Home, The Story, What comes before), Qualitative sections and Quantitative sections. Participants were asked to comment on how informative the introductory sections were; more specifically on whether the qualitative and quantitative sections helped them to understand more about data analysis techniques; whether the quizzes were a helpful way to revise what they had learned; usefulness of the links and resources provided; level of the learning object; and ease of navigation.

Participants were also asked to comment on any improvements they would like and whether they would use it again or recommend it to a friend. The evaluations revealed some bugs in the learning object which were quickly fixed.

Responses relating more to the content and display revealed that they thought the introductory sections were informative. Comments included:

"good use of headings to break up the text, allows the eye to quickly scan to the part you need...."

"not too involved, but a nice gentle introduction to what's involved. Quiz was useful".

Some commented that 'The Story' and 'What comes before' seemed a little out of order. In light of this 'The Story' page was removed and the text from this page (which describes two possible scenarios for data analysis) moved to the end of the first quiz, thus leading to either the qualitative or quantitative sections as appropriate.

Some commented that they were a little confused at first that external links did not open in a new window. The 'Notes on navigation' on the Home page now explains that external links did not open in a new window and it was necessary to use the Browser Back button to return to the learning object.

With regard to display of the quiz (this also applied to the qualitative and quantitative quizzes), an issue was raised with the quiz forward and back buttons which some felt were not prominent enough. In light of this an instruction for navigating through the quizzes has been clearly placed at the beginning of each quiz.

A common comment was that the links to resources and references were not as 'visible' as they could be, so easily missed. With this in mind, text has been added to the Home page with instructions on what the learning object involves and notes on navigation. Apart from the visibility issue, comments were positive:

"I liked this element of the program, particularly and used a lot of the links – would make me come back to this as a useful reference source".

All respondents found navigation easy, and some also provided suggestions for improvements (such as those relating to the quiz forward and back buttons). Overall, the respondents were enthusiastic about the learning object and said they would use it again and recommend to a friend, comments included:

"useful even outside social sciences for basic principles"

"plan to use asap!"

"very useful resource at this stage of the study"

"I liked it A LOT!"

CONCLUSIONS AND RECOMMENDATIONS

The changing nature of the student population in UK higher education, coupled with the advances in technology enhanced learning and personal experience in teaching students data analysis were instrumental in the formation of the conception of the Analyse This!!! learning object and it is hoped that it will prove to be a useful resource for students and staff alike, across many different subject disciplines and across different institutions.

In June 2008 Analyse This!!! was successfully launched as a re-purposable learning object for use by students and staff of LearnHigher partners and the wider academic community. We view this as the beginning of our work in this area and have recently been awarded further funding from LearnHigher to enable the development of a learning object on data collection approaches.

As with Analyse This!!! the theoretical and practical aspects of data collection are of a generic nature and are therefore transferable across subject areas and other higher education institutions. Consultations with colleagues have shown that varying degrees of customisation are preferable. Therefore, staff will be able to adapt (or re-purpose) it according to different staff and student requirements, levels, and subjects, and thus be able to cascade benefits down to students and other colleagues.

Further to this, consultations with students and staff over the use of Analyse This!!! raised a number of issues which we would like to explore further, such as increased interactivity through the use of screen shots, audio files and video clips. We will seek to address these in this learning object.

In addition, CERLIM has recently been funded funding from LearnHigher to undertaken work in the mobile learning area. One of the deliverables for this work will be the adaptation of Analyse This!!! for delivery via a mobile device – it is possible the data collection learning object could also be development in this way in the future.

ACKNOWLEDGMENT

- We would like to thank the staff and students at MMU who provided us with valuable feedback on the content of Analyse This!!! and also on the structure and usability of the online learning object.
- We are indebted to Eleanor Shember-Critchley who undertook the design of the learning object and who has provided the section on Technical requirements for re-purposing Analyse This!!!
- We are also indebted to Margaret Markland and Peter Brophy for their input into the pedagogical theories and e-learning sections. These sections have been informed by their work on the eMapps project [Online] http://www.cerlim.ac.uk/projects/emapps/index.php.
- We would also like to extend our gratitude to LearnHigher for funding the initial development of Analyse This!!!, its adaptation to the mobile learning environment and for the development of the second learning object on data collection.

REFERENCES

Afaneh, M., Basile, V., & Bennett, J. (2007). *E-Learning Concepts and Techniques*. Retrieved 1st December 2008 from http://iit.bloomu.edu/Spring2006_eBook_files/index.htm

Bernstein, B. (1996). *Pedagogy, symbolic control and identity. Theory, research, critique*. Maryland: Rowman Littlefield Published.

Boyle, T., Bradley, C., Chalk, P., Jones, R., Haynes, R., & Pickard, P. (2003). *Can learning objects contribute to pedagogical improvement in higher education: lessons from a case study?* Paper based on presentation given at CAL '2003, April 2003. Retrieved February 23, 2009 from http://www.londonmet.ac.uk/ltri/learningobjects/papers_pres/CAL_Obje cts_paper.doc

Buzzetto-More, N., & Pinhey, K. (2006). Guidelines and Standards for the Development of Fully Online Learning Objects. *Interdisciplinary Journal of Knowledge and Learning Objects, 2*, 95–104.

Callender, C., & Wilkinson, D. (2003). *2002/03 Student Income and Expenditure Survey, Research Report 487,* DfES, Nottingham, UK.

Dalziel, J. (2002) Reflections on the COLLIS (Collaborative Online Learning and Information Systems) demonstrator project and the "learning object lifecycle". In A. Williamson, C. Gunn & T. Clear (Eds.), *Winds of change in the sea of learning, Proceedings of ASCILITE 2002 Annual Conference*. Retrieved 23 February 2009 from http://www.unitec.ac.nz/ascilite/proceedings/papers/207.pdf

Gorard, S., Smith, E., May, H., Thomas, L., Adnett, N., & Slack, K. (2006). *Review of widening participation research: Addressing the barriers to participation in higher education.* Report to HEFCE by the University of Yorke, HEA and Institute for Access Studies, UK.

Haggis, T., & Pouget, M. (2002). Trying to be motivated: Perspectives on learning from younger students accessing higher education. *Teaching in Higher Education, 7*(3), 323–336.

Higgison, C. A. (Ed.). (2002). *Online Tutoring e-book.* Aberdeen, Scotland: Heriot-Watt University and The Robert Gordon University. Retrieved 1st December 2008 from http://otis.scotcit.ac.uk/onlinebook/

Hockings, C., Cooke, S., & Bowl, M. (2007). Academic engagement' within a widening participation context – a 3D analysis. *Teaching in Higher Education, 12*(5-6), 721–733. doi:10.1080/13562510701596323

Kelly, B., Phipps, L., & Swift, E. (2004). Developing A Holistic Approach For E-Learning Accessibility. *Canadian Journal of Learning and Technology, 30*(3, Autumn). Retrieved December 1st, 2008 from www.ukoln.ac.uk/web-focus/papers/cjtl-2004/

Laing, C., Chao, K., & Robinson, A. (2005). Managing the expectations of non-traditional students: a process of negotiation. *Journal of Further and Higher Education, 29*, 169–179. doi:10.1080/03098770500103572

Lammers, W., & Murphy, J. (2002). A profile of teaching techniques used in the university classroom. *Active Learning in Higher Education, 3*(1), 54–68. doi:10.1177/1469787402003001005

Lave, J., & Wenger, E. (1991). *Situated Learning: Legitimate Peripheral Participation.* Cambridge, UK: Cambridge University Press.

Livingston, J. (1997). *Metacognition: An overview.* Retreived December 1st, 2008 from http://webu2.upmf-grenoble.fr/sciedu/grangeat/EnsI/Meta/MetaLivingstonOverview.pdf

Pask, G. (1975). *Conversation, Cognition, and Learning*. New York: Elsevier.

Polsani, P. R. (2003). Use and abuse of reusable learning objects. *Journal of Digital Information, 3*(4). Retrieved 23 February 2009 from http://jodi.ecs.soton.ac.uk/Articles/v03/i04/Polsani/

Schön, D. (1987). *Educating the reflective practitioner.* San Francisco: Jossey-Bass.

Schuetze, H. G., & Slowey, M. (2000). Traditions and new directions in higher education. In H.G. Schuetze & M. Slowey (Eds.), *Higher Education and Lifelong Learners, International perspectives on change,* (pp.309-327). London: Routledge/Falmer.

Smith, R. S. (2004). *Guidelines for authors of learning objects* (p.8). The New Media Consortium. Retrieved December 1, 2008 from http://archive.nmc.org/guidelines/NMC%20LO%20Guidelines.pdf

Vygotsky, L. (1986). *Thought and Language* (original Russian edition 1934). Cambridge, MA: MIT Press.

Wenger, E. (1998). *Communities of Practice: learning, meaning and identity.* New York: Cambridge University Press.

Wiley, D. (2003). *Learning objects: difficulties and opportunities.* Retrieved February 23, 2009 from http://wiley.ed.usu.edu/docs/lo do.pdf

ENDNOTES

1 We are indebted to Margaret Markland and Peter Brophy for their input into the Pedagogical theories and e-learning sections of this chapter. These sections have been informed by their work on the eMapps project [Online] http://www.cerlim.ac.uk/projects/emapps/index.php.

2 We are indebted to Margaret Markland and Peter Brophy for their input into the Pedagogical theories and e-learning sections.

These sections have been informed by their work on the eMapps project [Online] http://www.cerlim.ac.uk/projects/emapps/index.php.

[3] (Mlearning Project, undertaken by the authors and Geoff Butters, funded by LearnHigher http://www.learnhigher.ac.uk/learningareas/mobilelearning/home.htm)

Chapter 3
Getting Teachers to Use New Technology by Just Giving Them Time:
A Case Study from the UK

Terry Haydn
University of East Anglia, UK

Roy Barton
University of East Anglia UK

EXECUTIVE SUMMARY

The chapter reports on a UK project which was designed to explore innovative ways of getting teachers to develop their use of new technology in subject teaching. The outcomes of this project suggest that in the area of developing teachers' use of ICT in subject teaching, simply providing support for teachers, in the form of time to explore the potential of ICT, to meet together to discuss ICT in subject groupings, and freedom to focus on their preferred ICT agendas, may be a more effective way forward than prescribing lists of required competences and providing generic 'training' type courses. This goes against the grain in an era characterised by 'top-down', centrally directed national strategies, high levels of accountability and auditing of teachers, and 'coverage' models of competence (Ball, 2003), but given the disappointingly sluggish and modest outcomes of such programmes, in the UK and elsewhere, such approaches may be worth exploring more extensively.

'Excessive directive methods of government that appear to treat front-line deliverers as unable to think for themselves, untrustworthy or incompetent, undermine the very motivation and adaptability on which real world success depends.... Driving through policies with an implicit assumption that the main players are the problem rather than the solution is usually a recipe for failure' (Performance and Innovation Unit, UK, 2001).

'The floggings will continue until morale improves' (Czar Alexander III of Russia, attrib.).

DOI: 10.4018/978-1-60566-942-7.ch003

INTRODUCTION

This chapter details the processes and outcomes of a small funded project in the UK which was designed to explore the potential of less directive and 'top-down' strategies for developing teachers' use of ICT and e-learning in their subject teaching. The basic premise was to give teachers dedicated time, and the opportunity to collaborate and discuss ICT and e-learning issues, in a way which allowed them complete freedom to choose what aspects of ICT and e-learning they wanted to explore and develop.

CONTEXT OF THE CASE STUDY

The UK is one of many countries in the developed world where politicians have invested considerable belief, pressure and funding in the potential of new technology to enhance educational outcomes. Politicians from all parties in the United Kingdom have been unequivocally positive about the part that new technology will play in enhancing educational outcomes, with Conservative Minister David Hunt stating that 'the nation which embraces technology most willingly and most effectively will be the winners in tomorrow's world' (Hunt, 1995), Labour Education Minister Charles Clarke (1999) elevating new technology above even literacy and numeracy in asserting that 'Familiarity with ICT is the most vital life skill for the generation now going through school', and former Prime Minister Tony Blair making a number of eulogistic speeches about educational transformation through technology use (see, for example, Blair, 1995, 1997).

This belief in the power of ICT to transform educational outcomes has resulted in considerable pressure on teachers in the United Kingdom to use information and communications technology (ICT) in their classrooms. The Education Ministry in the UK instigated biennial surveys to monitor teachers' use of ICT; school inspection reports criticised schools and teachers who were not using computers (Harrison, 2003), the teaching force was described by the ministry as 'a hurdle which needs to be overcome' in relation to developing the use of computers (quoted in Cohen, 1999) and a leading figure in the promotion of ICT in education asserted that 'in future, there will be two types of teacher, the IT literate and the retired' (Cochrane 1995).

In spite of these pressures, and in spite of substantial investment in ICT in schools, uptake in teachers' use of new technology has remained disappointingly sluggish in the UK (Nichol & Watson, 2003; Reynolds, Treharne, & Tripp, 2003; Selwyn, 2003). The ImpaCT 2 Report suggested that perhaps as many as 60% of teachers in the UK were making little or no use of computers in their day-to-day teaching (Harrison *et al.,* 2002). Research reports from outside the UK suggest that this problem is not limited to the UK, and that in spite of substantial financial investment in ICT in education worldwide, many teachers struggle to successfully integrate new technology into their teaching (Phillips, 2002; Zhao, Pugh, Sheldon & Byers, 2002; Zhao & Frank, 2003).

Learning from Mistakes Which Have Been Made

Although the UK government's commitment to the development of a technologically empowered and accomplished educational workforce has been steadfast and wholehearted, it has at times been misguided and ill-informed it its interventions and investments. Part of our research over the past decade has been in evaluating the impact of a range of government funded initiatives in education, and exploring the reasons why teachers and pre-service teachers do or do not use ICT in their teaching. There is some evidence to suggest that within the UK, many of the interventions, policies and investment in getting teachers to embed the use of ICT in classroom teaching have not been found to be helpful by teachers, or have even been counter-productive, and have had the effect of alienating teachers from the use of ICT (Haydn and Barton, 2007).

One early mistake was the failure to ask teachers what forms of investment in ICT they would find most helpful. The bulk of funding was ploughed into providing specialist ICT suites, where pupils would go for 'one-off' special occasion ICT lessons designed to develop pupils' basic ICT skills (described by one teacher as 'like dipping sheep'). Surveys of what forms of ICT investment was on the wish-list of practising teachers (see for instance, Haydn, 2004), revealed that most teachers wanted the projection facilities which would allow for ICT use as a regular component of lessons in day-to-day teaching so that computers could be used 'not as a special event, or to impress others, but naturally, when the need arises' (Ogborn, 2000: p. 26).

A major government initiative to improve the ICT capability of in-service teachers, the 'New Opportunities Fund' training programme was criticised as being unwieldy, over-prescriptive and unhelpfully focused on technological capability rather than pedagogical application and integration (Leask, 2002; Ofsted, 2002; Preston, 2004). The following comments from teachers involved in this project were not unrepresentative of teachers' responses to the initiative:

'A waste of money... and created a lot of resentment.'

'It was like a driving lesson that consisted only of learning the highway code but which had not actual driving tuition.'

'The training provided was patronising, unrealistic and painful in its delivery... The effect on staff morale was devastating. Nothing in recent years has done more to put teachers off using ICT.'

'I was tasked to deliver the training and implemented one course with 15 members of staff.... attendance dwindled within weeks to four or five and the course was never fully completed. I was unable to coax any other staff onto the course as they had heard of the experiences of others.'

'It was "done to us" in a way that left us deflated and angry. We needed time to debate and trial ideas – we did not have this. The sessions were geared around certain information which was going to be delivered come what may and usually this was not really linked to what we wanted to do. We persevered despite the training, not because of it.' (Haydn and Barton, 2007).

In an effort to add more rigour to pre-service teachers' education in the use of ICT, the government revised the specifications for competences which student teachers were required to possess before being allowed into the profession. These included 15 pages of competences related to ICT, with over 100 'micro-competences' identified as being essential (DfEE, 1998). This proved to be a nightmare in practice, and largely unworkable, and the specifications for ICT were quietly reduced (DfES, 2002), but not before many thousands of pre-service teachers had been made to feel that induction into the educational uses of ICT was an ordeal to be endured, rather than a useful and interesting opportunity to enhance their teaching repertoire.

Another 'punitive' innovation in the field of ICT policy for pre-service teachers was the introduction of an 'online basic skills test' in ICT, to be completed under time pressure at a nominated test center. A survey of student teachers' views on the test revealed that over 95% of them regarded the test as unhelpful. Although the students had reservations about various aspects of their ICT experiences on their course of training, the online test was the only element that evinced real anger. The following comments are representative of their views:

'Bloody waste of time.'

'Very, very, very unhelpful.'

'Insulting.'

'Pointless, pathetic and utterly ridiculous.'

'A waste of time: another hoop to jump through.'

'Unrelated to subject or to common sense.' (Barton and Haydn, 2006: 262).

(It is worth noting that at the time of writing, in spite of overwhelmingly negative feedback from students, tutors and employing institutions, the online basic skills test in ICT remains in place).

These were not the only mistakes and misjudgements that have been made. The idea that pre- and in-service teachers could learn to use ICT through online learning packages proved to be more problematic than policymakers had envisaged (Noss & Pachler, 1999, Preston, 2004), and many of the competence specifications and testing mechanisms for new teachers were found to be over bureaucratic and unhelpful (Barton & Haydn, 2007). Even well intentioned attempts to provide information brochures and specialist websites to provide ICT guidance proved ineffective, as teachers were simply overwhelmed with the volume of information that they were required to read. The volume of information relating to ICT was seen as excessive and unrealistic. In the words of one teacher involved in the project, *'Who writes this stuff? I don't have time to read through it all... they obviously don't have a clue about what teachers' lives are like.'* There was a general murmur of assent to this from other members of the group. In a sense, the last thing teachers needed was 'more stuff': they were already drowning in 'stuff'.

One of the characteristics of several government initiatives in ICT, both for pre-service and qualified teachers, has been a tendency to go for a 'coverage' model of ICT capability; that is to say, the idea that teachers needed to be expert *across the full range* of ICT applications – no matter what their subject specialism. In 1998, Anthea Millett, Chief Executive of the Teacher Training Agency in the UK argued that by spelling out more comprehensively than ever before the competences which trainee teachers would be obliged to possess before being licensed to teach, these new 'improved' Standards for the award of Qualified Teacher Status would ensure that the breadth of newly qualified teachers' competence would be higher than ever before (Millett, 1998). These competence models, the 'New Opportunities Fund' Programme for developing teachers' ability to use ICT in subject teaching, and the online basic skills test in ICT were all predicated on a 'coverage' mentality – teachers were to be 'trained' to use a wide variety of ICT applications so as to become 'completely equipped' in ICT. But to what extent do teachers need to be expert in all aspects of ICT? It is possible that teachers in different subject specialisms might (given the opportunity) choose to explore particular pedagogical possibilities in ICT in depth, and develop classroom applications related to these particular applications, rather than covering all aspects of new technology? What if you just gave them time to explore the potential of ICT to enhance teaching and learning in their subject? These were questions that we wished to explore in this case study.

The Genesis of the Research Project

After several years of researching factors influencing teachers' integration of new technology into their teaching, we wanted to see if we could learn from some of the mistakes which had been made in educational policy and ICT. As well as the findings from our own studies, and those of other researchers in this field, many of the teachers and lecturers in our own initial teacher education partnership had expressed concern about the increasing prescription of government initiatives relating to ICT, with the proliferation of strategy documents and the possible atrophy of teacher initiative, originality and imagination in approaches to developing ICT use.

Funding from the Department of Culture, Museums and Sport (DCMS) as part of its 'Creative Partnerships' programme enabled us to undertake a two year action research project which aimed to give teachers in particular subject areas time to develop their own ideas for enhancing teaching and learning and time to meet up to share their ideas and initiatives. The underpinning hypothesis behind the project was that recent government initiatives in education in England have added to teachers' overall workload (see, for example Smithers and Robinson, 2000, NFER, 2000, Cockburn and Haydn, 2004) and resulted in a lack of time and opportunity for teachers to develop their own agendas and ideas for the use of ICT in the classroom.

The overarching aims of the project were to use the funding to provide time for teachers to develop their ideas for integrating ICT into their subject teaching, to be able to 'try things out' in areas of their subject which were of interest to them, and to provide time for them to meet together to discuss and share their ideas with other teachers.

Research Design

The research design was predicated on the proposition that most teachers want to teach their subject well, that they are at least open-minded and interested in exploring the potential of ICT, and that one of the barriers to the development of ICT in subject teaching is lack of time (Barton & Haydn, 2007; Zhao & Frank, 2004). The project was designed to give teachers 'dedicated time' with which to explore their ICT agendas, both in terms of time on their own/within their own department, and also in subject groupings with colleagues from other schools.

There was a conscious attempt to avoid a 'top-down' prescriptive approach and a 'coverage' mentality (in the sense of attempting to address all ICT applications which might be relevant to participants). It was also felt to be important to let participants choose what facets of ICT they wanted to explore. Although this might lead to a degree of replication or overlap, the project rested on the teachers being able to work on whatever they wished. We also wanted to avoid an 'audit' or 'target-setting' culture. The research activity was to be conducted in separate subject groups – one of history teachers, the other of science teachers. A key element of the research design was to ensure that there was provision for teachers to have dedicated time within their own school *and* to discuss their work with other teachers of the same subject. The overall approach was based on an action research model which would allow teachers to reflect on their practice and explore possibilities (Elliott, 1991), rather than a 'training' model such as the NOF scheme which had preceded this project (Leask, 2002; Ofsted, 2002; Preston, 2004). One of the research questions to be tested by the project was whether it was possible for teachers to generate their own development in the use of ICT in subject teaching, without recourse to external inputs, training, and national strategy based approaches.

Research Process

It should be stressed that because the funding available for the study was modest (around 40,000 US Dollars), this was a small-scale project. The funding for the project supported the formation of two groups of secondary teachers in science and history. The choice of subjects reflected the curriculum specialisms of the university tutors involved, and all the teachers involved in the project were members of the regional initial teacher education (ITE) partnership. Teachers were invited to participate in the project, which was described as focusing on developing the creative use of ICT in subject teaching.

The framework for the project activities involved one day 'research workshops' at the start and at the end of the project and one day of supply cover to be taken at any time in between the two research days, so that teachers had some dedicated time to work on their own ideas for developing their department's use of ICT.

The composition of both groups was 'mixed ability' in terms of ICT expertise. Some already had a high level of technological expertise in ICT, and as well as being subject teachers, were ICT coordinators for their schools. Others were not particularly accomplished in terms of technological expertise but were heads of department interested in exploring the potential of ICT, and some were self-acknowledged ICT novices who were keen to learn from colleagues who were more expert in ICT.

In all, 37 teachers were involved in the project. Over 80% of those involved in the first phase of the project opted to continue their involvement into the second year.

Some time at the start of the initial research workshops was spent discussing the action research approach which we intended to adopt. Participants then explained 'where they were up to' in terms of facilities, staff expertise, interest and recent use of ICT, including some discussion of the factors which they felt were either conducive or unhelpful to the development of ICT use in the department. Although this was time consuming, and reduced the time available for 'hands-on' practical demonstration of the forms of ICT related activity that had already been developed, it gave some indication of the current interests and aspirations of the participants in the area of ICT in subject teaching.

The aim was that during the course of the subsequent school year, the members of the two groups would work on a specific development of their use of ICT to support their teaching. This was to be followed by a second research worshop to be held towards the end of the next school year, where they would report back on ideas, experiences and reflections on the project. The meetings concluded with a brief discussion on the format of the second research day and email addresses were exchanged, in the hope that there would be communication between those involved between the two research workshops, and so that teachers could work collaboratively if they wished to do so.

In terms of ICT development, as might be expected given the broad (and perhaps vague) remit which teachers had been given, teachers worked on a wide range of ICT applications to see how they might improve teaching and learning in their department. Some teachers and groups (some teachers had chosen to work collaboratively with other departments) had explored quite sophisticated ICT agendas, others had worked to improve the benefits which might be derived from basic generic applications such as word processing and PowerPoint.

Amongst the history group, three teachers had developed departmental websites, two had focused on exploring how to make best use of interactive whiteboards, and a cluster had formed 'to find ways of making PowerPoint less boring'. Two teachers had learned how to use Macromedia Flash, and other projects included the development of themed collections of images on particular historical topics, materials which might develop pupils' understanding of the reliability of the internet as a source of information about the past, and materials to develop pupils' democratic literacy. Another department had focused on the development of web templates which enabled pupils to make their own web pages without wasting too much time on technical issues, and another had focused on the use of databases and word processing exercises to develop and test historical understanding.

The science teachers reported on using ICT to address pupil misconceptions at Key Stage 3 (11-14 year olds), bringing applied science to life, on-line testing to prepare pupils for science exams, developing pupils' skills of scientific education using ICT and new approaches to running an out of hours science club, and the development of science department websites. As with the history group, the project had facilitated the development of some quite sophisticated departmental websites. Exploration of data logging demonstrated the fact that science teachers and history teachers sometimes have very different agendas in ICT.

Second phase funding made it possible to continue the project into a second year, with the same pattern of group meetings early and late in the year, and supply cover to support activity between meetings. By the end of the second phase of the project, over 30 of the 37 participants had reported on some form of development or activity in their departmental use of ICT. Again rather than looking for some objective measure of the quality of the individual end products, our main focus was to explore the extent to which engagement in the project had impacted on teachers' commitment to developing the use of ICT in their subject teaching. We felt it was significant that the teachers were already using these resources in their teaching, impacting directly on their pedagogy. There was no significant disparity between the science and history cohorts, in terms of the proportion of teachers producing some form of finished product or activity in ICT, with 13 out of 16 science teachers and 17 out of 21 history teachers presenting some form of development in ICT in the end of year workshops.

OUTCOMES

Evaluation of the project was conducted partly through discussion groups in the concluding sessions of the research workshops, and partly through e-mails and subsequent phone or face-to-face conversations with participants. There was ample evidence from the project outcomes that many of the teachers involved invested many hours of work in the project, above and beyond the days of supply cover which were funded by the project.

However, it is important to stress that not all the aims of the project were fulfilled. The hope that teachers would stay in close touch by email and phone over the course of the year to keep up to date, share ideas and map developments, proved to be unrealistic. Beyond a handful of phone calls and emails this just didn't happen. At the end of year workshop, several participants acknowledged that it was difficult to keep up a sustained collaborative research agenda in the face of the many other demands on their time. Compared to the exigencies of examination classes and departmental responsibilities, this project was clearly a luxury item, and in some schools, cover was not possible even though funding was available. The sheer 'business' of teachers' lives hampered our project in the same way that it has impeded recent government initiatives in ICT, although there was some evidence to suggest that because of the sense of ownership and autonomy involved in the project (and perhaps because of a sense of collegiality with colleagues), teachers did try to commit to the project to at least some degree. The grounds for this tentative assertion rest to some extent on the fact that at the end of the two year project, almost all the teachers involved attended the end of project workshop, and reported some positive gains in terms of their subject department's use of new technology.

The project was not uniformly successful; not all the teachers involved made progress with their declared area of interest, and good intentions had not always translated into action. At the end of year research workshops, there were a few teachers who acknowledged that they had not completed the work that they had intended to. One of them reported *'I'm afraid I don't have a great deal to offer. We were supposed to work on updating the revision website but as with many great intentions it didn't end up happening.'* However, over 85% of the participants felt that they had accomplished something worthwhile from involvement in the project, and in some cases, reported that the work had a transformational effect on their practice. Sometimes this was in the form of a particular resource, such as a departmental website, in other cases it was a radical change in the use of new technology, the ways in which planning for learning was organised, or the degree of autonomy and responsibility accorded to pupils.

In some cases, the impact on departmental use of ICT was quite dramatic, with the development of highly successful classroom activities, collections of resources and departmental or revision oriented websites. Some of the participants have gone on to achieve national prominence in the field of ICT, and have been involved in national and international research and development programmes.

There also appeared to be important process and attitudinal issues involved. Participants reported that the project had been more favourably received than centrally directed government training schemes for ICT, and that teacher attitudes to ICT, and their willingness to experiment with new technology had improved considerably.

The evaluations suggested that the participants involved had enjoyed being involved in the project and considered it worthwhile, in spite of the many other demands on their time. Some teachers 'confessed' to not getting as far as they had hoped with their research but still felt that there had been some positive aspects to having been involved in the project:

'The day gave you time to think and it was very enjoyable... there was a nice atmosphere.. I like the combination of having some time with other teachers to just share ideas – and frustrations.'

'The end of term was taken up with the organsation of the Battlefields trip. We were hoping to spend time on it towards the end of term but I'm not sure. I would love to have the opportunity to try and do something like this at my new school so if anything ever comes up let me know.'

'We've not followed up things as much as we hoped but I am meeting up with J. (from another school involved in the project), using one of the supply days.... It was helpful and I'm glad I got involved.'

Transcripts of the evaluations suggested that teachers welcomed the opportunity to talk about pedagogy with their peers, they enjoyed having time to reflect and explore ideas and recent developments which have the potential to influence the ways in which they teach their subject, and they enjoyed the chance to be creative in their approaches to subject pedagogy. This was apparent from the end of session discussions (teachers valued both the chance to meet up together *and* to have time to develop their ideas further within their own departments), and from the fact that nearly all of the teachers involved in phase 1 of the project had wanted to be involved in phase 2.

The end of workshop discussions included (unsolicited) comments about other recent experiences of ICT courses, and in particular, unfavourable comparisons were made with participants' experiences of New Opportunities Fund (NOF) training in the use of ICT.

POTENTIAL SIGNIFICANCE OF THIS CASE STUDY

We are aware of the limitations of this project, in terms of sample size and geographical specificity (the project took place within one 'county' authority in the UK). However, there are some tentative hypotheses which might be explored further in other school systems and contexts.

- **Possible advantages of moving away from 'coverage' models of ICT competence**

The outcomes of the project suggest that allowing teachers to pursue particular facets of ICT in some depth may be more productive than putting them through 'general' training courses. As one teacher remarked, '*It's better to just do one thing that makes a real difference to your teaching... then you'll go on to do other things as well.'*

The attempt to break ICT down into a list of technical competences, which has been a feature of curriculum specifications in teacher education in the UK (see, for example, DfEE, 1998), has tended to reduce learning about ICT into a chore and a burden, rather than something that is intrinsically interesting, and which can lead to exciting practical outcomes when a teaching artefact or activity is produced rather than the ticking off of a technical competence. In the words of Dickinson *et al.* (2001: p. xiii).

In training, beginner teachers are monitored on their achievement of 'standards'. These are discrete 'outcomes' statements, that closely resemble a long-discredited behavioural objectives model, and are so numerous as to be unworkable. The danger is that a system of this kind produces mechanical, rule bound assessment, in which monitoring against discrete statements supplants teaching towards understanding. A merely 'accounting' assessment against such standards can mean that real understanding of complex practices essential for effective teaching in the long term is discounted in favour of simplistic and low-level short term procedures.

One of the most important lessons to learn from UK experience of trying to get teachers and pre-service teachers to use ICT in their teaching is the abject and decisive failure of 'coverage' models of ICT competence: the idea that the more competences stipulated, the better teachers will be.

- **Possible advantages of letting teachers choose which facets of ICT they wish to explore**

Another of the factors which we believe had a positive effect on the outcomes was that teachers chose their own learning agenda in ICT and had ownership and control of their work. The government's own Performance and Innovation Unit pointed out that 'driving through policies with an assumption that the main players are the problem rather than the solution is usually a recipe for failure' (Performance and Innovation Unit, 2001). Many of the interventions in ICT policy which did more harm than good (Haydn and Barton, 2007) derived from a 'low trust' rationale which was based on the assumption that unless subjected to pressure, 'training' and testing, teachers and pre-service teachers would not engage positively with new technology. Our project proceeded on the assumption that most teachers want to teach well, and are interested in anything which might help them to do this. Another lesson to be learned from the recent history of ICT in teacher education in the UK is that there are advantages to be derived from engaging in dialogue with the teaching force about which forms of investment in ICT they would find most helpful. Recent research on technology planning and funding in K-12 public schools in Florida notes 'a significant increase in parent, administrator, teacher and student involvement in the technology planning process' (Ritzhaupt *et al.*: p. 1). It might be noted that lack of consultation with stakeholders in the UK is not limited to ICT policy: the government's 'Building schools for the future' project has been criticised for not asking teachers or pupils what they want from a school building, with some schools having been built without any form of leisure space for pupils because of this (Davies, 2008). Not consulting closely with those directly involved is an obvious form of unintelligent policy making. The idea that teachers are the problem rather than the solution to ICT integration is unhelpful, given that in the long run, it teachers who will have to make high quality ICT integration happen.

The experience of this project suggests that there may be some advantages in letting teachers have at least some say in determining their own learning agendas for ICT, and that allowing them to pursue particular areas of interest and enthusiasm may be more beneficial than attempting 'across the board' training, which may be broader but more shallow in effect. Although this may result in some 'black holes' in their areas of expertise, there would appear to be significant advantages in terms of their motivation, engagement and sense of professionalism.

We believe that the absence of a 'performativity' climate of targets, audits and testing (Ball, 2003; Elliott, 2001) also had a positive influence on teachers' commitment to the project. The fact that over 80% of teachers involved in the project chose to continue in the second year extension to the project, and the overwhelmingly positive comments about participation in the project are in stark contrast to teachers' responses to the £200 million 'New Opportunities Fund' government programme for teacher education in ICT (Leask, 2002, Ofsted, 2002, Preston, 2004, Haydn and Barton, 2007).

- **Possible advantages to developing more collaborative models of teacher development in ICT**

One of the strongest strands in the evaluation of the project was the enthusiasm for working collaboratively with other teachers from the same subject discipline. Although teachers valued having some time to work on their own to develop their ideas and capabilities in ICT, in the evaluation stage of the project, they were at pains to stress that they valued the combination of time to get on with their own ICT priorities, and time to come together and share ideas with others.

Several teachers talked about the impetus to try things out which derived from talking together, looking at *'other people's stuff'* and discussing ideas, examples and possibilities. In spite of the absence of intermediate meetings or contact, the combination of some time for teachers to just get on with it and work on their own, and some time to get together and share ideas appeared to have worked well. All the teachers who had taken part in the project had enjoyed being involved, exploring creative approaches to ICT use, and coming together to present their work and look at the work of others. Some teachers said that they had been 'inspired' by talking and working with colleagues.

This preference for working collaboratively echoes the sentiments expressed by teachers in other recent studies in the UK (see, for example, Cordingley *et al.*, 2004, Barton and Haydn, 2006).

- **Possible advantages of giving teachers more time to explore the use of ICT in subject teaching**

As long ago as 1975, Professor Lawrence Stenhouse argued that 'the most serious impediment to the development of teachers as researchers – and indeed as artists in teaching – is quite simply shortage of time' (Stenhouse, 1975: 111). The outcomes of this project suggest that in the area of developing teachers' use of ICT in subject teaching, simply providing support for teachers, in the form of time to explore the potential of ICT, to meet together to discuss ICT in subject groupings, and freedom to focus on their preferred ICT agendas, may be a more effective way forward than prescribing lists of required competences and providing generic 'training' type courses. This goes against the grain in an era characterised by 'top-down', centrally directed national strategies, high levels of accountability and auditing of teachers (Ball, 2003).

Although the amounts of time made available for teachers to explore ICT were quite modest, they spoke of this time as (in the words of one teacher) *'an incredible luxury'*. One of several comments which suggested that the project had transformed departmental use of ICT came from a teacher who had used the time to set up a departmental revision website:

'Our school is well resourced in terms of equipment, but the funding bought us time. We feel that time is one of the crucial components teachers lack because of the extensive workload. We were able to take three days to plan and design the (web) sites, something that would have been otherwise impossible without financial backing. The web site has been a big success, it has had a massive impact on the department and the pupils and it has led us on to develop other things.'

It should be emphasised that it is not primarily a question of simply increasing the *amount* of time that teachers spend deliberating on the integration and development of ICT into their subject teaching, it is rather a matter of improving the *quality of thinking and discussion* that is a product of the time spent reflecting on ICT use. The outcomes of this project suggest that the quality of that thinking and discussion is likely to be enhanced if teachers have some control over the facets of ICT that they choose to focus on. This applies not just to the ways in which teachers choose to explore particular ICT applications, but their thinking about the whole concept of e-learning – how they might improve their website or use of websites, how effectively they use the institution's virtual learning environment, whether to explore the use of wikis, blogs or other Web 2.0 applications and so on. The key factor is that they are likely to make the best decisions about what elements of e-learning to prioritise, and as it is their choice, they are more likely to sustain their commitment to developing their practice in the area chosen.

A MODEST PROPOSAL

The past few years have seen a modest resurgence of interest in the idea of 'teacher-centred' professional development in the United Kingdom, with initiatives such as the Networked Learning Communities initiative (NCSL, 2004), the 'Best Practice Research Scholarships' sponsored by the Department for Education and Skills (DfES), and the idea that teachers across schools will work collaboratively to improve teaching and learning outcomes (DfES, 2005a). However, in terms of overall investment in continuing professional development for teachers in England, the emphasis has been on centrally directed programmes and 'strategies' which are 'rolled out' and 'delivered' across the system as a whole (see, for instance, DfES, 2005b, DfES, 2005c). 'Bottom-up', teacher led professional development remains a very marginal and poorly funded strand of education policy. Given the disappointingly sluggish and meagre outcomes of big government programmes to promote teachers' use of ICT, and the modest but encouraging outcomes from projects such as this, the potential of giving teachers time to think about integrating ICT into their practice, and to talk with subject colleagues about developing their use of ICT, is an approach which may be worth exploring more extensively. It is possible that simply providing teachers with time to think, and to talk to each other, is a comparatively cost-effective form of developing a technologically empowered teaching force.

REFERENCES

Ball, S.J (2003). The teacher's soul and the terrors of performativity. *Journal of Education Policy, 18*(2), 215–228. doi:10.1080/0268093022000043065

Barton, R., & Haydn, T. (2006). Trainee teachers and 'impact' learning. A study of trainees' views on what helps them to use ICT effectively in their subject teaching. *Journal of Computer Assisted Learning*, (22): 257–272. doi:10.1111/j.1365-2729.2006.00175.x

Blair, T. (1995). A week in politics. *Channel, 4*(February), 18.

Blair, T. (1997). *Foreword in Connecting the learning society*. London: DfEE.

Clarke, C. (1999). Quoted in *The Guardian*, March 9.

Cochrane, P. (1995). Quoted in *The Times Educational Supplement,* 23 June.

Cockburn, A., & Haydn, T. (2004). *Recruiting and retaining teachers: understanding why teachers teach.* London: RoutledgeFalmer.

Cohen, M. (1999). Turning teacher into a computer. *The Guardian*, October 12.

Cordingley, P., Bell, M., Rundell, B., Evans, D., & Curtis, A. (2004). How does collaborative continuing professional development for teachers of the 5-16 age range affect teaching and learning? EPPI Centre, Institute of Education, University of London. Retrieved December 6, 2008 from http://eppi.ioe.ac.uk/cms/Default.aspx?tabid=133&language=en-US

Davies, C. (2008). Teachers denied a say in school design. *The Observer*, September 21.

Department for Education and Skills. (2005a). *White Paper: Higher standards, better schools for all*. London: DfES.

Department for Education and Skills. (2005b). *Secondary National Strategy for school improvement: pedagogy and practice, teaching and learning in secondary schools*. London: DfES.

Department for Education and Skills. (2005c). *Secondary National Strategy for school improvement: leading and coordinating CPD in secondary schools*. London: DfES.

DfEE (Department for Education and Employment). (1998). *Teaching: high status, high standards. Requirements for courses of initial teacher training, Circular 4/98*. London: DfEE.

DfES (Department for Education and Skills). (2002). *Qualifying to teach: professional standards for Qualified Teacher Status and requirements for initial teacher training.* London: DfES.

Dickinson, A. Gordon, P. & Lee, P. (2001). *Raising standards in history education,* London: Woburn Press.

Elliott, J. (1991). *Action research for educational change.* Buckingham, UK: Open University Press.

Elliott, J. (2001). Characteristics of performative cultures: their central paradoxes and limitations as resources for educational reform. In D. Gleeson & C. Husbands (Eds.) *The Performing School*, (pp. 192-209). RoutledgeFalmer: London and New York.

Harrison, C., Comber, C., Fisher, T., Haw, K., Lewin, C., Lunzer, E., et al. (2002). *ImpaCT2: The impact of ICT on pupil learning and attainment.* Coventry, UK: BECTa.

Harrison, S. (2003). The use of ICT for history teaching: slow growth, a few green shoots. Findings of HMI inspection 1999-2001. In T. Haydn and C. Counsell (Eds.) *History, ICT and learning in the secondary school* (pp. 38-51). London: RoutledgeFalmer.

Haydn, T. (2004). *The use of ICT in history teaching in England and Wales, 1970-2003*, PhD thesis, University of London, London.

Haydn, T., & Barton, R. (2007). 'First do no harm': factors influencing teachers' ability and willingness to use ICT in their subject teaching. *Computers & Education*, *51*(1), 439–447. doi:10.1016/j.compedu.2007.06.001

Hunt, D. (1995, March 27). Quoted in 'Stemming the decline in science.' *The Guardian*.

Leask, M. (2002). *Training for teachers and school librarians in the use of ICT*. London: TTA.

Millett, A. (1998)... *Quoted in The Times Educational*, (May), 22.

National College of School Leadership. (2004). *Networked Learning Communities: learning about learning networks*. Cranfield, UK: NCSL.

National Foundation for Educational Research (NFER). (2000). *Who would be a teacher? A review of the factors motivating and demotivating prospective and practising teachers*. Slough, UK: NFER.

Nichol, J., & Watson, K. (2003). Editorial: Rhetoric and reality - the present and future of ICT in education. *British Journal of Educational Technology, 34*(1), 1–6.

Noss, R., & Pachler, N. (1999). The challenge of new technologies: doing old things in a new way or doing new things? In P. Mortimore (Ed.) *Understanding pedagogy and its impact on learning,* (pp. 195-211). London: Paul Chapman.

Ofsted (Office for Standards in Education). (2002). ICT *in schools: the impact of government initiatives*. London: Ofsted.

Ogborn, J. (2000). Shaping the future: revitalising physics Education. In A. Morris (Ed.), *Post-16 physics for tomorrow.* Bristol: Institute of Physics.

Performance and Innovation Unit (UK). (2001). *Better policy delivery and design*. London: DfEE.

Phillips, J. (2002). The dinosaurs didn't see it coming, but historians had better: computer aided activities in the history classroom. *History Computer Review, 18*(1), 27–36.

Preston, C. (2004). *Learning to use ICT in classrooms; teachers' and trainees' perspectives: an evaluation of the English NOF training programme 1999-2003*. London: MirandaNet and the TTA.

Reynolds, D., Treharne, D., & Tripp, H. (2003). ICT: the hopes and the reality. *British Journal of Educational Technology, 34*(2), 151–167. doi:10.1111/1467-8535.00317

Ritzhaupt, A., Hohlfeld, T., Barron, A., & Kemker, K. (2008). Trends in technology planning and funding in Florida K-12 public schools. *International Journal of Education Policy and Leadership, 3*(8), 1–17.

Selwyn, N. (2003). Exploring patterns of computer use in schools. *Welsh Journal of Education, 12*(1), 75–97.

Smithers, A., & Robinson, P. (2000). *Attracting teachers: past patterns, present policies, future prospects*. Liverpool, UK: Carmichael.

Stenhouse, L. (1975). *An introduction to curriculum research and development.* London: Heinemann.

Zhao, Y., & Frank, K. (2003). Factors affecting technology use in schools: an ecological perspective. *American Educational Research Journal, 40*(4), 807–840. doi:10.3102/00028312040004807

Zhao, Y., Pugh, K., Sheldon, S., & Byers, J. (2002). Conditions for classroom technology innovations. *Teachers College Record, 104*(3), 482–515. doi:10.1111/1467-9620.00170

Chapter 4
Dealing with Affective Needs in E-Learning:
Contrasting Two Cases, in Two Cultures

Yi-Ching Jean Chiu
Wenzao Ursuline College of Languages, Taiwan

John Cowan
Edinburgh Napier University, Scotland

EXECUTIVE SUMMARY

This chapter presents and contrasts descriptions of two cases of online affective support provided to support students engaged in higher level learning tasks. The cases are set in different cultures, centre upon different intended learning outcomes, and follow different tutorial styles. One (Eastern) tutor acted as a "shepherd leader" in response to needs arising in the Confucian Heritage Culture as the teacher promoted critical thinking, according to the Western model. The other (Western) tutor provided Rogerian facilitation of reflective learning journals, kept by students seeking to develop personal and professional capabilities. In both styles, affective support features strongly. The cultural and pedagogical comparisons between the cases have proved useful to the writers. These distinctions together with the similarities between the two online styles emerge in the comparisons.

BACKGROUND

Following the energetic and thorough consideration of the cognitive domain by Bloom and his colleagues (Bloom et al, 1956), the natural desire to move on to deal similarly with the affective domain ran into considerable and well-documented problems (Krathwohl et al, 1964). The result was that, for the next 40 years, affective outcomes in higher education received scant attention. Indeed a sampling of popular texts used in the training of UK university teachers (Cowan, 2005), found that only two made mention of the affective domain or of affective outcomes. One devoted a marginal sentence to this topic (MacDonald, 1997, in Boud et al, 1997); the other (Heywood, 2000) referred to the almost unique practices of Alverno College (Mentkowski, 2000).

Recent publications addressing the importance of affective outcomes (Robinson & Katalushi, 2005) concentrate usefully and worthily on the hitherto neglected and important area of values and ethics

DOI: 10.4018/978-1-60566-942-7.ch004

(Barnett, 2003). But direct consideration of affective learning outcomes and learning needs, and of tutorial approaches to respond appropriately and effectively to them, are difficult to unearth (Cowan, 2005, p 161). Affective outcomes having been neglected in this way and to this extent, affective learning needs (and supportive teaching) feature only in writings which report pastoral concerns and their resolution. Fry et al (2000), while including a generous chapter on support and guidance by Wallace (2000), make guidance, rather than the fundamental support of learning and development, the selected focus of their advice. Biggs and Tang (2007) index no mentions of "affective", "needs" or "support". Eysenck and Piper (1987), in the closing pages of an authoritative although early text (Richardson et al, 1987), made the telling comment that "cognitive psychologists rarely consider motivational or emotional factors at all, a factor which one of these writers had recently bemoaned."

However, interest in affective needs has recently awakened, or re-awakened. Even as the final draft of this chapter was being refined by its writers, Huyton's (2009) paper "support and development needs of HE tutors engaged in the emotion work associated with supporting students" cites Beard's work in helping students to "develop a better understanding of the energies and challenges involved in coming to terms with studying" (Beard et al, 2007, p250). While Huyton (2009) concentrates on the emotional well-being and support of tutors, she usefully reminds readers of the earlier work by Earwaker (1992), stressing, as do the present writers, that learning support should be based on a pedagogy which recognises and takes account of the effects on learners of personal change. Earwaker (2009) drew a firm distinction between counselling and Rogers' (1980) position of providing learning support which involves both the emotional and the cognitive, and the act of professional counselling.

The present writers remain committed to Rogers' ideal of unified learning in which the cognitive, experiential and affective are melded (Rogers, 1980). Therefore they do not pursue here the concept of "therapeutic pedagogy" (Ecclestone, 2004: p118). Rather do they offer their cases to this anthology aware that they have been engaging with a relatively neglected topic and challenge.

Circumstances brought the two writers together as strangers, when John was reviewing a paper by Jean (Chiu, 2009) prior to publication. They found common ground, and began to correspond electronically, contrasting the cultures in which they supported their students' development in e-learning and online interactions. It became clear that they shared concerns regarding their students' affective needs, although following somewhat different approaches to providing support. Jean's pedagogy centred on shepherd leadership within a learning community (Chiu, 2007); John followed a Rogerian approach (Cowan, 2004) on a distinctly individual basis.

This chapter is a joint response to the request for cases dealing with e-learning. The writers are both firmly committed to the view that effectiveness in the provision of e-learning depends to a great extent on the prompt identification, and effective resolution, of students' affective needs. They therefore relate their contrasting experiences in two distinct cases, identify the issues and lessons emergent from them, and report their responses, present and intended. Each narrative account is assembled under sub-headings following the same sequence as is used in the comparison. in accordance with the somewhat detailed sub-headings which will be used in the comparative section, "ANALYSIS OF THIS COMPARISON", to enable the reader to compare the writers' experiences. The sub-sections for each case are titled accordingly.

SETTING THE STAGE (JEAN'S CASE) DEVELOPING CRITICAL THINKING IN AN EFL CLASS IN TAIWAN

The Context and the Cultural Setting

The Confucian Heritage Culture (CHC) is characterized by a social norm which is not readily supportive of the public verbalizing of thoughts, or of challenging others face-to-face. Yet both of these overt cognitive activities are, of course, important in the practice of critical thinking. In the paper entitled "Breaking the culture of silence," Akindele and Trennepohl (2008) - who were writing of a Japanese context not to be confused with Botswana in Africa - found that 92% of their CHC students perceived that the western communicative approach "broke the culture of silence," which forced quiet students out of their comfort zone. More than half (63%) approved the approach, leaving 37% opposed the approach, preferring exams and quiet, individual, learning.

Jean's teaching approach evolved gradually, as she pursued her online facilitation of verbalized critical thinking according to the Western model, for Taiwanese students of English as a Foreign Language (EFL). From 1999 to 2006, she tried different learning and teaching approaches, ranging from online newsgroups to E-course discussions of media related issues. The ensuing interactions were frequent though lacking evidence of critical thinking; and the overload of consequent e-mail messages turned her towards investigating other approaches and software (Chiu, 2001).

In view of the continuing dearth of thorough critical thinking within her students' interactions, Jean wondered (Chiu, 2002) about the potential of online reading and personal journalling for the development of her students' critical reading skills. Jean's subsequent doctoral study and teaching innovations (Chiu, 2006) therefore attempted to explore, culturally and pedagogically, suitable blended models to promote critical verbalization by her EFL students.

CASE DESCRIPTION

The Underlying Pedagogy and Rationale for Jean's Case

Jean was now following the culturally appropriate practices commended by McCormick and Davenport (2004). This they had termed "shepherd leadership", following the practice of shepherds in the field, who build up affectionate relationships with their flock, knowing each one on an individual basis. Academic shepherd leaders, similarly, will know their individual students' backgrounds, and their behaviour patterns. They will reach out to any "lost" (silent) students by cognitive modelling, and recruit disciples from their class to exercise secondary leadership.

Resolving affective needs has a vital role to play in the development of critical thinking for students from a CHC background. This cultural upbringing prepares them to defer, to refrain from expressing disagreement or challenges in public or formal situations, and especially so in exchanges with a teacher. However, Merryfield (2003) found that engagement in online interactions can assist non-Anglo-Saxon students to break through "the veil of silence". Hence Jean opted to nurture critical thinking (mainly) online and to foster affective needs by adopting the role of a shepherd leader (McCormick & Davenport, 2004). She therefore chose that she would not be the teacher of the course, but rather would be the "shepherd leader facilitator" (Chiu, 2007), aiming to minimize the periods of CHC-originating silence, which occasioned failure to engage with critical thinking.

The Basic Challenge for the Students

The students in this class were confronted by four major challenges (Chiu, 2009):

- To free themselves from CHC customs, practices and inhibitions;
- To think critically, according to Western norms;
- To feel free to question, disagree and criticise – and to express themselves accordingly;
- And, of course, to first identify valid grounds for questioning, disagreeing or criticising.

The Reasons for the Students' Silence

Jean already knew that she would encounter various reasons rendering her students unwilling to "speak out" on line (Sun, 2000; Merryfield, 2003). These included:

- The inhibiting influence of Confucian Heritage Culture (CHC);
- Difficulties in seeing "critical thinking" as potentially constructive;
- Embarrassment with regard to exchanges involving loss of "face";
- Low self-efficacy;
- Lack of confidence in risk taking;
- Lack of practice in verbalising thoughts and questioning;
- Worries about English grammar and writing style;
- Becoming silent after being embarrassingly engaged (either as challenger or challenged) in a confrontation (Merryfield, 2003)

The Main Activity for the Students

In Jean's programme, the e-learning interaction patterns within a CHC context embodied a three-phase process, which she has described according to the metaphor of nurturing growth in a garden: budding, blossoming and fruiting (Figure 1). She illustrates this figure here with examples taken from a recent presentation of her programme.

Budding: The first of seven e-learning discussions, planned to occur during one semester and commencing in week 7, was designed to allow students to become familiar with the online environment, while they were "budding". It was also the warm-up stage used to cultivate student-student and student-teacher interactions – while moving from the face-to-face situation towards an online environment. This phase concentrated on providing affective, cognitive, technical and pedagogical preparation (Chiu, 2008) to help to minimise the teacher's authority, and to maximise the impact of the e-learning environment, in order to make the "budding" of autonomous critical thinking more possible.

In the second online discussion, a face-to-face class debate was carried forward. One student posted a message on her debate stance. The negative side debater followed suit, but was challenged by Jean for lack of evidence in her claims. That student did not reply, but a teammate volunteered to present evidence to support her. Thus teammates helped one another in finding and presenting credible evidence to validate viewpoints. This budding phase helped the students to prepare themselves online for the next phase.

Blossoming: After the warm-up stage, the participants moved on to engage actively in critical discussion of controversial issues. The phase of blossoming transmuted at times to overt confrontation. Prompted by reminders, face-to-face grouping dynamics transported directly online. A student who had previously been identified as a "silent girl", broke the cultural wall (Chiu, 2009) in week 10, to question the assumptions behind

Figure 1. Diagram of the process of the cross-cultural online experience

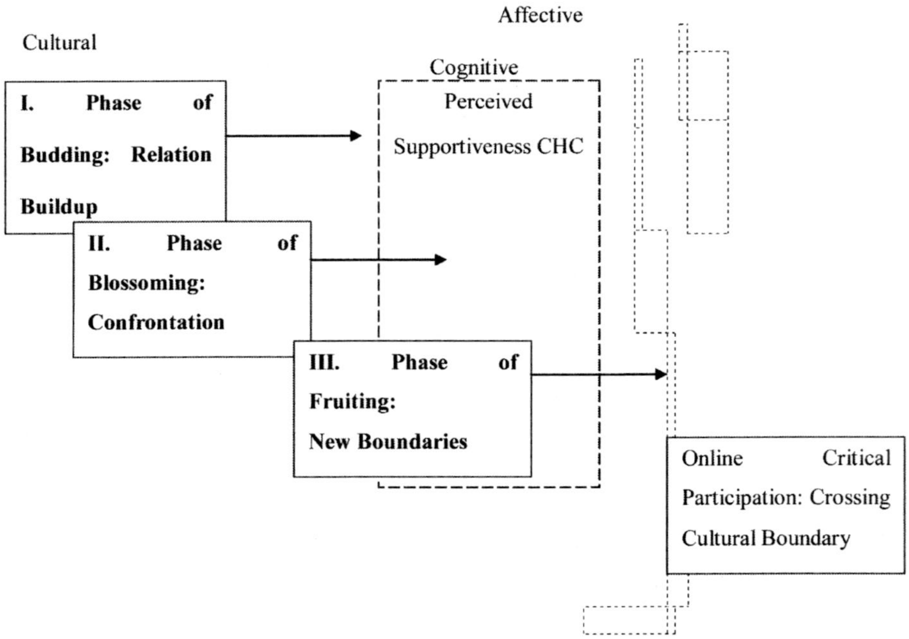

an issue. Three quiet students cross-referenced one another. Additionally, a fourth member of the group firmly refuted her group mate's stance. In this stage, the informal and courteous face-to-face small group dynamics seemed to have been transformed, with appropriate cognitive support and encouragement, to being "critical".

One sharp online confrontation led to a call for "Peace", by which a mediating student sought to promote cultural harmony in accordance with CHC norms. A follow-up message then offered affective support for the challenged one, with an appreciative response. Students had now dared to challenge one another, had risked violating the CHC "wall", and had even challenged the facilitator. These were thus acts of blossoming, leaping ahead in the public expression of confrontation. However (and typically), in this incident challenger and challenger withdrew from subsequent exchanges for different cultural reasons. Each required the energetic support of their shepherd leader to overcome their cultural embarrassment, to regain their confidence and, only after some time and facilitative effort, to again find willingness to participate.

Fruiting: After confrontation, the final phase of fruiting culminated and reached fulfilment in the full e-learning discussion experience. Affective messages had smoothly followed the previous three discussions (Chiu, 2009). A quiet student now broke her silence, and in effect challenged a senior who was more outspoken, both in class and online. However, in this her first challenging contribution, the quiet student found it was more empowering to speak appropriately to someone who was already in line with her viewpoint than to directly confront an opposing senior. This oblique confrontation adjusted the cultural boundary from open disagreement towards indirect disagreement, avoiding unpleasantness and with only limited embarrassment. This normalized the confrontation as part of a democratic e-learning culture, as a result of the cognitive, affective, and pedagogical support and leadership which had been offered.

Jean had found that, if their critical thinking was to be effectively developed, her students would need:

- cognitive, affective, pedagogical and technical support;
- face-to-face small group support before the online interactions;
- modified (preparatory) debates in face-to-face situations;
- personal, often face-to-face, encouragement;
- a friendly teacher's "shepherd leadership" throughout (Chiu, 2007).

The Students' Affective Needs

Affective needs consequently arose for Jean's students, due to their CHC upbringing which had established commendable values and desirable behaviour patterns apparently incompatible with engaging actively in Western critical thinking. In breaking from this precedent and the pattern of cultural expectations, the students needed support and encouragement – constantly, and from the outset. Hence the focus for the supportive activity of shepherd leadership concentrated primarily on affective needs and anxieties, supporting students in resolving these needs in pursuit of the determined cognitive goals of the programme.

The Basic Challenge for the Tutor

Educators in the West (Kneser, Pilkington, & Treasure-Jones, 2001; Pilkington, 2001) identified the desirability for moderating tutors in e-learning discussions to use Socratic strategies, and to play the role of the devil's advocate as frequently as possible (Walker, 2004). They found teachers' online "challenging" and "probing" to be effective in eliciting students' further clarification of their arguments and thinking. However Jean had then to discover as a matter of some importance whether or not this confrontational model would fit into, and promote, a worthwhile and vital e-learning discussion in a CHC culture, where probing, challenge and even arguing in formal classes are alien.

The shepherd leader can influence students in the long term through her relationship with them, by cognitively modelling the type of thinking and interaction into which she seeks to lead them. Cognitive modelling is a gradual process, which is most effective when it brings about modification of thinking and behaviour by example, and by "immersing" students in a different practice, in this case precluding fragmented adoption of the Western framework.

In so doing, the tutor-shepherd can in due course lead some of her flock to take up secondary leadership on their own part. She can do this by encouraging and training student leaders to participate actively, and thus to influence other students by their example. As already mentioned, Merryfield (2003) found online interactions by student leaders an effective way to help remove the "veil of silence" - particularly on the part of students who are not native English speakers. An active student leader can then help the rest of the class to dismantle cultural and interpersonal barriers to challenging different thoughts online (Merryfield, 2003).

Facilitating Consideration of Weaknesses in Students' Critical Thinking

The facilitator's content-centred facilitation took the form of clear but gentle reminders, incorporating prompts relating to such questions as:

What is the stance/opinion of the author of the article you quote?

What is the assumption behind this stance? Is the evidence credible? Can it be double checked with other sources? What is the counter opinion?..... and your evaluation, after logical thinking? Before you state your stance, present facts or post the extra links which you found online to build up your understanding about this complex issue.

Such prompts, which were not all asked in one communication and were always expressed in mild terms, nevertheless provided reminders of the need to critically evaluate the sources, the stance and the assumptions of each student contributor.

Declaration of Criteria and Standards

The above style of facilitation and cognitive modelling implicitly and gradually conveyed criteria and standards, in respect of the need to present facts and to cite sources, and to consider these objectives, before coming to, and expressing, a conclusion (Yu & Chou, 2004).

The Nature of the Tutor/ Student Relationships

As Watkins and Biggs (2001, p282) convincingly describe, the role of the CHC teacher goes beyond that of being a lecturer and an authority in the classroom. It extends to the moral role of a caring "parent" – which is why students feel a "collectivist obligation to behave within the socially accepted ways." For Jean, in an informal role, this cultural expectation implied her placing of stress on informal contacts (Ho, 2001), providing support and encouragement orally or by supportive smiles, and generally by building up strong and warm personal relationships, to prepare students for cognitive challenge (Walker, 2004).

The Pedagogical Style of the Tutor's Initiatives

Jean's actions were thus primarily pro-active. She initiated and actively guided from the outset, modelled behaviour, trained, (Yu & Chou, 2004), and reached out for critical thinking development (Kneser et al., 2001; Pilkington, 2001).

The Students' Interactions with Peers

Interaction with peers was an integral and predominant part of the students' activity, productive or otherwise (Merryfield, 2003).

The Focus for the Tutor

Jean had deliberately abstained from a teaching role. Although many of her interjections related to the students' handling of the module and task content, she took her primary function as being the provision of affective and cognitive support (Chiu, 2009).

CURRENT CHALLENGES

The next immediate challenge for Jean is to build in face-to-face and online discussions, and foster students' critical thinking according to the Western norm, provided online with modelling by a British academic recruited to this collaboration. Students will learn to cross their cultural boundaries with the affective support provided by the tutor and the British academic, prior to the critical thinking activity.

SETTING THE STAGE (JOHN'S CASE) DEVELOPING REFLECTIVE THINKING IN A WESTERN CONTEXT

The Context and the Cultural Setting

John's classes have included a majority of students of Scottish origin, with a few English students and (later) some from the Middle and Far East.

John has long been committed to the promotion of student-centred learning and the development of generic capabilities (Cowan, 1987). He pioneered large-scale independence in undergraduate learning in the UK (Cowan et al, 1973). He offered autonomy to students, first in their choice and use of a suitable learning method and rate of study (Butts et al, 1976), then in determining the content of parts of their studies (Cowan, 1978), and finally in self-assessment (Cowan, 1984; Boyd & Cowan, 1986; Cowan, 1988). More recently he has concentrated on the promotion of reflective practice (Cowan, 2006a), as a means of purposefully developing personal and professional capabilities.

John has depended upon several types of reflective activity. The first (then without e-learning) involved learning journals concentrating on reflection-on-action (Schön, 1987). This was a radical programme, seeking thorough development of interdisciplinary capabilities at first year undergraduate level (Cowan, 1987). It established the pattern for what he subsequently developed on-line. Following social-constructivist workshops focused on key generic abilities, each week his students reflected in a personal journal on their answers to the question "What have I learnt about learning, or thought about thinking, which should make me more effective next week than I was last week?" Each weekly journal was handed in, as hard copy, for speedy (and confidential) facilitative comment by John or a colleague, who nudged or challenged the journallers to consider further, and then to make their own decisions.

Thereafter John went on to facilitate students' electronic reflective journals in a blended learning degree programme (Weedon & Cowan, 2001, 2002). These journals called for what John has called "reflection-for-action" (Cowan, 2006a). Each week, students studying an Enquiry Skills module were expected to write reflectively, asking themselves how best to tackle the next cognitive challenge in their self-directed Enquiry project, and trying to answer as they wrote in "stream of consciousness" writing. Many students increasingly devoted almost as much of their online journalling time to relevant affective needs as they did to the metacognitive process analysis - which concentrated on thinking about how to tackle the various aspects of enquiring. Once again, John and colleagues pinpointed the need for consideration of affective and cognitive issues, but did not suggest solutions.

CASE DESCRIPTION

The Underlying Pedagogy and Rationale

John expected his students to initially find reflection a strange and demanding activity, and one with outcomes which might not emerge for them in the short term. For that reason, he and his colleagues devoted a great deal of effort to introductory activities, using appropriate examples (Cowan, 2006a), to ease and encourage the students' induction.

This pedagogy is constructivist (Cowan, 2006a, p.46, Fig 4.1), though preliminary group work can be structured (Cowan, 2006a, p.49, Fig 4.2) to be social-constructivist (Cowan, 2004; Francis & Cowan, 2007). In the examples in this case, however, John followed a primarily constructivist pedagogy, which usually passed through several iterations of ascending impact (Cowan, 2006a, p. 53, Fig 4.3). In such personal reflection, social-constructivist interaction other than with the tutor is necessarily slight and inappropriate.

The Basic Challenge for the Students

Journallers were soon quite frank, privately, about their needs and feelings - regarding their lack of confidence in dealing with unfamiliar course tasks, their difficulties in identifying possible options, their resentment at times on being asked to take responsibility for tasks which they saw as their teachers' responsibilities, and their unwillingness to reveal their assumed inadequacy to teaching staff, or even to fellow-students.

The Reasons for the Students' Reticence

John's students were reluctant from the outset to reveal to anyone else the fine detail of their highly personal cognitive thinking, their articulation of that, and their success – or lack of it – in applying their consequent generalisations. Hence he opted to promote constructive reflection in personal (and almost private) journals, which were not to be assessed and whose content remained strictly confidential. Following Gibbs (1988), he suggested that writers might begin by writing about their feelings in reaction to the recent experience or forthcoming task which was to feature in their reflections.

The content of many of the journals became intimately personal – in an academic sense. Students analysed their satisfaction with approaches which had developed effectively for them and which they were applying elsewhere, to good effect. Many entries were specifically affective in origin, stemming from the consequences of poor choices and difficult decisions, and admitting reluctance to discuss these problems with peers or even with tutors. Increasingly, in addition, most journals became metacognitive in nature, as writers self-reliantly sought, analysed and actively experimented with their own solutions. Open frankness in such matters is not characteristic in overt conversations between Scottish students or with their tutors.

It was always notably rare for a student to mention or respond to a comment on a journal entry, whether cognitive or affective in focus. It seemed to be firmly understood and accepted by both parties that journal content and tutor's comments remained personal and confidential, both ways. This confidentiality (and accompanying reticence) was akin to the constrained communication in a small community when a villager meets their parish priest midweek, and neither will make reference to what was said when the priest was partly hidden, and they had spoken together in the confessional.

The Main Activity for the Students

Thus John's aim for his students was for them to develop and put to good use effective metacognitive skills through self-questioning; and thereby to enable them to enhance their higher level cognitive and interpersonal capabilities. He relied on facilitated and highly personal reflective thinking to lead to the development of relevant personal and professional capabilities of transferable usefulness and value.

The Students' Affective Needs

The affective needs which became apparent as these experiences progressed resonated with findings in John's Open University activities (George, 1996; George et al, 2004). The journallers needed to:

- establish a sound estimate of their abilities, and develop realistic self-efficacy;
- benchmark their study experiences, or discover whether a task was proving depressingly difficult because it was indeed difficult, or because they lacked the intellectual grasp with which to tackle it (Cowan, 2007);
- norm-reference their progress;
- avoid exposing themselves to frank and hurtful comments from peers or tutors.

These affective needs emerged for the individual concerned from the cognitive and interpersonal demands of the course. They were only sometimes, and almost grudgingly, admitted face-to-face to a tutor or peers. And yet they were needs with which students seemed able to cope to their eventual satisfaction in the semi-anonymous and non-judgemental context of their facilitated personal journalling.

The Basic Challenge for the Tutor

From these online experiences, John formulated three subjective impressions. The first was that the intimacy and frankness between journallers and commenter seems radically greater, and hence more effective, when this is their only working relationship with each other. John has discussed this deliberately virtual relationship (electronically) with former students. He has been told how much easier it is to be frank with yourself "when you don't feel that someone you know is looking over your shoulder". The somewhat dispassionate commenter is then "just a name at the foot of a computer screen".

Next, John has noted that open-ended feedback through routine university channels has often testified to the fact that the responding student had "trusted" John, during their year of contact. Given what Rogers (1961) and, later, Brookfield (1990) had written about the importance of students being able to trust their tutor, this aspect of the relationship seems an important affective e-learning need (and variable) which merits further attention (Cowan, 2006b), seeking especially to discover the features of online relationships which promote trust.

Finally, John has found that his students can feel the pressures of being solitary, even if they meet others face-to-face in class. This was apparent at times in the social content of the friendly e-mail cover notes which were increasingly sent with the journals as the attachments. John has found it useful (and appreciated) to display limited congruence (usually within his own cover notes) with the students' volunteered feelings.

Facilitating Consideration of Weaknesses in Students' Thinking

John and his colleagues followed a Rogerian approach (Rogers, 1961, 1969, 1980). They set out to empathise implicitly with the students' analyses of challenges and methods, and their expressions of feelings; and they endeavoured to accept them all with unconditional positive regard. To provide an effective framework for cognitive development, they relied on carefully planned course activities to initiate reflective thinking, linked to their blunt and very directed facilitative comments, albeit on Socratic lines, as questions worthy of consideration. Facilitation with regard to the associated affective needs was thus mostly indirect, but tangible.

Declaration of Criteria and Standards

John soon decided (Cowan, 2006a, p22) that every time he called upon a student to share their personal reflections with him, he should thereafter share with them his own current reflections, drafted to the same remit. This arrangement has been the subject of much volunteered appreciation from students. It probably constitutes cognitive modelling; but, for John, it is primarily an attempt to establish an equitable relationship with those whom he asks to expose their intimate thinking to him.

Criteria and standards were thus made explicit in the commenters' facilitative questions, and were then also implicit in John's own shared journalling. Thus criteria and standards were, in effect, being modelled collegially, having been tabled explicitly and for discussion at the outset. It was left to the students to determine and adopt these for themselves, and to apply them to their own reflective thinking.

The Nature of the Tutor/ Student Relationships

Initially, and for a long time thereafter, commenting was detached and impersonal. In some cases, associated informal exchanges began to develop within the cover notes which sent or returned reflections as attachments. Consequently, by the end of a year of contact, a few significant virtual relationships had developed and were to continue on almost personal terms, for years after students had completed the programme which brought them together with their commenter.

The Pedagogical Style of the Tutor's Initiatives

In contrast to the comparable account in Jean's pro-active case, John and his fellow commenters followed Rogers' principles and re-active practice (Cowan, 2006a, pp 200-203). With unconditional positive regard for the subject matter of students' journals, they empathised with the thoughts, priorities and feelings recorded by students. On a few occasions they volunteered congruence through a personal example or response. However, they firmly used questions or comments to point out inconsistencies in thinking, omissions from considered factors or possible outcomes, and neglect of implications. They did not offer suggestions about how the journallers should respond; they simply tried to highlight instances where the journaller's thinking might become more rigorous and searching. Subsequent evaluations have shown that students perceived the style, overall, as caring and supportive – despite the overt concentration on developing rigorous reflective thinking.

The Students' Interactions with Peers

To the knowledge of the commenters, the students had no relevant interactions with their peers. In the early days, before electronic working in a blended format and when the handwritten journals were returned physically to the students' pigeon holes, 90% of the class swarmed around, seized their journals, and stood aside on their own to read the comments quietly, without any observable conversation between students immediately thereafter.

The Focus for the Tutor

The commenting tutor sought above all to promote pertinent self-questioning on the part of each student. When the learning experience had been successful, it led to situations in which student journallers (rightly) declared that they had learnt to self-question. "John, I don't think I need you any more. I know the questions I still need to think about."

CURRENT CHALLENGES

In the face of ever-increasing numbers, the next challenge for John is to find ways to influence progress of relationships which will be discerned as personal in accordance with the effective practice described here, yet with numbers which preclude the arrangements already described.

ANALYSIS OF THIS COMPARISON OF THE TWO CASES

The writers have here reported their involvement with affective needs arising in respect of different educational aims, in different settings, in different discipline areas, and leading to different approaches to the teaching challenges. It seems useful now to highlight both similarities and comparisons. The writers follow a helpful suggestion from a reviewer, in presenting that comparison in tabular form. The sequence used is in accordance with the two case accounts, but the short summaries given are often derived with the corresponding item in these sections.

CONCLUSIONS AND FURTHER CHALLENGES

As the table of comparisons suggests, there are common grounds in the two cases of student affective and cognitive needs across cultural boundaries; yet these may call for contrasting proactive-reactive approaches within different cultural contexts.

Table 1.

Feature	Jean's Case	John's Case
The context and cultural setting	Developing critical thinking in an EFL class in Taiwan	Facilitating development of generic abilities, in Scotland
The underlying pedagogy and rationale	Shepherd leadership	Rogerian facilitation of reflective writing
The basic challenge for the students	To think critically according to predominantly Western norms	To probe thinking metacognitively, and apply the outcomes to professional development
The reasons for the students' reticence	Cultural unacceptability of expressing disagreement and challenging peers and the teacher	Embarrassment when expressing feelings, and awkwardness with the strangeness of metacognitive thinking
Main activity for students	Contributing in English to group discussions calling for critical thinking	Constructive personal reflection on experiences and challenges
The students' affective needs	Support and encouragement, especially when breaching CHC norms	Objectively judging self, and confronting that judgement without embarrassment
The basic challenge for the tutor	Students' reticence to speak out and publicly express disagreement in class	Students' reticence to think and work deeply and personally, and to commit thoughts and feelings to writing
Engaging with weaknesses in students' thinking	Early in the process; gentle prompting and individual encouragement	Early in the process; explicit though non-judgemental identification of questions to consider
Declaration of criteria and standards	Emerge as experience progresses,	Explicit from the induction, and indirectly from John's shared journals
The nature of the tutor/student relationships	Warm, supportive, and encouraging person-to-person relationships, often face-to-face	Detached, almost anonymous, facilitation, helping the student to be the best that they can be
The pedagogical style of the tutor's initiatives	Pro-active: guiding, modelling, training, reaching out, usually on an informal basis	Reactive: facilitating formatively on the basis of student's submitted and explicit reflections
The students' interactions with peers	Frequent, and central to the method	Infrequent, and incidental
The focus for the tutor	Affective Providing affective support, and facilitating critical thinking	Reactive and metacognitive Prompting deep thinking and self-questioning. Affective support implicit in empathy.
The next challenge for the tutor	Enhanced effectiveness in achieving desired critical outcomes, with some Western input	Making an already effective process cost-effective with larger numbers of students

The underlying pedagogies and rationales which have been described in the two cases in this chapter are:

- Shepherd leadership, for CHC students (Chiu, 2009)
- Rogerian facilitation of reflective writing, for western students (Cowan, 2006b)

The pedagogical style of the tutors' initiatives in these cases has focussed on being:

- Pro-active: in guiding, modelling and reaching out to CHC students on an informal basis (Akindele & Trennepohl, 2008; Brookfield, 1990);
- Re-active: facilitating on the basis of the students' submitted and explicitly private reflections (Cowan, 2006a).

In these reported cases, then, the nature of the tutor/student relationships has proved effective when it has been either

- Affectively supportive, and encouraging of person-to-person relationships which are often face-to-face, for non-Anglo-Saxon Students \(Merryfield, 2003; George, 1996; George et al, 2004; Chiu, 2009), or
- Detached and explicit facilitation according to induction and tutor-modelling (Cowan, 2006a), for western students, with embedded and implicit responsiveness to affective needs (Rogers, 1961, 1969, 1980). .

The writers have thus come to see that their approaches are complementary – and not contradictory. So they conclude and recommend that:

- **In activities designed to develop higher level abilities, affective needs are significant for student learning and development.**

Affective needs should therefore feature prominently on the agenda for those who aspire to provide effective student support.

- **The origin of many affective needs is to be found in the students' prior experiences.**

Recognition of cultural factors and responses to them, should be reflected in chosen supportive approaches. This will call for thought and ingenuity, by students as well as teachers, in dealing with individuality and especially in cross-cultural groups of international students, with their very varied affective needs (Merryfield, 2003; Watkins & Biggs, 2001).

- **An approach embodying empathy with, and unconditional positive regard for, student feelings, is effective in resolving affective needs.**

Empathy with, and unconditional positive regard for, student feelings, apparently provides an important key to unlocking and resolving affective needs. Developing empathy, as indicated in both cases, is easier said than done. It can thus usefully feature as a priority in staff development, as well as in educational research. Effort should also be devoted to future action research enquiries which may identify the effectiveness of such interventions, and their impact on the resolution of affective needs.

- **The writers' approaches as described in their two cases are complementary, cater for different groupings of affective needs, and are both important for learners.**

Both approaches should somehow receive attention in every learning and teaching situation. In the writers' cases, this will prompt Jean to further explore the potential of her current one-to-one interactions to cater for affective needs; and will encourage John to seek a collaborator, to undertake the shepherd leader role in his classes.

REFERENCES

Akindele, D., & Trennepohl, B. (2008). Breaking the culture of silence: Teaching writing and oral presentation skills to Botswana University students. *Language, Culture and Curriculum, 21*(2), 154–166.

Barnett, R. (2003). *Beyond All Reason: Living with ideology in the university.* Buckingham, UK: Open University Press.

Beard, C., Clegg, S., & Smith, K. (2007). Acknowledging the affective in higher education. *British Educational Research Journal, 33*(2), 235–252. doi:10.1080/01411920701208415

Biggs, J., & Tang, C. (2007). *Teaching for Quality Learning at University*. Maidenhead, UK: Open University Press.

Bloom, B. S., Engelhart, M. D., Furst, E. J., Hill, W. H., & Krathwohl, D. R. (1956). *Taxonomy of Educational Objectives – Handbook I: Cognitive Domain*. London: Longmans Green.

Boud, D., & Feletti, G. (Eds.). (1997). *The Challenge of Problem-based Learning*, (2nd Ed. pp. 93-102). London: Kogan Page.

Boyd, H. R., & Cowan, J. (1986). The case for self-assessment based on recent studies of student learning. *Assessment & Evaluation in Higher Education, 10*(3), 225–236.

Brookfield, S. D. (1990). *The Skilful Teacher*. San Francisco: Jossey Bass.

Butts, D., Cowan, J., & Bingham, E. G. (1976). *Learning systems, a case study*. Glasgow: Jordanhill College of Education.

Chiu, Y. C. (2001, June). *Online support for journalistic news course*. Paper presented at Developing College Specialization Seminar, Wenzao, Kaohsiung, Taiwan.

Chiu, Y. C. (2002, November). Exploring the learning effectiveness of internet news reading. In *Selected Proceedings of the 4th Pacific Asian Conference, and 11th English Teachers' Association Annual Conference,* Crane, Taipei.

Chiu, Y. C. (2006). *Exploring student and teacher interactions for critical thinking in face-to-face and online environments in an EFL course in Taiwan*. Unpublished doctoral dissertation, National Australian Catholic University, Melbourne, Australia.

Chiu, Y. C. (2007, July). Using the Shepherd Metaphor in Crossing Cultural Boundary Online for CHC Students. In *Proceedings of the 2007 E-Learning IADIS Conference, Lisbon, Portugal, International Association for the Development of the Information Society.*

Chiu, Y. C. (2008). Technical and Pedagogical Support for EFL Online Interactions. *Journal of Foreign Language Instruction, 2*(1), 83–102.

Chiu, Y. C. (2009). Facilitating Asian students' critical thinking in online discussions. *British Journal of Educational Technology, 40*(1), 42–57. doi:10.1111/j.1467-8535.2008.00898.x

Cowan, J. (1978). Freedom in selection of course content: a case study of a course without a syllabus. *Studies in Higher Education, 3*(2), 139–148. doi:10.1080/03075077812331376199

Cowan, J. (1984). *Learning contract design; a lecturer's perspective*. Royal Society of Arts - Occasional Paper No 7, London, Royal Society of Arts

Cowan, J. (1987). *Education for Capability*. Unpublished higher doctoral dissertation, Heriot-Watt University, Edinburgh.

Cowan, J. (1988). Struggling with self-assessment. In D. Boud (Ed.), *Student Autonomy in Learning*, (pp.192-210). London: Kogan Page.

Cowan, J. (2004). Beyond reflection: where next for curricula which concentrate on abilities? In C. Baillie & I. Moore (Eds.), *Effective Learning and Teaching in Engineering* (pp 202-19). Abingdon, UK: RoutledgeFalmer.

Cowan, J. (2005) Atrophy of the affect in academia. In S. Robinson & C. Katalushi. (Eds). *Values in Higher Education.* Leeds, UK: Aureus.

Cowan, J. (2006a). *On Becoming an Innovative University Teacher* (2nd Ed.). Maidenhead, UK: Open University Press.

Cowan, J. (2006b). Introduction. In J. O'Donoghue (Ed.), *Technology Supported Learning and Teaching*, (pp. 1-14). Hershey, PA: Information Science Publishing.

Cowan, J. (2007). Quoted at p. 62 in E. J. Burge, (Ed.), *Flexible Higher Education*. Maidenhead, UK: Open University Press.

Cowan, J., Morton, J., & Bolton, A. (1973). An experimental learning unit for structural engineering studies. *The Structural Engineer, 51*(9), 337–339.

Earwaker, J. (1992). *Helping and Supporting Students*. Buckingham, UK: SRHE and Open University Press.

Ecclestone, K. (2004). Learning or Therapy? The Demoralization of Higher Education. *British Journal of Educational Studies, 52*(2). doi:10.1111/j.1467-8527.2004.00258.x

Eysenck, M. W., & Piper, D. W. (1987). In J. T. E. Richardson, M. W. Eysenck, & D. W. Piper, (Eds), *Student Learning* (p. 215). Milton Keynes, UK: Society for Research into Higher Education and Open University Press.

Francis, H., & Cowan, J. (2007). Fostering an action-research dynamic amongst student practitioners. *Journal of European Industrial Training, 32*(5), 336–346. doi:10.1108/03090590810877067

Fry, H., Ketteridge, S., & Marshall, S. (2000). *A Handbook for Teaching and Learning in Higher Education*. London: Kogan Page.

George, J. W. (1996). Action Research for Quality Development in Rural Education in Scotland. *Journal of Research in Rural Education, 12*(2), 76–82.

George, J. W., Cowan, J., Hewitt, L. A., & Cannell, P. (2004). Failure dances to the tune of insecurity: affective issues in the assessment and evaluation of access learning. *Journal of Access Policy and Practice, 1*(2), 119–133.

Gibbs, G. (1988). *Learning by Doing: A Guide to Teaching and Learning Methods*. Birmingham, UK: SCED.

Heywood, J. (2000). *Assessment in Higher Education.* London: Jessica Kingsley Publishers.

Ho, I. T. (2001). Are Chinese teachers authoritative? In J. B. Biggs & D. A. Watkins (Eds.), *Teaching the Chinese learner: Psychological and pedagogical perspectives*. (pp. 99-114). Hong Kong: Comparative education research centre, University of Hong Kong.

Huyton, J.L. (2009). Significant Personal Disclosure: exploring the support and development needs of HE tutors engaged in the emotion work associated with supporting students. *Journal of Learning Development in Higher Education,* (1).

Kneser, C., Pilkington, R., & Treasure-Jones, T. (2001). The tutor's role: An investigation of the power of exchange structure analysis to identify different roles in CMC seminars. *International Journal of Artificial Intelligence in Education, 12*, 63–84.

Krathwohl, D., Bloom, B. S., & Mersia, D. B. (1964). Taxonomy of Educational Objectives. In *Handbook 2: Affective Domain*. London: Longmans Green.

MacDonald, P. J. (1997). In D. Boud & G. Feletti, (Eds.), *The Challenge of Problem-based Learning*, (2nd ed., pp. 93-102). London: Kogan Page.

McCormick, B., & Davenport, D. (2004). *Shepherd leadership; wisdom for leaders from Psalm 23*. (Y.C. Geh, trans.). Taipei: Apocalypse Press.

Mentkowski, M., et al. (2000). *Learning that Lasts*. San Francisco: Jossey Bass

Merryfield, M. (2003). *Like a veil: Cross-cultural experiential learning online*. Retrieved Nov 26, 2007, from http://citejournal.org/vol3/iss2/socialstudies/article1.cfm

Pilkington, R. (2001). Analysing educational dialogue interaction: Towards models that support learning. *International Journal of Artificial Intelligence in Education, 12*, 1–7.

Richardson, J. T. E., Eysenck, M. W., & Piper, D. W. (1987) *Student Learning*. Milton Keynes, UK: Society for Research into Higher Education and Open University Press.

Robinson, S., & Katalushi, C. (Eds.). (2005). *Values in Higher Education*. Leeds, UK: Aureus.

Rogers, C. R. (1961). *On Becoming a Person*. London: Constable.

Rogers, C. R. (1969). *Freedom to learn; a view of what education might become*. Columbus, OH: Merrill.

Rogers, C. R. (1980). *A Way of Being*. Boston: Houghton Mifflin

Schön, D. A. (1987). *Educating the Reflective Practitioner*. San Francisco: Jossey-Bass.

Walker, S. A. (2004). Socratic strategies and devil's advocacy in synchronous CMC debate. *Journal of Computer Assisted Learning, 20*, 172–182. doi:10.1111/j.1365-2729.2004.00082.x

Wallace, J. (2000). Supporting and guiding students. In H. Fry, S. Ketteridge, & S. Marshall. (Eds). *A Handbook for Teaching and Learning in Higher Education*. London: Kogan Page.

Watkins, D. A., & Biggs, J. B. (2001). Insights into teaching the Chinese learner. In D. A. Watkins & J. B. Biggs (Eds.), *Teaching the Chinese learner: Psychological and pedagogical perspectives* (pp. 277-300). Hong Kong: Comparative Education Research Centre, University of Hong Kong.

Weedon, E. M., & Cowan, J. (2001). Commenting Electronically on Students' Reflective Learning Journals. In *Proceedings, ISL conference*, Edinburgh, (pp. 134-42). Oxford, UK: Oxford Centre for Staff and Learning Development.

Weedon, E. M., & Cowan, J. (2002). The Kolb Cycle, reflection and all that what is new? In *Proceedings, ISL conference*, Brussels, (pp 97-108). Oxford, UK: Oxford Centre for Staff and Learning Development.

Yu, Y. L., & Chou, S. G. (2004). Integrating writing with reading in EFL writing classroom with metacognitive approach. *Selected papers from Thirteenth International Symposium and Book Fair on English Teaching, 1*, 708-719.

Chapter 5
A Cyber-Apple for the Teacher:
A Case Study of Anti-Hegemonic Adult Education Practices in a Cyber-Education Environment

Mark Federman
Ontario Institute for Studies in Education, University of Toronto, Canada

Marilyn Laiken
Ontario Institute for Studies in Education, University of Toronto, Canada

EXECUTIVE SUMMARY

In an age seemingly defined by near-ubiquitous access to Internet-based communication, there is little wonder that adult educators are turning to online, distance education as a means to reach their participants. In the traditional academy, post-secondary institutions increasingly include online courses and programs as elements, or comprising the entirety, of both undergraduate and graduate degrees (Allen & Seaman, 2006). Even in the realm of non-formal adult education, "hacktivism[1]" has become one of the most effective mechanisms through which engagement for social change – especially on a global scale – occurs (Day, 2004; Ganesh, Zoller & Cheney, 2005). Ironically, rather than truly integrating the philosophy of emancipatory and transformative adult education, cyber-education environments as typically implemented throughout the academy, overwhelmingly – if unwittingly – reproduce and reinforce the hegemony of traditional teacher-pupil power relations. By examining the mechanism of hegemony, and its pervasive presence in contemporary pedagogical technologies, this chapter will demonstrate how organized power is maintained through these mechanisms. In contrast, a case will be offered that demonstrates how engaged intellectuals can reconstruct the cyber-education environment in order to challenge the pretensions of entrenched academic power, and manifest adult education principles. In particular, the case will explore how the many years of research on how adults learn can be applied with the use of technology, so that the cyber learning milieu is as dynamic, personal and collaborative as the physical presence classroom context can be in the hands of a skilled adult educator.

DOI: 10.4018/978-1-60566-942-7.ch005

THE RISE OF CYBER-EDUCATION

Online[2], distance education has grown to comprise a significant fraction of post-secondary, institutional education. In a survey of 2,200 post-secondary institutions in the United States for the academic year 2005-2006 (Allen & Seaman, 2006), 3.2 million students were taking at least one online course. This represents approximately 15% to 20% of total enrolment, a 39% growth year-over-year – more than double the number added in the previous year. Almost all (96%) of large institutions, defined as having more than 15,000 enrolled students, offer online courses, with two-thirds of them offering at least one fully online degree program. Among institutions offering graduate-level, research-oriented degrees, more than 80% have online offerings.

With online courses and degrees gaining credibility and acceptability among both faculty and employers (Allen & Seaman, 2006), it is not surprising that nearly 60% of post-secondary Chief Academic Officers consider that online distance education is not only important, but "critical" to the long-term strategy of their institution. The reasons for such a strategic emphasis are not difficult to fathom. Post-secondary (and especially graduate) enrolment is increasing among working adults in a societal environment that actively promotes so-called lifelong learning. Distance education enables learners to gain access to education and advanced degrees without significantly compromising income, thereby creating a potentially large, lucrative market. Additionally, institutions are no longer limited to geographically-bounded markets consisting of a local population proximate to the physical institution. Distance education programs offered online – what we refer to as *cyber-education* – create a potential student population from which to draw that is geographically unlimited (Hiltz & Turoff, 2005). Moreover, once the initial investment in technical infrastructure, instructor training, and course management templates has been made, the marginal financial and instructor-time cost for additional cyber-education courses and programs is no more than that for adding conventional face-to-face programs of equivalent quality (Allen & Seaman, 2006).

André Grace (1998) describes how, in the post-war period, the focus of adult education emphasized post-industrial skills, knowledge, and planning, particularly in scientific and technological areas. This instrumental focus diminished the relative importance of social and cultural education in favour of utilitarianism. Despite – or perhaps because of – a rhetoric that nominally values notions like the distinctive nature of the adult learner in an environment of self-directed, goal-oriented, "facilitated" learning, adult education as contextualized by utilitarian knowledge acquisition is inextricably linked with a knowledge hierarchy that is, not coincidentally, foundational for the traditional academy (Federman, 2005). However, such an instrumental practice represents not adult education, but a "counterfeit" of adult education practice, at least according to Thomas Heaney; that is:

Practices in which knowledge and skill are transferred, in which the assumed superior knowledge and skill of the educator dominate the learning environment, in which the task is to impart knowledge that is already given, and in which learning is assessed in relation to the normative expectations of others. (Baptiste & Heaney, 1996).

Heaney's description of the counterfeit version corresponds closely to the characteristics of utilitarian and instrumental "lifelong learning," that focuses on "worker performativity in a global economy in which knowledge is commodifed and information literacy is valued" (Grace, 2004, n.p.). Indeed, the discourse of lifelong learning as framed by Grace establishes a normative expectation throughout contemporary society that would tend to favour cyber-education of a form that delivers commodified knowledge from an individual of presumed superior knowledge and skill, to a subordinate cohort, through a mechanism that particularly values information literacy. The apparent willingness of adult learners to seek such cyber-education in ever-increasing numbers suggests the existence of a pervasive, cultural hegemony with respect to adult learning in cyber-environments.

GRAMSCI'S THEORY OF HEGEMONY AS (ADULT) EDUCATION

Antonio Gramsci's basic notion of hegemony observes that people are not ruled by force, but by ideas – specifically, the ideas of the ruling, or dominant, class. Gramsci observes that political rule occurs largely with the consent of those ruled, "a consent which is secured by the diffusion and popularization of the world view of the ruling class" (Bates, 1975, p. 32). From the limited vantage point of his prison cell under Mussolini's fascist rule, Gramsci saw that the masses subordinated their rightful (Marxist) revolutionary ambitions not only to the economic oppression of the ruling class, but more to its *cultural* domination. Whereas Marx believed that the ruling class exerted its control via force and coercion, Gramsci maintained that even the most repressive states could not sustain their control exclusively through force and coercion. There had to be a subtle form of *ideological control* – shared values, attitudes and beliefs – that permeated society, to which the masses tacitly consented, providing the ruling class's support and legitimacy. "To the extent that this prevailing consciousness is internalised by the population it becomes part of what is generally called 'common sense' so that the philosophy, culture and morality of the ruling elite comes to appear as the natural order of things" (Burke, 1999, ¶9).

Gramsci sees two vital and opposing roles for intellectuals with respect to both establishing and maintaining the hegemony, and resisting it. *Traditional intellectuals* comprise people who are normally thought of as intellectuals, such as those in the academy, who typically align themselves with the ideological tenets of the dominant class and are therefore instrumental in their reproduction. *Organic intellectuals* are thought of as those professionals and other people of influence – including some who might also be traditional intellectuals – who could emerge from the working class to lead an active resistance to the established hegemony. Just as Lenin believed that a vanguard of the Communist Party leadership must initiate and lead the masses to emancipation through revolution, Gramsci believed that a vanguard of organic intellectuals was required to break down and break through the domination of the popular discourse that is firmly embedded in a culture, precluding both the general perception of domination, and its opposition in revolution (Lears, 1985; Mumby 1997; Burke, 1999).

According to Dennis Mumby (1997), it is the organic intellectuals that could combine theory and practice in such a way so as to enable and encourage ordinary people to attain a *critical consciousness* and reflexivity with respect to everyday occurrences and circumstances. Gramsci calls such critical practice a *philosophy of praxis* that moves philosophy away from the exclusive realm of ideas, and action away from the exclusive realm of uncritical everyday experience. "Gramsci situates the philosophy of praxis not as a body of ideas that confronts other philosophies, but as a collective activity with a material force, the goal of which is to engage the social forces and their ideologies that produce common sense (i.e., unreflective) conceptions of the world" (p. 350).

For Mumby, the *hegemonic struggle* describes a dialectical tension concerning people's ability to locate their everyday experience within an explicitly conceived discursive context, in order to achieve an awareness that ultimately moves them to action. When hegemony is interpreted in a functionalist ground (as is mostly the case according to Mumby), people have as their only option to blindly and unquestioningly succumb to a dominant ideology. This interpretation precludes the awakening of awareness through critique, and the possibility of an emergent counter-hegemony other than in a crisis. However, Gramsci's philosophy of praxis necessitates acknowledging actors' recognition of the contradictory tensions that exist in their hegemonic social dynamics, maintaining a continual, if latent, possibility for resistance, action and change. Frank Youngman sums up the hegemonic struggle this way:

Gramsci argued that the struggle for hegemony takes place in all spheres of cultural and intellectual life, and of social practice. He believed that there was a need for the working class to develop a new world-view which would undermine the legitimacy of ruling ideas and create an oppositional culture, a counter-hegemony. He concluded that socialist political activity should create intellectual leadership within the working class to develop a counter-hegemony through which that class could achieve its own emancipation." (Youngman, 2000, p. 18)

Youngman's "intellectual leadership" comprises the core of true adult educators. Mumby echoes these thoughts, describing a "process of struggle rather than an existing state of consensual domination that is continually produced and reproduced. ... Gramsci's philosophy of praxis recognizes *both* the possibilities for social change and the tenacity of the dominant hegemony that resists such change" (1997, p. 365-366; emphasis in original).

In keeping with the theme of struggle for social change, Grace (1998) advocates that adult educators adopt the vocabulary of social revolution and struggle against repressive and dominant power to counter the instrumentalism and individualism that seems to be embedded in popular, hegemonic, adult learning discourses, such as lifelong learning. He calls for a "Theory of Adult Learning Community" that incorporates equalization of roles among educators and learners, a focus on knowledge production rather than knowledge consumption, the incorporation of praxis in areas of collective change, constructivist approaches to identity and meaning making, politicization of the learning environment, and explicit recognition of the effects of culture and power. These principles would be intended to equitably address all of the instrumental, social and cultural concerns that surround instantiations of adult education.

Given the increasing pervasiveness of cyber-education environments in post-secondary – and especially graduate – institutions, one is moved to ask whether the particular implementations are expressing the principles of adult education, or whether they are further entrenching a traditional hegemony of teacher dominance, and knowledge as consumable property with a primary focus on content and instrumentality. Perhaps more important to consider, if more subtle, is the question of whether specific technological implementations themselves are more conducive to either supporting hegemony or promoting a counter-hegemony. Put more simply, we ask, is technology political?

EXAMPLE: THE BIAS OF HYPERLINKS

danah boyd (sic) has recently completed a Ph.D. at the School of Information (SIMS) at the University of California, Berkeley, and is a social media researcher with Microsoft Research. She investigated the gendered nature of hyperlinks among weblogs, and particularly, the gendered effects of networking patterns of hyperlinks on imputed authority and power granted by algorithmic authorities such as Google and Technorati (boyd, 2005). In a survey of 600 weblogs, including the so-called Technorati 100 (Technorati, 2006), boyd discovered that there is a correlation between link structure and content type, and that content type demonstrates a clear gender split, with male bloggers tending to favour, and create, blogs containing more links. Further, she found that when bloggers link to another blog, it is more likely to be a blog whose author is of the same gender as the linker. This, she says, is consistent with gender affinity in social networks. "Men tend to have large numbers of weak ties and women tend to have fewer, but stronger ties. This means that in traditional social networks, men tend to know far more people but not nearly as intimately as those women know" (boyd, 2005). Men, therefore, tend to link promiscuously, whereas women tend to refrain from linking until there is a stronger relationship established. boyd describes the implications of this behaviour: "all links are created equal. All relationships are not. Treating everything like a consistent weak tie is quantity over quality and in social networks, that means male over female."

Algorithms (like Google and Technorati) that claim to be neutral proxies for knowledge authority, influence or prestige, are actually, and very effectively, measuring the available link structure [and generally favouring blogs written by males over those written by females]. The difficulty is that there is nothing consistent whatsoever with that link structure. There are disparate norms, varied uses of links and linking artifacts controlled by external sources (like the hosting company). There is power in defining the norms, but one should question whether our companies or collectives should define them. By squishing everyone into the same rule set so that something can be measured, the people behind an algorithm are exerting authority and power, not of the collective, but of their biased view of what should be. *This is inherently why there's nothing neutral about an algorithm.* (boyd, 2005; emphasis added)

THE HISTORY AND POLITICS OF THE THREADED FORUM

The vast majority of so-called learning management systems or course management systems in use for cyber-education are designed around the *threaded forum* as the primary means of interaction among both instructors and students. To gain a perspective on the pervasiveness of the threaded forum, consider the following items from three distinct countries. After the merger of the two, then market-leading course management vendors, Blackboard and WebCT – both of which predominantly use threaded forums – the merged company is estimated to hold 75% of the market for such systems (Poftak, 2006). A study of cyber-education platforms in use throughout Switzerland undertaken by the University of Fribourg showed that across 25 post-secondary institutions, over 95% of students enrolled in cyber-education courses used a threaded forum-based system (University of Fribourg, 2006). The University of Toronto has standardized on the Blackboard system for all of its distance education, or hybrid physical/cyber, offerings. Even the Ontario Institute for Studies in Education at the University of Toronto – one of the world's largest graduate faculties of education – had, until the recent transition to Blackboard, almost exclusively used (and exclusively supported) its home-grown Web Knowledge Forum system that relies on threaded forums as the primary venue of interaction. These illustrative examples suggest that there might be a valid pedagogical[3] reason for such an overwhelming choice of format. In fact, the reason is more historical in nature.

In 1979, Tom Truscott and Jim Ellis, two graduate computer science students at Duke University, wanted to create a mechanism whereby the users of the university's Unix system could share information with users at other universities. At the time, connecting to the nascent ARPANET was financially infeasible, as the cost to join was approximately $100,000. The two students created a jury-rigged system that used home-built autodial modems, automated scripts, and the Unix-to-Unix-Copy-Program to connect, and exchange files with, computers at other universities. Files are synchronized and placed in order according to their intrinsic date-time stamp, and the list of new, topically related files are linked, or *threaded*, together into *threaded forums*, also known as *newsgroups*. As news of this capability spread among the academic community, more universities and systems participated in what eventually became known as Usenet News. The threaded forum, newsgroup style became the commonly accepted cultural norm for both the precursor networks to, and the contemporary, Internet (Hauben, Hauben & Hauben, 1994/1997). The form's hegemonic dominance became clear after the inception of the World Wide Web in the early 1990s, as the threaded forum was quickly implemented in, and dominated, interactive web-pages – despite the fact that the original technical reasons for the particular form had long since disappeared. It seems that today, most cyber-education environments use threaded forums because of techno-cultural inertia and user familiarity (Klemm, 2002), and the hegemonic notion that forums are the "natural" way to provide interaction among multiple users.

On the other hand, software developer Joel Spolsky deliberately selected the online threaded forum to provide spaces of discussion and support for his clients. He articulates the reasons for this choice based on the specific effects of threaded discussion forum. Spolsky notes that the threaded forum tends to focus the discussion of a particular topic. It does this by forcing the reader to read through the discussion as it has evolved so that the context becomes very well established in the reader's mind before s/he is moved to respond. In doing so, the forum tends to create a fairly (topic-) controlled environment in each thread. In the threaded forum, all comment posts are created equal. There is no mechanism, aside from a limited editorial role sometimes played by a forum moderator, to apply judgement or allow particularly insightful or useful comments to "rise to the top," both metaphorically and literally. Thus, in the threaded forum, there is a subtle exercise of cognitive control that is characteristically linear and transactional (Spolsky, 2003).

The problematics of a threaded-forum cyber-education environment have been identified among other pedagogically-focused researchers. Serce and Yildirim (2006) observe that in an online course they studied, only 34% of postings were student-student interactions (representing only 40% of the participating students). The majority of students – and their interactions – were transactional exchanges between student and teacher. Matthew Thomas (2002) found that student threads quickly became divergent, with a significant decrease in the chance of a given message being read in relation to its "depth" in the thread. Over half of student postings in his study receive no response and instead terminate their particular branch of a topical thread. Rather than participating in anything that resembles true discussion and collaborative knowledge creation, students effectively state their own opinion for the benefit of the instructor (not to mention the benefit of their marks that are assigned by the instructor), without drawing from the collective knowledge and views of their cyber-classmates, as often happens in face-to-face engagement in a well-run physical classroom, especially at the post-secondary level. Thomas (2002) sums up these observations by stating:

There was little on-going development and exchange of ideas in any of the discussion themes. Rather, the disjointed and fragmented individual contributions were abstracted in space and time from other students' contributions. ... This incoherent structure of the discussion threads is not compatible with a truly conversational mode of learning. From this analysis it is evident that the virtual learning space of the online discussion forum does not promote the interactive dialogue of conversation, but rather leads students towards poorly interrelated monologues. (Thomas, 2002, p. 360-361; emphasis added)

As William Klemm observes, this is a situation circumscribed by the technology itself:

Threaded-topic design typically requires the cumbersome process of opening and closing many messages. There is no way for students to create in-context links from within a given message or to insert text or multimedia into any jointly prepared document, because there are no jointly prepared documents. ... Indeed, "discussion" is probably the wrong word to use for this activity, because posted messages are more like monologues. (Klemm, 2002, ¶2)

Based on the relatively few available critiques of the threaded forum as technical implementation of cyber-education, there seems to be a pattern of pedagogical philosophy and politics emerging from the choice of implementation itself, that contradicts Grace's (1998) Theory of Adult Learning Community, mentioned previously. In particular, the *de facto* focus on instrumentality, the structural isolation of each students' ideas (Klemm, 2002), and the implicit emphasis on interacting with the instructor rather than learner peers, starkly contrasts with the language of collaboration, dialogue, conversation and interaction that pervades the cyber-education discourse (Lapadat, 2004; Pelz, 2004; Haythornthwaite, 2006). Indeed, the specific implementation choices for the overwhelming majority of cyber-education students seem to manifest and reify Youngman's contention that, "education does not diminish inequality but, in fact, serves to reproduce the class structure of capitalist society from generation to generation... Schools not only reproduce capitalist social hierarchies but also reproduce the contradictions inherent in those hierarchies" (2000, p. 154). Additionally, the question of leadership has only begun to be explored (Bourhis, Dube & Jacob, 2005), and opens up many questions – in particular, the aspect which is the focus of the current study – that of structured versus emergent leadership in on-line learning communities.

"BEST PRACTICES" AND CYBER-EDUCATION AS HEGEMONY

In accepting the 2003 Sloan Consortium award for excellence in online teaching, Bill Pelz, a professor at Herkimer County Community College, a State University of New York campus, shares his "three principles of effective online pedagogy" (Pelz, 2004): "let the students do the work," "interactivity is the heart and soul of asynchronous learning," and "strive for presence." In all of Pelz's online classes, threaded forum technology is used to create the spaces for interaction within the cyber-education environment. Students "doing the work," according to Pelz, requires each student to initiate a topic thread and "lead" the ensuing postings that comprise the putative discussion. In this way, the traditional, didactic teacher-pupil interaction is reproduced among the various threads, with each student in turn taking on the role of (dominant) teacher/leader over her or his (subordinate) student colleagues. For situations in which there are specific homework problems, "students submit their solutions to the professor, and then check their answers against the [professor-provided answer] key. They discuss with one another the errors they made, then suggest their grade for the assignment to the professor" (p. 36).

These, and other similar dynamics described in the Pelz article maintain and replicate the traditional classroom hegemony, while merely transferring *workload* from professor to student (teacher workload being a major issue in online, threaded-forum style courses) rather than actually transferring *power*. The hegemonic dominance of professor power is replicated via the assigned exercise in the forum structure and thereby reinforced among the subordinated students. While some power (e.g., the "power" to initiate a discussion thread) is transferred, the professor retains the ultimate power as arbiter, as he scores the individual posts and participation relative to a professor-imposed rubric.

Pelz encourages interactivity, that is, active participation in the forums, through the use of assigned marks. Even when he creates an opportunity for collaborative construction of knowledge that might counter the hegemonic structure of the cyber-environment, that opportunity is quashed: what are called "collaborative papers" in Pelz's context are actually individually researched and written, traditional papers, merely posted to a threaded forum, available for comment, with the author taking the traditional role of teacher/leader.

"Presence," a characteristic apparently to be strived for by the students, comprises social presence (merely "showing up" for the online class through postings), cognitive presence, and teaching presence. The latter pertains to facilitation and instructional aspects to be conducted by the students assuming a teacher-role on the threaded forums, and reproduces the cultural hegemony of the traditional classroom. Cognitive presence relates to the quality of each post, about which Pelz instructs, "treat each discussion submission as if it is an answer to a test question" (2004, p. 42). Were there an organic intellectual among the students they might be moved to question the inconsistency and contradiction between the first principle of letting the student do the (marking) work – however nominal that might be – and the not-so tacit reminder of the omni-presence of coercive mark-granting authority, what Gramsci describes as the agent of direct domination (Bates, 1975), in the presence of the students' consensual subordination.

Pelz includes "two cardinal rules of discussions," mandating that forum posts must comprise "relevant and new information," and must have "an appropriate subject line" (2004, p. 45). On the surface, these would seem like reasonable requirements. However, they are requirements that are imposed by the threaded forum technology itself that have the effect of reinforcing the cultural domination of the teacher. Thomas observes that:

It would be more appropriate to conceptualise students' messages as data stored for potential access by other students, rather than contributions to an on-going dialogue. In this way the online discussion forum promoted an individualistic mode of learning rather than an interactive mode. (Thomas, 2002, p. 362)

Because each posting *can* be individually and discretely attributed, there seems to be a compulsion to treat it as if it *must* be marked, in Pelz's words, "as if it was an answer to a test question" (2004, p. 42). Although this might nominally compare to the teacher's evaluation of students' traditional classroom participation, during which the instructor forms an emergent impression of who contributes and who does not, the control-enabling elements inherent in the technology change the social environment of a physical presence discussion into a disciplinary environment of continual evaluation *à la* Foucault (1980). Such change harshly reinforces the dominant position of the teacher, and the reproduction of that role in each student as they individually are charged to lead so-called discussion threads. Pelz's "cardinal rules" clearly serve the purpose of "uncover[ing] the mechanisms whereby social inequality is reproduced and legitimated on one hand, and resisted on the other" (Youngman, 2000, p. 35).

Pelz is not alone in this practice. Judith Lapadat, presenting at the Canadian Society for the Study of Education at the "Learneds" conference in 2000, proudly observes that participation in her class online forums was above her minimum expectation. Why?

One reason for this was that I built specific expectations for participation into the course design. I explicitly informed class participants that they were expected to contribute thoughtful remarks to the discussion each week by the posting deadline, to present their ongoing work online, and to provide feedback to each other; and that *participation would be graded on the basis of both quantity and quality.* (Lapadat, 2004, p. 243; emphasis added)

The politics introduced through the threaded forum technology, coupled with the ever-present threat of direct, coercive domination, creates a hegemony within the cyber-education environment that not only reinforces the pretensions of the traditional power of the classroom teacher. In their willing subordination, the students enthusiastically learn to reproduce the teacher's power, directed at each other, thread by thread, post by post.

A COUNTER-HEGEMONIC IMPLEMENTATION OF PELZ'S THREE PRINCIPLES

From the Gramscian principle of philosophy of praxis, an engaged intellectual – even traditional intellectuals, as we must categorize ourselves – can combine Pelz's recommendations for practice, and combine it with more emancipatory theories of adult education, such as those recommended by writers like Grace (1998), and Baptiste and Heaney (1996). The result would be an instantiation of praxis that might inform the selection of technology and specific implementation design for a cyber-education environment that is consistent with adult education principles. Thomas observes:

The challenge is for interface design which promotes a more coherent structure and true many-to-many interaction in the virtual learning space. ... While the online discussion forum has become a ubiquitous element of Internet-supported flexible delivery of education, it is apparent that it might not be the best technology to support the interactive and collaborative processes essential to a conversational model of learning. These new developments must involve the redesign of both the technological support tools and curriculum structures to support collaborative learning processes. Accordingly, such innovation would emphasise the implementation of learning tasks that promote collaborative engagement towards knowledge development and problem solving. It is perhaps this route that may prove to be the most productive means of realising truly conversational modes of learning, given the inherent problems involved in traditional online discussion. (Thomas, 2002, p. 364)

- ***Principle 1: Let the students do the work.*** In a counter-hegemonic design, students would create entire conversations in a non-threaded format on the various course content themes. While the course designer would likely choose the themes for the early part of the course in order to provide some structure to the new environment, students would actively be encouraged to deviate from the pre-chosen themes if newer themes emerge amidst their mutual engagement. Through the design and technical implementation of the environment, conversations would be encouraged to flow and emerge according to the interests and explorations of the participants, without the restrictions of threads, subject lines, or coerced requirement to constantly consider each post as the answer to a test question. To this end, students would self-evaluate for the participation component of their final grade, possibly guided by a student-created rubric.
- ***Principle 2: Interactivity is the heart and soul of asynchronous learning.*** Students would interact freely throughout a hyperlink-enabled environment without the stricture of linear forums. Topical reflections would be distributed throughout the environment, connected to wherever might

be appropriate, irrespective of the topic of the week. Participants would be guided by design to interact among many areas of the environment in relatively unstructured ways in order to enable a collaborative, community structure from which normative patterns of behaviour and governance would emerge specific to the particular cohort of course participants.

- ***Principle 3: Strive for presence.*** Social, cognitive and teaching presence would be facilitated by enabling students to engage actively with each other's postings – including interjecting and changing another participant's writing. This replicates – as closely as is possible in an asynchronous environment – patterns of conventional, active, engaged, face-to-face conversation, dialogue, and discourse. Social presence is enhanced not by creating a segregated set of social forum threads – often called by names like "Back Porch," "Café," or "Living Room" to highlight their directed function as a venue for more sociable postings – as is the case in almost all threaded forum cyber-education environments. Rather, a concept of "leaving footprints in someone else's garden" by cross-posting to others' learning journals or profile pages can serve as strong "social glue" among the learning community. Indeed, this style of engagement emulates the strong social presence enablers of contemporary social networking sites, such as MySpace, Facebook, Friendster, Orkut, and the like.

Finally, teaching presence in an adult education environment is more than instrumental reproduction of didactic functions. Adult education principles call for the incorporation of critically framed lived experience into knowledge construction, in contrast to Pelz's contention that comments based on personal experience are of average to minimal value (2004, p. 43). The course curriculum should be designed to encourage students to augment and enhance the designer-provided material with critical reflections of their embodied experience. Additionally, participants would be specifically encouraged to reflect on the processes of the course, to critique its structure and dynamics, and offer suggestions for its modification and improvement throughout the term of the course. Moreover, the course designer and facilitator(s) would engage with participants in lively, albeit mostly asynchronous, conversation about these critiques and suggestions and take appropriate action to modify the educative environment as appropriate and possible. In this way, "both adult educator and adult educatee are jointly and unrepealably educated, changing both the immediate environment in which the education occurs, and the larger social environment to which each contributes, and in which each lives." (Federman, 2006b).

THE DESIGN OF AN ADULT EDUCATION, CYBER-EDUCATION ENVIRONMENT

These counter-hegemonic versions of Pelz's three principles are integrated in an experimental cyber-education environment design used for a distance education course in the 2006 fall semester in the Adult Education master's degree program at OISE/UT. The core implementation technology for this course is a wiki: "a type of Web site that allows the visitors themselves to easily add, remove, and otherwise edit and change some available content, sometimes without the need for registration. This ease of interaction and operation makes a wiki an effective tool for collaborative authoring" (Wikipedia, 2006). Perhaps more important is the context that provides the focus of the case study – an OISE course entitled: *Organizations and the Adult Educator: Historical and Theoretical Perspectives on Organization Development* (AEC 1141). The classroom version of the course was designed by one of the authors and a team of colleagues, to explore the concepts of organization change and development as it evolved over the past century. Its major purpose was to enable adult educators to understand the historical and theoretical roots of these concepts so that they would have a context in which to develop the skills and expertise needed as organizational change professionals. With this grounding, the learner was expected to be able to benefit from other more skills-based courses, as well as to more fully appreciate the origins of many contemporary workplace practices.

The course developers used the metaphor of "unrolling the ribbon" to express the genesis, flow and evolution of the ideas underlying the concept of Organization Development (OD). The metaphor implies a slow and gradual transition, in which ideas are retained and transformed, but never lost or discarded. We called each section of the course a "Block" – mirroring the evolution of OD thinking through the 20th century. In this resource-based course, students were expected to cover a substantial amount of material on their own, between classes, and then participate in a collaborative knowledge-building design during the three-hour class sessions. The readings were organized to reproduce the chronology of the development of the field of OD, from its genesis as a reaction to early management thought, with its repressive use of Scientific Management and hierarchy (Taylor, 1911; Weber, 1947; Fayol, 1960), through the era of "T-groups" (Miles, 1970), and Socio-technical Systems Design (Cherns, 1976), to a current-day focus on organizational change within the context of a globalized world (Laiken, 2002), including contemporary practices of enhancing organizational learning (Senge, 1990) and non-hierarchical leadership approaches (Semler, 1994).

The original course was designed to embody specific principles of adult learning – a key one being the role of the instructor as a facilitator of knowledge building, as opposed to being the primary source of information and expertise. Thus, this course is particularly well-suited to study the translation from physical to cyber-presence, because the majority of the learning occurs in interactions among a community of learners. One of the authors is the designer of the online environment, and one of the course's co-instructors. His own experience of taking the course as a student yielded the insight that, be it in physical presence or as cyber-education, the course is intended to effect personal transformation for each of the course participants.

The transformation occurs through the students' own evolving understanding of organization development as the theory, history, and shared experiences are assimilated and embodied. The individually experienced process of transformation parallels the history of management theory through the various blocks. Each week, course participants are provided with various readings and encouraged to form conversation groups to discuss one of four suggested themes, or to create a theme of their own. The conversation groups subsequently gather in a general plenary session to share each group's collective experiences and learning relative to the weekly topic. Reproducing the transformative effects of face-to-face, conversational interactions that occur across varied group topologies represents a significant challenge within an online, distance education context. Nonetheless, this was the primary design objective of the cyber version of the course.

As a secondary design objective, fully participating in the course (including reading and preparation time) in the cyber incarnation should require no more time or effort on the parts of either the students or instructors than participating in physical presence. Additionally, and introducing a critical element to the design, a final key objective was to incorporate appropriate principles of adult education into the design and dynamics of the online offering. Grace's (1998) Theory of Adult Learning Community is used as a rubric for judging the successful enactment of critical adult education precepts in cyber-presence. Grace's theory comprises six principles: equalization of roles among educators and learners; a focus on knowledge production rather than knowledge consumption; the incorporation of praxis in areas of collective change; constructivist approaches to identity and meaning making; politicization of the learning environment; and explicit recognition of the effects of culture and power.

TECHNICAL IMPLEMENTATION OF THE CYBER-EDUCATION ENVIRONMENT

The course was implemented in a wiki software environment using the hosted PBWiki service, augmented by Skype voice-over-IP telephony. Unlike other common technological platforms currently being used for on-line, distance learning, which are overwhelmingly based on threaded forums (Poftak, 2006; University of Fribourg, 2006), a wiki platform is particularly well suited to the development of community, as the participants can both effect and affect the evolution of the learning environment itself. The key question then becomes, how much, or little, leadership structure is needed to help participants, as quickly as possible, take control of their own learning milieu and work together interdependently? In the Fall of 2006, fourteen students and two co-facilitator-instructors embarked on discovering the answer.

As almost all of the participants, including one of the instructors, were relative novices in using this sort of technology, the first two weeks prior to any course content being introduced were spent in technological acclimatization. Course participants were invited to participate in specific, very structured exercises designed to acclimatize them to both the instrumentality of the technical environment, and the radically different cognitive and social environment enabled by the wiki. For example, in the following excerpt taken from the actual course wiki (Federman, 2006a), these initial exercises variously encourage participants to:

1. Go play in the SandBox [a wiki page on which to experiment with wiki formatting codes];
2. Explore this wiki environment to find your bearings and begin to construct a mental picture of the architecture of the learning environment. If you find a page that you want to change, change it!
3. "Leave footprints in someone else's garden," that is, to change and add to [other people's] posts ... because of prior experience in other environments, and sometimes because the posted material feels almost like a garden planted by another person, participants are reluctant to change or add to what has been posted; [and perhaps most important of all,]
4. Don't Panic!

Although the week-to-week pace of the course was pre-set, students were advised that they could change any other aspect of the course content or process, having full control of the wiki environment. As in the classroom version of the course, the participation component of the grade – 40% of the final mark – was self, or small student-group evaluated to remove any instructor-coercive aspect to online participation and contribution. The rubric suggested for this evaluation was developed over time by previous groups of students enrolled in the physical presence versions of this course, with this cyber-environment group invited to either adopt the rubric as is, modify it as they individually or collectively deemed appropriate, or develop a new rubric. The remaining 60% of the grade would be assessed by one of the instructors, based on the final essay. Students were also offered a "no-fault" assessment of an essay draft by the non-marking instructor to minimize the instructor-student power differential inherent in an all-or-nothing essay assignment, for such a substantial portion of the grade[4]. Weekly activities involved reading from a prescribed set of materials (which could be changed or augmented), self-assigning small groups to work on suggested "probes" (or additional, student-proposed "probes"), participating in other's small groups, and reflecting on process in individual learning journals and on a feedback page. A mid-course check-up – a concept taken directly from the face-to-face version of the course as a way of reducing the power differential between instructor and students – specifically asked for critique and suggestions, many of which were immediately implemented through design and process changes. There was also a peer-evaluated dialogue exercise (Bohm, Factor & Garrett, 1991) conducted via Skype, and a final feedback session prior to marks being submitted.

STRUCTURE VS. EMERGENCE IN THE CYBER-EDUCATION ENVIRONMENT

One of the key questions for the instructors – and the polarity held and examined by the authors of this paper – concerns the degree to which the instructor should provide explicit leadership and guidance as the course begins. A "Situational Leadership" approach to teaching (Hersey, Blanchard & Natemeyer, 1979) suggests that more structured guidance should be provided to participants early in the course, progressively reducing the instructor's involvement as the students become increasingly self-directed. This enables a team to painlessly develop their own ability to self-direct in order to effectively perform a task. Indeed, some of the students who had previously participated in another OISE course on Developing and Leading High Performing Teams (AEC1107), which subscribes to this model, especially sought this approach as an effective way in which to develop a task-focused team (Tuckman, 1965; Laiken, 1998). The instructor's role in that situation was to initially provide not only the technical and subject matter structure of the course, but to facilitate the evolving social structure as well. The instructor then progressively took a "back seat", until members had incorporated effective team behaviour into their own repertoire and could work interdependently with each other.

The opposing polarity, used in the design of AEC1141, held that the ultimate objective of the course was to enable participants to learn the values and practices of OD by experiencing them, from the start, without instructor intervention. The intention was to replicate one of the effects of the face-to-face course experience, in which this transformation is accomplished not through the mere acquisition of historical and theoretical material, but through the embodied experience of making sense of lived organization dynamics via the class-as-organization. In this situation, if social norms were facilitated by the instructor, s/he would impose a potentially stifling power dynamic on the transformative potential of the class. As Tony Huzzard notes in the context of organizational learning, "power is articulated through social practices that produce the "truths" that make up our self-concepts and the institutions in which our selves are embedded. Power is not exercised by sovereign individuals, but is located in social practices and the relationships on which such practices are built" (2004, p. 352). He distinguishes between "sensemaking," which, in the context of a group (organization), is an act of dominant individual(s) creating meaning; and "sensegiving": ... acts [that] mobilize other actors and thereby establish collective activity in the emergent community of practice" (p. 356).

If the instructors in AEC1141 were to have constrained the discussions of behavioural norms by pre-establishing them, it would have imposed a particular type of sensemaking on the class. One of the students who had taken AEC1107 offered a suggestion to help fill the structural vacuum created by the instructors: "I'm wondering if it makes sense for us (all of us) to negotiate some expectations/norms for using the Wiki environment before we get started with the task portions of our course? ... We really need to build a team to do this successfully." And another weighed in, saying: "My suggestion is that we set some norms at the start of this learning journey to assist us in our work together and help us develop as a cohesive team."

True to the experiential learning intentions of the course design, the instructors neither resisted nor specifically encouraged this particular discourse, having established a premise that the students could introduce any dynamic they chose in the context of creating their environment of interactions. This concept, that is fundamental to the ethos of the wiki environment, and was deliberately chosen as a design criterion for the AEC1141 cyber-environment, created a tension with the expectations for instructor-led direction that most students brought with them to the course. Responding to the suggestion to establish norms, one student offered, "perhaps this is something we can approach Mark about." In response, Mark gently queried, "why would you approach Mark about this since you all now (collectively) control the environment? Perhaps this conversation is part of a reflection on process (hint, hint)."

The tensions (with the potential for learning about power and control in the workplace) that this approach instigated are reflected in the following comments and responses (Federman, 2006c):

...any empowerment to us from my perspective is limited. Yes we could have taken initiative in some areas but why and for whom and who says that that is what is best. Who's [sic] needs would have been met?"

Another student immediately responded:

If we really had control and [were] empowered, we would have changed many of the readings ... the weekly probes ... the final assignment, etc. ... Had we tried this, would we have been successful? Probably not ... so we weren't really in positions of power or control.

To this, the instructor replied, "Actually, probably 'yes' for most changes." One more student, realizing the intended learning objective, made the following insightful and key observation:

We were only limited by our own notions re: what we could and could not do within this environment. Mark, for his part, certainly did not limit us. I think a lot of the ill feeling or criticism may stem from us wanting, on some deep level to be directed. Perhaps, as alluded to above, this is simply a product of our own conditioning within education settings over the years.

One specific way in which the struggle over establishing behavioural norms for the group manifest was related to the question of whether or not to sign one's name to wiki page contributions – an issue that remained unresolved throughout the course. One student finally became aware of the learning experience that was being created amidst the seeming disorientation and chaos: "I wonder if there is more at play here than simply the need to know who you are communicating with." Had behavioural norms been explicitly imposed – such as an instruction to sign a contribution, or not – the students would not have had the opportunity to experience figuring out the (admittedly random) extant culture in which they found themselves, or encouraging the creation of a healthy culture, which are both critical aspects of work in OD. From the instructor's perspective, pre-establishing a normative culture would have been disrespectful to the reality that the culture of each class will be different, depending on the experiences, knowledge and context of each individual participant.

Enabling the participants to negotiate the tension between structure and emergence turned out to be one of the most important parts of the organization development learning experience. The learning from this research was that this differed from the intention of the High Performing Teams course, in which the team needed to form quickly and effectively in order to accomplish a specific task (a community-serving project). In AEC1141, a discursive environment in which the participants were actively encouraged to challenge both the instructors' choices and their own decisions, enabled the exploration of issues of power and control – not unlike the T-groups[5] of the 1970's. Some students came to the realization that any limitations on group norms and community culture were self-imposed by the class participants, and could have been changed at any time. This notion carries over to observing and understanding pre-existing normative behaviours and expectations that comprise organizational culture – the context of the OD professional. As one student commented in their final reflection on the course, "it's important to realize that part of the learning of the course is in dealing with the difficulty and frustration, and reflecting on the process of change and transformation that it entails" (Federman, 2006c).

The question of whether the culture of on-line adult learning environments should initially be more structured by the instructor, or be allowed to emerge in response to the interpersonal dynamics of each collective of participants cannot be resolved through the experience of one course. However, it seems clear that the decision needs to be made in relation to both the curricular and pedagogical purposes of the course. In AEC1141, the intention is to enable an exploration of leadership in relation to issues of power and control. As suggested by Grace (1998), a course's intention might be to politicize the learning environment by explicitly recognizing the effects of culture and power. If this is either the curricular or pedagogical intent of the course, then the educational benefit of a non-directive environment clearly is illustrated by these final participant reflections (Federman, 2006c):

I still struggle with the initial lack of structure... Perhaps this is merely a matter of degrees... how much structure is enough. Or perhaps merely style. Certainly for me, a large degree of structure and design would drive me nuts and stifle creativity.

I feel there has been room for true exploration, mistakes, reflection, etc. which for me is essential in a learning environment. ... Our group culture ... seems to have been formed by everybody's behaviour and input. ... By providing us with lots of avenues to interact, he [the instructor] facilitated our collaborative community development.

I am sharing the feelings of lack of structure, but am also enjoying what I see as freedom to explore. Watching myself and everybody coming to terms with it is a great experience in itself.

For this particular exploration, wiki technology and the design and architecture elements of the course environment were specifically chosen as much for the their ability to force an explicit experience and realization of the collaborative and emancipatory potential of the various elements, as for their conduciveness to create a pedagogical counter-hegemony. The challenge is one at which not all students seem to be successful. Some find it extremely difficult to break from the hegemonic expectation that the teacher is in charge, and that the guidance and direction offered are merely suggestions. However, as Mumby astutely identifies:

Hegemony always involves struggle over systems of meaning and the processes by which social reality is framed. While meanings may be temporarily fixed and certain interpretations hold sway, there is constant slippage between discourses and meanings, such that alternative and competing definitions of the world arise. (Mumby, 1997, p. 364)

The social reality of pedagogical hegemony pervades cyber-education environments through the politics of the technical implementation as much as through the desire to maintain a dominant status quo among the elites of the academy. With radical and subversive implementations based on a philosophy of praxis, students' awareness of the actual dynamics of cyber-education can be awoken. Formal education that is reproduced in traditional, threaded forum cyber-education environments may be seen as a systemic means to maintain hegemony in the academy. Alternative implementations that emphasize active, collaborative construction of knowledge, and developing critical self- and social- awareness can enable the emergence of counter-hegemony that directly challenges the pretensions of organized academic power.

REFERENCES

Allen, I. E., & Seaman, J. (2006). *Making the grade: Online education in the United States, 2006*. Needham, MA: The Sloan Consortium

Ally, M. (2004). Foundations of educational theory for online learning. In Anderson, T. & Elloumi, F. (eds.). *Theory and practice of online learning*. Athabasca University (online book). Retrieved November 10, 2007, from http://cde.athabascau.ca/online book/index.html

Baptiste, I., & Heaney, T. (1996). *The political construction of adult education*. Paper presented at the Midwest Research-to-Practice Conference in Adult, Continuing, and Community Education, October 17-19, 1996, Lincoln, NE. Retrieved on November 23, 2006 from http://www.nl.edu/academics/cas/ace/facultypapers/TomHeaney_IanBaptiste_Construction.cfm

Bates, T. R. (1975). Gramsci and the Theory of Hegemony. *Journal of the History of Ideas*, *36*(2), 351–366. doi:10.2307/2708933

Bohm, D., Factor, D., & Garrett, P. (1991). *Dialogue: A proposal*. Retrieved April 24, 2006, from http://www.infed.org/archives/e-texts/bohm_dialogue.htm

Bourhis, A., Dube, L., & Jacob, R. (2005). The success of virtual communities of practice: The leadership factor. [Available online at www.ejkm.com]. *Electronic Journal of Knowledge Management*, *3*(1), 23–24.

Boyd, D. (2005). *The bias of hyperlinks*. Retrieved December 3, 2008, from http://www.zephoria.org/thoughts/archives/2005/08/07/the_biases_of_links.html

Burke, B. (1999, 2005). *Antonio Gramsci and informal education*. Retrieved December 3, 2008, from http://www.infed.org/thinkers/et-gram.htm

Cherns, A. (1976). The principles of sociotechnical design. *Human Relations*, *29*(8), 783–792. doi:10.1177/001872677602900806

Dahlberg, L. (2001). Computer-mediated communication and the public sphere: A critical analysis, journal of computer-mediated communication.

Day, R. J. F. (2004). From hegemony to affinity: The political logic of the newest social movements. *Cultural Studies*, *18*(5), 716–748. doi:10.1080/0950238042000260360

Fayol, H. (1960). General principles of management. In H.F. Merrill (Ed.), *Classics in management* (pp. 216-224). New York: American Management Association.

Federman, M. (2005). *Why Johnny and Janey can't read, and why Mr. and Ms. Smith can't teach: The challenge of multiple media literacies in a tumultuous time*. Paper presented at the University of Toronto Senior Alumni Association Lecture Series, Toronto, Canada, November 2005.

Federman, M. (2006a). *Online1141/EntryClass*. Retrieved on December 3, 2008 from http://online1141.pbwiki.com/EntryClass

Federman, M. (2006b). *What is the (next) message: Reflections of an adult educator - part 5.* Retrieved December 3, 2008 from http://whatisthemessage.blogspot.com/2006/10/reflections-of-adult-educator-part-5.html

Federman, M. (2006c). *Online1141/ExitClass.* Retrieved on December 3, 2008 from http://online1141.pbwiki.com/EntryClass

Foucault, M. (1980). *Power/Knowledge.* New York: Pantheon Books.

Ganesh, S., Zoller, H., & Cheney, G. (2005). Transforming resistance, broadening our boundaries: Critical organizational communication meets globalization from below. *Communication Monographs, 72*(2), 169–191. doi:10.1080/03637750500111872

Grace, A. P. (1998). Parameters, pedagogy and possibilities in changing times. In S. M. Scott, B. Spencer, & A. M. Thomas, (Eds.), *Learning for life: Canadian readings in adult education* (pp. 114-24). Toronto, Canada: Thompson Educational Publishing.

Grace, A. P. (2004). *Mediating lifelong learning in Canada in inclusive contexts.* Paper presented at the 3rd International Lifelong Learning Conference, Central Queensland University, Yeppoon, Central Queensland, Australia.

Hara, N. and R. Kling (2002). *IT supports for communities of practice: An empirically-based framework.* Center for Social Informatics, 2003.

Hauben, M., Hauben, R., & Hauben, J. (1994, 1996). *Netizens: On the history and impact of USENET and the internet.* Retrieved December 3, 2008, from http://www.columbia.edu/~rh120/

Haythornthwaite, C. (2002). Building social networks via computer networks: Creating and sustaining distributed learning communities. In K.A. Renninger & W. Shumar, (eds.). *Building virtual communities: Learning and change in cyberspace,* (pp. 159-190). Cambridge, UK: Cambridge University Press.

Haythornthwaite, C. (2006). Facilitating collaboration in online learning. *Journal of Asynchronous Learning Networks, 10*(1), 7–24.

Hersey, P., Blanchard, K. H., & Natemeyer, W. E. (1979). Situational leadership, perception, and the impact of power. *Group & Organization Studies, 4*(4), 418–428. doi:10.1177/105960117900400404

Hiltz, S. R., & Turoff, M. (2005). Education goes digital: The evolution of online learning and the revolution in higher education. *Communications of the ACM, 48*(10), 59–65. doi:10.1145/1089107.1089139

Huzzard, T. (2004). Communities of domination? Reconceptualising organisational learning and power. *Journal of Workplace Learning: Employee Counselling Today, 16*(6), 350–361.

Johnson, C. M. (2001). A survey of current research on online communities of practice. *The Internet and Higher Education, 4*(1), 45–60. doi:10.1016/S1096-7516(01)00047-1

Klemm, W. R. (2002). *Extending the pedagogy of threaded-topic discussions.* Retrieved December 3, 2008, from http://technologysource.org/article/extending_the_pedagogy_of_threadedtopic_discussions/

Knowles, M. S. (1980). *The modern practice of adult education: From pedagogy to andragogy.* Englewood Cliffs: Prentice Hall/Cambridge.

Laiken, M. (1998). *The Anatomy of High Performing Teams: A Leader's Handbook (3rd Edition).* Toronto: University of Toronto Press.

Laiken, M. (2002). Models of organizational learning: paradoxes and best practices in the post-industrial workplace. *Organizational Development Journal, 21*(1, Fall), 8-19.

Lapadat, J. C. (2004). Online teaching: creating text-based environments for collaborative thinking. *The Alberta Journal of Educational Research, 50*(3), 236–251.

Lave, J., & Wenger, E. (1991). *Situated learning: Legitimate peripheral participation*. Cambridge, UK: Cambridge University Press

Lears, T. J. J. (1985). The Concept of Cultural Hegemony: Problems and Possibilities. *The American Historical Review, 90*(3), 567–593. doi:10.2307/1860957

Miles, M. B. (1970). *Learning to work in groups: a program guide for educational leaders*. New York: Teachers College Press, Teachers College, Columbia University.

Mumby, D. K. (1997). The problem of hegemony: Rereading Gramsci for organizational communication studies. *Western Journal of Communication, 61*(4), 343–375.

Pelz, B. (2004). (My) three principles of effective online pedagogy. *Journal of Asynchronous Learning Networks, 8*(3), 33–46.

Poftak, A. (2006). Blackboard-WebCT merger draws concern. *Technology & Learning, 26*(6), 6.

Samuel, A. (2004). *Hacktivism and the future of political participation.* Unpublished Ph.D. dissertation, Harvard University, Cambridge, MA.

Semler, R. (1994). Why my former employees still work for me. *Harvard Business Review, 72*(1), 64–74.

Senge, P. (1990). *The Fifth Discipline: The Art & Practice of The Learning Organization*. New York: Doubleday.

Serce, F. C., & Yildirim, S. (2006). A web-based synchronous collaborative review tool: A case study of an on-line graduate course. *Educational Technology & Society, 9*(6), 166–177.

Spolsky, J. (2003, 2003/03/03). *Building communities with software*. Retrieved November 4, 2006, from http://www.joelonsoftware.com/articles/BuildingCommunitieswithSo.html

Taylor, F. W. (1911). *The principles of scientific management*. New York: Harper.

Technorati. (2008). *Technorati: Popular blogs.* Retrieved December 3, 2008 from http://www.technorati.com/pop/blogs/

Thomas, M. J. (2002). Learning within incoherent structures: the space of online discussion forums. *Journal of Computer Assisted Learning, 18*(3), 351–366. doi:10.1046/j.0266-4909.2002.03800.x

Tuckman, B. (1965). Developmental Sequence in Small Groups. *Psychological Bulletin, 63*(6). doi:10.1037/h0022100

University of Fribourg. (2006). *Platform use - 2005*. Retrieved December 3, 2008 from http://www.edutech.ch/lms/inst-platforms.php

Weber, M. (1947). *The theory of social and economic organizations*, (T. Parsons (Ed.), A.M. Henderson & T. Parsons, Trans.). New York: Free Press.

Wikipedia. (2008). *Wiki*. Retrieved on December 3, 2008 from http://en.wikipedia.org/wiki/Wiki

Youngman, F. (2000). *The political economy of adult education and development*. New York: Zed Books Ltd.

ENDNOTES

1 A portmanteau word comprised of "hacking" and "activism," defined as "the nonviolent

use of illegal or legally ambiguous digital tools in pursuit of political ends" (Samuel, 2004).

2 "Online" courses and programs are taken to mean those in which at least 80% of the material is delivered via the Internet, in most cases with no face-to-face meetings (Allen & Seaman, 2006).

3 The references to "pedagogy" throughout this chapter are not intended to distinguish between a teacher-oriented approach to education that is considered appropriate for younger learners, in contrast to an "andragogical" (Knowles, 1980), learner-centred approach that Knowles considers more appropriate for older learners. Both authors of this chapter agree that learner-centred, embodied, and experiential adult-education approaches, encompassing both formal and informal learning, are universally appropriate irrespective of the learners' – or instructors' – ages. In our usage, we are seeking to reclaim "pedagogy" in the context of enlightened and reflective practitioners.

4 The face-to-face course had two marked essays, the first offered to provide feedback early in the course, to help students become familiar with the writing expectations of the instructor, and to help the instructor gain a sense of the students' written work.

5 T-groups were a form of organizational training in which the facilitator/instructor provided little or no guidance with respect to either the content or process of the course. The students/participants were to independently determine their collective objective, process and necessary materials as an experiential learning exercise in team building and collaboration.

Section 2
Addressing Various Divides in E-Learning

Chapter 6
Cultural Implications of E-Learning Access (& Divides): Teaching an Intercultural Communication Course Online

Pauline Hope Cheong
Arizona State University, USA

Judith N. Martin
Arizona State University, USA

EXECUTIVE SUMMARY

This chapter presents a case study of developing and teaching an intercultural communication (IC) course online, within the context of a department in a large research University in the U.S. In so doing, the authors discuss a broadened and recursive model of cultural access and divides in E-learning. Expanding on van Dijk's (2005) framework, the authors present several ways in which their IC course attempts to address multiple pathways of E-learning access, including motivational, material, skills and usage access. They describe both the successes and challenges of meeting the goals of e-learning access with specific examples of the content, activities, assignments, pedagogical strategies, and student assessment in this online course. Finally, they identify challenges of this e-learning at the micro and macro level context—in the course, university writ large and in the communication discipline.

INTRODUCTION: UNDERSTANDING E-LEARNING ACCESS

Increasing numbers of students are enrolling in online courses in institutions worldwide (Allen & Seaman, 2007). The internationalization of higher education and popularity of applying a global approach to education is increasingly facilitated by the use of communication technologies in E-learning (Burbules, 2000). An important topic in new media use is the issue of access, related to the 'digital divide' or the technological chasm between information haves and have-nots which is of concern among policy makers and educators both in the United States and abroad (Warschauer, 2003). As van Djik (2005) and others suggest, the digital divide is more than just a question of access to computer software, but rather includes technocapital on many levels including: mental, material, skills, and usage.

DOI: 10.4018/978-1-60566-942-7.ch006

Furthermore, in many places in the world, technocapital is in competition with basic necessities of life (Olaniran & Agnello, 2008).

There is a research gap in considering the cultural implications of E-learning access and (on the flip side, divides). In particular, there are cultural aspects of socio-technical divides that tend to be overlooked in E-learning (Ess & Sudweeks, 2005; Schwartzman, 2007). E-learning technologies should not be considered *fait accompli* but recursively constructed and defined by a host of psychological, social and political influences and actors (Dutton, Cheong & Park, 2004a). In this paper, we present a case study of developing and teaching an intercultural communication (IC) course online, within the context of a department (subsequently referred to as 'the department') in a large research University ('the University') in the U.S. In doing so, we discuss a broadened and recursive model of cultural access in E-learning, to encompass access to communication technologies, information, people and services associated with online pedagogy. This paper has theoretical and practical implications for educators' curricula design and implementation of E-learning courses.

BACKGROUND AND SETTING THE STAGE

We begin by first discussing the backdrop and circumstances leading to this IC course development. The online course was first delivered as a face-to-face course (Elements of Intercultural Communication), an integral part of the department's offerings - a popular major choice among undergraduates. The course is designed "as an introduction to the basic concepts, principles, and skills for improving communication across racial, ethnic and cultural differences." The course has been offered each semester since Fall 2006. The course is always enrolled to the maximum capacity (30 students) and there is usually a waiting list of students. While the course is designed as an introductory level course, most of the students are juniors and seniors, a few sophomores and occasionally one or two freshmen. This is due to the fact that courses in this department are impacted in general and students are not able to enroll in required courses until late in their college career. The students in this online course have a variety of majors. Most are business or pre-business majors, a few communication students and the remainder represent a range of majors including pre civil engineering, anthropology, music, art, biotechnology. The cultural backgrounds of the student usually reflect the make up of the University (approximately 65% white, 35% international students, and ethnic and racial minorities)—the largest minority group is Latino and the smallest American Indian.

Spurred by the following factors, this course was first developed in 2004 and offered Fall semester 2006. At the time of the course's inception, the University was undergoing a period of rapid expansion and growth, and the leadership was promoting a philosophy of "access, excellence and impact". Like many academic institutions, the University saw potential income in online courses The department had been at the forefront of online courses, offering a very successful course since 1996 and was eager to expand its offerings. Consonant with its mission, from 2000-2005 the University provided monetary incentives and technology assistance to faculty interested in developing new online courses. The initial proposal to put the existing IC course online included the following rationale: "Our undergraduate students will benefit as we will be able to expand our offerings of IC to students who are not traditional students, who may be off campus students. Secondly, our graduate students and faculty who teach this course will benefit by having a competitive edge in this aspect of curriculum delivery. Finally, from a scholarly point of view, gaining expertise in this new area of communication research will enhance our visibility in the field." In this light, this online course can be seen to be part of the overall curriculum infrastructure to open up opportunities for students to access intercultural courses and complete their degree.

An interview with the Director of Online Programs revealed the extent of the University support for online programs, as well as information about the administration philosophy, culture and practices.[1] She reported that the numbers of online courses are steadily increasing and gave four reasons for the University support: 1) The University is focused on access and bringing education to more people. As she noted "Online degree programs are a way to bring a University education to those who may not be able to physically make it to a campus"; 2) students want online courses, so the University is meeting student demand by offering increasing online course options; 3) the University believes that online courses can be as *effective* for student learning and in large classes can be *more efficient* than face-to-face classes (due to technological advances, students in courses with high enrollment of several hundred students receive the same lecture material from the professor as in face-to-face class, can have *more* interaction with each other and the professor and *more* access to supplemental materials like online quizzes, video examples, learning activities). In addition, there are better retention checks—a major priority for the University, e.g. when an assignment is missed, there is prompt communication and follow-up with the student by the professor. There is also a new "lockdown browser" in development which will provide better control for online exams (cheating is a huge problem in a 500 student class); and 4) today's students learn better with new media than students in the past because they are comfortable with technology.

She described the specific ways in which the University offers support. For example, her office offers technical as well as pedagogical assistance for existing online courses and those in development, and thus serves as a type of e-learning quality control. While it is not mandatory for instructors to work with her office staff, most do take advantage of the available expertise. She and her staff work with faculty who already teach online but want to improve the design and delivery of their courses—e.g. incorporate the latest technology in testing, discussion board and lecture delivery.

While the macro organizational culture was supportive of E-learning, and believes that e-learning can be as effective as traditional instruction, there were and continue to be a number of challenges. In a recent review of online programs, Bejerano (2008) notes that (1) e-learning is not for everyone, that some college students need the physical presence of instructors and other students to socially and intellectually integrate and adapt to the college experience, (2) success in online courses require discipline which not all students possess, and 3) there is little evidence of higher-order learning, e.g. evaluation and synthesis, in online instruction.

For these and other reasons, some faculty were at first hesitant to transfer the IC course online. They knew that developing and teaching online courses takes an enormous commitment. Early estimates were that it takes 500 hours to put an existing course online (Santovec, 2003), and teaching online is as time-consuming as teaching a traditional course (Akintunde, 2006; Carnes, Awang & Marlow, 2003; Sieber, 2005; Young, 2005). Also, some faculty perceived that an online intercultural communication course could not replicate the richness of experiential learning that takes place in the face-to-face course. Many of the department's students are white, middle class, with little intercultural experience, and the course was intended to deliver theoretical concepts and provide exposure to intercultural experiences. However, as online courses became more prevalent and it was apparent that students (and faculty) live increasingly mediated lives, this IC course, as explained below, was developed to incorporate meaningful experiential learning activities.

CASE DESCRIPTION: EXAMINING MULTIPLE LAYERS OF E-LEARNING ACCESS

There exists a need to address the dialectics of access in terms of technology adoption to minimize the disparities in Internet use for education (Natriello, 2001). Dominant discourse in digital divide research is grounded in a functionalist approach, focusing on the binary nature between the technologically rich and poor (Warschauer, 2003). The lion's share of digital divide research consists of quantitative data analyses with various studies concluding that socio-demographic factors like age, gender, and race play a significant part in access disparities in terms of Internet adoption. However, leading conceptualizations of digital inclusion rarely consider the social embeddedness of the Internet and the dynamic interactions between technological and social inequalities (Selwyn, 2004). According to van Dijk (2005), digital access to information technology should be theorized as cumulative and recursive, including motivational, material, skills and usage access. Past survey studies employ measures of Internet use to reveal summary data differences in access but quantitative data may enshroud how cultural vectors dynamically affect the situated experiences of certain populations, including how college students grapple with various layers of the E-learning access.

In the next section, we present several ways in which our IC course attempts to address multiple pathways of E-learning access. As we do not want to essentialize instructors, administrators or students' behavior to group membership in our presentation, we note that various dialectics characterize peoples' relationships with E-learning technologies, (Martin & Nakayuma, 1999) that may simultaneously engender digital bridges and divides. A consideration of these related and oppositional logics illustrate how E-learning in this IC course case study may bridge multiple layers of access, albeit incompletely within the context of the larger department and University.

Material Access

As Mitchell (1999) argues, bridging the digital divide entails access to electronic appliances as "equitable access" requires access to fast digital connections, the affordable appliance, user-friendly software and the skills and motivation to learn and benefit from the new technologies (Mitchell, 1999, p. 151). In our case, several efforts have been made on the University and department level to provide parity in students' connections with regards to access to a panoply of digital resources to fulfill their educational goals.

The role of social institutions, including the University, is significant in shaping technological access (Kerckhoof, 1995), as the "political will" of governing organizations can help turn past divides into present and future digital opportunities (Koss, 2001). In recent years, the University has advanced several steps to build physical and technological infrastructure to create a wired and wireless campus. Computer laboratories in multiple buildings and student residential halls on campus provide free computing facilities. To provide equal access in computer facilities across campus for qualified students with disabilities, the University ensures federally mandated physical and program access for students with disabilities in all computer laboratory facilities commensurate with the general student population. Furthermore, the University provides wireless coverage throughout the campus. A 'mobile initiative 1:1' has been launched in partnership with Apple, Dell and Verizon Wireless. This initiative provides discounts on laptop purchases and also provides opportunities for students from financially disadvantaged households to apply for a $500 (locally endowed) scholarship toward a new laptop purchase. Additionally, operating software is provided gratis by the University as students can download popular applications like Microsoft Word, SPSS, Photoshop, Microsoft Access, Acrobat Professional, and Microsoft Powerpoint, needed to complete their coursework. On the departmental level, there are several computers available to students. To the extent that differences in family resources and environment play a part in students' using computers to benefit their academic pursuits (Attewell, 2001), the University's plan to provide various channels of digital connections to students serves as an entry point for them to gain technical access to electronic appliances and software to log on to course materials.

Every student is provided access to the University's virtual E-learning environment and electronic course management system, Blackboard, where course materials for online courses are housed. This IC course is hosted on Blackboard which is integrated into the students' personalized start page from which they can access their University email, receive messages from various University offices (e.g. library, messages about registration, etc), and a number of other Internet resources. As soon as they register, this course is immediately listed on their personal start page and they click on the course listing which takes them to the course website.

E-learning allows students to experience a sense of continuous dialogue – they can post anytime; discussions and insight are not confined to the classroom. A student may send a paper or post a message after school hours, a unique characteristic of online instruction. However, it should be noted that gaps may still exist in terms of technical access, for example, in online connectivity. Regular, if not monthly maintenance may disrupt online services. Dutton, Cheong & Park (2004b) highlight in their case study that technological glitches on electronic course management systems like Blackboard represent a more substantial barrier to Internet use than anticipated by students and faculty, including slow response times and trouble uploading course materials. A recent Pew Internet and American life survey, entitled 'When technology fails' also points to technical limitations that are a facet of digital divides as 39% of American adults surveyed with desktop or laptop computers reported to have had their machines not work properly at some time in the previous year (Horrigan & Jones, 2008).

Therefore, it is noteworthy that technical problems may still exist despite advances in courseware upgrades and the wiring of classrooms. When the course was first offered, the electronic course system was frequently unavailable for periods of time due to technological failure. This created anxiety for the students if the breakdowns happened to occur when an assignment was due. Course exams have also been the source of some problems since a technical glitch can cause an exam to close before students have completed it. The course policy is that once an exam is started it cannot be reopened (to prevent students from opening the test, copying it and then taking the exam later). When a test is prematurely closed (because of technical breakdowns or less legitimate reasons) the instructor makes a judgment as to whether to open the exam again.

In order to deal with the current occasional technological glitch and the challenges for first time online students, the IC instructor is readily available and open to student contact—particularly at the beginning of the semester and the University maintains a 24- hour help desk where students can call, email or chat electronically with help-desk personnel if they have technical problems. In addition, the University maintains a website with useful information—including a tutorial for first time students, which walks them through the details and logistics of their course website. This information about the website, helpline, and instructor contact information is posted prominently in the IC course syllabus, and students are instructed who to contact for various problems they might encounter (the help desk for technical problems, the instructor for answers to questions about grading, etc.). On the IC course syllabus, students are advised to submit their assignments early before the stipulated deadline and to always keep a copy of their assignments until they have received their grades. The IC course also allows for one missed assignment to give students a little leeway if they are not used to online courses, or if they encounter technical problems which might prevent them from submitting an assignment on time.

Motivational and Mental Access

Besides access to physical electronic gadgetry and the Internet, intellectual access is another dimension of E-learning. Stanley (2003) in an ethnographic research among marginalized populations highlights how individuals' "self-concept", "fear", and the "relevance" of the Internet interfere with their motivation to engage and thus potentially benefit from electronic technologies. With regard to E-learning, prior studies on student motivation and their attitudes toward online pedagogy highlight connections between learning style preferences and particular cultural groups (Aragon, Johnson & Shaik, 2003). For instance, one way of framing learning styles is according to field independence/dependence (Witkin, Moore, Goodenough, & Cox, 1977). Field independent learners prefer to work alone, with narrow focus; they impose structure on environment and have self-defined goals. In contrast, field-dependent learners rely more on the context for clues about information, they prefer structure and experience environment more holistically and globally; they are also interested in people and learn better in more social settings. For example, Mestre (2006) notes that African American students tend to prefer experiential learning and minimal structure, while Native Americans and Latinos prefer relational, social learning. All three groups tend to be more field-dependent learners while Asian Americans and white males tend to be field independent learners. Notwithstanding the complexity of online communication (Hewling, 2005), these insights raise significant issues for E-learning instructors to address in order to encourage fruitful participation for students of diverse educational and cultural backgrounds.

This IC course was designed to take into account various learning styles so as to maximize students' motivational access. Typically, online learning situations stress logical, text-based, passive learning, more suited to those students who prefer abstract conceptualization and reflection, as well as field-independent learning (Mestre 2006). Thus, in addition to information delivery, this IC course was designed to provide opportunities for experiential learning practices via several course assignments that prompt students to apply IC theories to real life situations. For example, the initial assignment asks students to create and share their cultural profile: Where they grew up, their language background, any intercultural experiences they've had (e.g overseas travel or study, and family members and friends from different religious, ethnic, national cultures). This exercise helps students and instructor to get to know each other in a more personal way, which is especially important for relational learners who learn best through interpersonal connections. A second assignment that connects real life to intercultural concepts is the family history interviews. Students are instructed to investigate their family's immigration history by interviewing the oldest members of their family (Martin & Davis, 2001). They write a paper summarizing this history and then post discussion messages ("After writing your family history paper, what did you learn about your family that might help you understand today's immigrants better?"). Through this exercise students learn from their own and others' histories about the complexities of intercultural communication in an immigration context. A third assignment asks students to interview someone in an intercultural relationship, to identify and describe the benefits and challenges of their relationship, and to connect their findings to concepts discussed in the course materials. Results of this assignment also form the basis for a discussion board forum where students are asked to discuss among themselves what they learned from those interviews that they might apply to their own relationships. Students seem to learn a lot from this experience that they can apply to their own lives. As one student reported:

It was great to learn more about [intercultural relationships] because I have many friends who are in intercultural relationships and I cannot always relate. I feel that learning in depth about intercultural relationships can definitely be applied to my own life. The interviews that we conducted as well as discussion posts that I was able to read from my fellow classmates has given me some insight that I can apply to the important people in my own life who are currently in an intercultural relationship.

Furthermore, various aspects of the course design caters to what Mestre (2006) identified as the "global learners" and "millennium learners"; new, emerging types of learners who are ethnically diverse, used to digital environments, used to multi-tasking, visual learning and interactivity via mobile computing and online social networking tools. Online courses may provide increased opportunities for interactivity where there is more student-student and teacher-student interaction than in traditional classrooms (Akintunde, 2006; Merryfield, 2001).

Many E-learning studies suggest that the most effective discussions are student-centered where the role of instructor is not as authority but as facilitator (Dennen, 2005; Kelly, Ponton, & Rovai, 2007), however there is little agreement on the specifics of optimum instructor participation (Mandernach, Gonzalez & Garrett, 2006). This IC course provides very structured, specific questions to encourage student reflection on their own knowledge and opinions (e.g. often students must complete a prior activity before posting messages to the discussion board). The IC instructor participates little during the actual discussion but provides a weekly summary of each discussion board, highlighting the main points in the discussion, often quoting particular students' messages, and reinforcing class norms of respectful, thoughtful discussion posts. In this way, students learn quickly that the instructor consistently monitors the discussions and that their contributions play a significant role in what they learn and the grade they earn. However, sometimes students remark that they would like more consistent instructor input during discussions. This remains a challenge for the course instructor to know how to achieve optimal input, given the variety of student personal preferences and learning styles.

The course evaluation is designed to encourage honest and open discussions as students earn credit/no credit points for the discussion section of the course. In order to receive credit, the students must: 1) post messages on different days during the unit, so that a *discussion* takes place and it's not just every student posting two messages on the last day of the unit; 2) address the questions listed in the assignment, 3) demonstrate that they have completed a prior activity, if required, 4) make a substantive contribution that moves the discussion forward (they can't just say "I agree") and 5) be respectful of others in their postings. On the syllabus, it is stated that "Some of the topics we cover in this course can lead to emotionally intense discussions. One of the goals of the course is to help us become more aware of how people from different backgrounds think and experience life. This learning cannot occur when people feel threatened or defensive. If postings are deemed by me to be hostile or demeaning to others, the message will be removed and the sender will not receive credit".

Research shows that students who are too shy or anxious to talk in face-to-face classes may feel freer to speak up in online discussion (Merryfield, 2001; Thompson & Ku, 2005). This IC course was designed to encourage equitable power dynamics by structuring discussion assignments so as "move the center," creating a space wherein all learners feel they are the center of instruction. Merryfield (2001) describes how international students and students of color are often marginalized in classroom discussions (even online), while white U. S. students dominate the discussion. She describes how class interaction changed after she "shifted the center" in her online teacher education course by requiring that each student post the exact same number of discussion messages: "I found a lack of dominance of any one group in initiating new ideas.....there is no obvious pattern of one group of teachers (students) controlling the discourse or silencing others. Nor did I find evidence that people chose to interact with others like themselves" (p. 295). In this IC course, the instructor maintains an awareness of students who may feel marginalized (e.g., non-native English speakers, ethnic/racial minorities, gay students) and attempts to make a legitimate space for them in class discussion, particularly in the early days of the course. This is done by quoting *these* students in the instructor discussion summaries and/or explicitly reinforcing the opinions/feelings that these students express during discussions so that not one group of students seems to control the discourse.

While online pedagogy does not easily afford opportunities for skill practice (Doo, 2006), in this IC course, students have been asked to engage in experiential activities to gain intercultural experience and develop intercultural communication skills. For example, in one assignment, students experientially explore the role of nonverbal communication in prejudicial thinking. Using Breeze Plug-in technology, they close their eyes and listen to the audio recording of their instructor telling them to image a U. S. student and Japanese student interacting in front of the student union. Then they are asked to open their eyes and describe the U. S. student very specifically (how tall, what color eyes, hair, appearance, clothing etc)—in a written paragraph. They then look at their description and are asked to write about who they did NOT see as U. S. American (e.g., Latino/a, disabled person, old person, heavy person?) and then write about and later post their ideas on the discussion board about the implications of intercultural interactions (e.g., if we only "see" certain people as "Americans" what does that say about how we interact with people we meet)? This exercise has proven to be very effective and has led many majority (and some minority) group students to insights about how ingrained and pervasive prejudice and discrimination are. As one student reported:

I have learned how easily I and everyone else stereotype people. I realized this from the assignment that we did when you had us visualize two people from different cultures having a conversation. I was surprise at the fact I did that. ...After that assignment I realized that I need to be careful on how I view others before I get to know them. Now, before I go and talk to someone I make sure I do not prejudge what they are going to be like based on their outer appearance. .... I believe I am a better communicator with those that are culturally different from me because I keep an open mind and do not prejudge how I think they should act.

Students in this IC course are also required to participate in two different virtual collaborations, one with other classmates and one with students in a similar course in an overseas University. Recognizing that there may be cultural differences in students' online collaborative behaviors (Kim & Bonk, 2002), students are asked to reflect on their own participating and cultural learning—particularly in their collaboration with overseas students. One written assignment asks the students to analyze the intercultural communication that took place in their online collaboration—relating their own success and challenges in terms of what they have learned in the course about intercultural skills and effectiveness. In addition, the IC instructor maintains an awareness of cultural differences that may impact the performances of students and on occasion meets with or maintains email contact with students who have particular challenges in meeting course expectations.

In sum, this IC course provides motivational and mental E-learning access in the aforementioned ways, yet it should be noted that divides may continue to exist for some students, dependent on their backgrounds and experiences. Several scholars have highlighted the potential impact of different communication styles in online communication, using the frameworks of E. T. Hall (1959, 1966; low/high context communication style) and communication styles reflecting Hofstede's (1980, 1997) values framework (individualism/collectivism, power distance, masculinity/femininity, high/low uncertainty avoidance). For instance, Olaniran (2001) notes that conversations between low and high-context communicators may be difficult online as the low-context communicator may be comfortable being direct about feelings and opinions, whereas the high-context communicator might feel rather constrained by computer-mediated communication.

In this course, several strategies are used to motivate all students. If a student seems to "disappear" from discussion fora or fails to turn in more than one assignment, the instructor contacts the student, asks if everything is alright. However, it is a challenge to assist students when they disappear from an online course. As Stanford-Bowers (2008) reminds us, there are many factors beyond the instructors' control that affect student persistence in online courses, and often require attention to minute details that are sometimes overlooked or taken for granted.

Approximately 50% of the evaluation points are credit/no credit points, which also helps motivate all students, regardless of communication style or prior knowledge. They just need to show up and participate. Furthermore, students participate actively in evaluation; they are asked to evaluate themselves and others in their virtual team assignment. They assign points to themselves and others and the Instructor does not add any additional evaluation. Therefore, students know from the beginning that there will be consequences if they do not actively contribute to the course project.

Technological Skills Access

Another related area of mental access relates to the students' online skills. According to Warschauer (2003), technological "literacy" involves the development of relevant understandings of devices, content, skills, understanding, and social support in order to engage in meaningful pedagogical practices. Literacy also links to a "second- level" digital divide in terms of online skills as past experiments show that there exists a considerable variance in the way and time in which individuals access and find information that they need online (Hargattai, 2002). Lack of technical skills and assistance is noted as a barrier to persistence in online courses (Stanford-Bowers, 2008). Cheong's research among Asian college students also highlight another chasm differentiating highly versus lowly skilled Internet users, in terms of their daily computer and Internet problem solving behaviors which has implications for users' productivity and potential benefits that can be reaped from the Internet. Contrary to popular conceptualizations of Asian youths as a cohort of technically savvy experts, findings from survey and interview data show considerable variance in participants' Internet expertise and problem solving behaviors, with some demonstrating limited knowledge of Internet use and awareness of troubleshooting strategies (Cheong, 2008). Among American adults, findings show that about half the population surveyed needed help from others to set up new devices or to show them how they function (Horrigan & Jones, 2008), illustrating the significance of computing skills for literacy access and divides.

Among the steps taken on the University level to increase computer literacy and improve students' online skills was the formation of the E-learning network in 2001. The technology-based learning and research unit, as part of the College of Education, partnered with Cisco systems to develop online networking and IT curriculum. Students can earn credit from the University by completing modules on topics related to hardware and networking called the "basic fundamentals approach technology" online courses that include graphics, video, hands-on virtual labs and stimulations.

As noted earlier, students in the IC course are provided technical help from the University helpdesk on occasions when they have problems accessing the discussion board, or submitting their papers to the course "digital dropbox". In each instance students contacted the technical staff and were given assistance. In one case a student taking an exam in a computer lab insisted that the system has inexplicably closed his exam and that the action was seen by a lab attendant. The instructor contacted the lab attendant who verified that the action had occurred as described by the student and the exam was reset.

As people in many professions find themselves increasingly working in globally distributed environments (Connaughton & Shuffler, 2007), this IC course requires students to work in virtual teams and complete an assignment where they develop a powerpoint presentation on a particular relevant topic, which is then posted on the course website. This assignment is designed to provide students with transferable technical skills which should serve them well in future professional situations, particularly in contexts where they are working with culturally diverse teams. Each virtual team has their own communication "center" on the course website, where they can post documents, exchange emails etc. The course also provides exposure to different kinds of Internet resources and interactions. For example, students learn to post messages on a discussion board, learn when to start new discussions on the fora, use a digital drop box to submit assignments, and listen to Breeze powerpoint presentations with audio posted by the instructor.

Content Access

A fourth facet of E-learning access is the availability of content appropriate to students as past content analyses of websites through the most popular Internet portals revealed the dearth of inadequate content for traditionally disadvantaged communities; newer immigrants and ethnic minorities in the U.S. (Lazarus & Mora, 2000). In this IC course, the assignments cover a wide range of cultural knowledge; many are topics that all students can relate to regardless of ethnic/racial background (e.g., cultural dimensions of nonverbal communication, conflict, relationships), and in addition some topics (e.g. code-switching, multicultural identities, managing intercultural transitions), tap particular knowledge and expertise held by students not generally privileged in many University classes—those who are bi/multilingual, those with minority cultural backgrounds, those who have direct experiences with prejudice and discrimination, and those who have learned to navigate different cultures in their everyday life.

Merryfield (2003) points out that "online technologies increase the depth of study and the meaningfulness of academic content" (p. 162), in part because students have time to think and reflect in asynchronous interactions and also because writing, in contrast with spoken discussion, encourages deeper, more thoughtful analysis. This IC course provides structured online intercultural experiences for students to reflect upon in undergraduate courses, particularly where there is little cultural diversity among the students. Online dialogue can be an effective way of engaging students in discussion about race and ethnicity, specifically "the use of an online platform can facilitate one's learning about 'others' in a more engaged, open and accommodating manner, which goes beyond the traditional classroom teaching and learning of intercultural communication" (Kanata & Martin, 2007, p. 1).

The course also provides access to students overseas as one assignment pairs each IC course student with a student from a similar course in another university in either Europe or Asia. Each student pair conducts an "ethnographic" research project on cultural differences/similarities of a communication practice (e.g. workplace conflict, Internet relationships). Each student interviews 5 of their friends on the topic, the two students then compare their ethnographic data, draw some conclusions about cultural similarities/differences between the two cultural groups and speculates on the reasons for their findings, using course materials to guide their analyses and conclusion. A final assignment asks the students to reflect on their own intercultural behavior with the overseas student, what they learned from the experience and what they can take from the experience that may help them in their future work and social relations. One student describes what he learned in the virtual collaboration assignment:

The Virtual Collaboration experience was very rewarding... I learned several important things that I will use in the future when working on projects such as this. Firstly, language barriers and time zone differences can increase the time it takes to complete common tasks. Realistic timelines need to allow extra time for the completion of tasks due to these factors. In addition, proper planning and frequent communication is essential. Clarification on issues can often take 24 hours or more to resolve. However, the unique perspective that collaboration with people from differing cultures produces is often a better result than either is capable of alone.

CURRENT CHALLENGES

This chapter has addressed multiple access/divide dialectics, in the case of an online IC course, in order to illuminate the multifaceted nature of E-learning access. While the delivery of this IC course seems to have met many of the challenges of access, a few challenges remain. Starting at the most local level, concerning material access—while the University has addressed this issue in many ways, there are still some students who cannot afford home Internet access and many nontraditional students have difficulty accessing the University computers because of child-care issues, work schedules, or lack of reliable transportation. Concerning mental access, while a great deal of effort is extended to motivate all students and to adequately meet the challenges of students' various learning styles and culture-specific communication styles, this course could be altered to incorporate more student input and suggestions concerning this goal. Similarly with issues of content access, it is possible that the content could be altered to be more explicitly inclusive of cultural issues, i.e., while cultural communication issues of race, ethnicity, age, and religion are fairly explicitly addressed, issues concerning disabilities are not.

At a more macro level, administrative challenges remain. As noted, while the course seems successful and desired, issues at the departmental level remain to prevent increasing access—i.e. adding more sections. The extant and looming economic crisis in the U.S. has also recently impacted the University in various ways (e.g. mandatory revertment of funds to the State) and may lead to the implementation of more bottom-line and cost-cutting philosophies to manage future technical and administrative support for E-learning. There are also challenges at the University level, described by the Director of Online Programs quoted earlier. One challenge is the evaluation of the effectiveness of existing courses which led to a recent initiative "Midterm Course Design Survey" where students were asked to evaluate the design of and materials in online courses, what technologies were most beneficial, and which activities were most helpful in student learning. These results will be useful in helping faculty and administration to improve the quality of online courses. A second challenge is to get the word out to faculty that technological and pedagogical support exists. She noted that while most instructors are open to teaching courses online, there are some who are resistant. Finding ways to persuade faculty to become involved in E-learning remains a challenge.

Finally, a word about the broader disciplinary context. It is important to context this case study in the larger scholarly endeavors of cultural and communication research. Historically, intercultural communication scholars have focused on face-to-face encounters and there is a dearth of research investigating the relationship between culture and mediated communication; theory lags substantially behind practice. Everyday, millions of people communicate online with culturally different others in social networking sites (SNS), on blogs, through email, in virtual teams—and communication scholars have only begun to scratch the proverbial surface of knowledge. Concomitantly, most intercultural communication courses are offered in face-to-face contexts. There has been some recent interest in online instruction, some institutions are supportive, but as we have shown in this paper, the issues of culture and access in E-learning processes are complex and multi-layered.

REFERENCES

Akintunde, M. (2006). Diversity.Com: Teaching an online course on white racism and multiculturalism. *Multicultural Perspectives, 8*(2), 35–45. doi:10.1207/s15327892mcp0802_7

Allen, I. E., & Seaman, J. (2007). *Online nation: Five years of growth in online learning*. Needham, MA: The Sloan Consortium. Retrieved September 22, 2008 from http://www.sloan-c.org/publications/survey/pdf/online_nation.pdf

Aragon, S. R., Johnson, S. D., & Shaik, N. (2003). The influence of learning style preferences on student success in online versus face-to-face environments. *American Journal of Distance Education, 16*(4), 227–244. doi:10.1207/S15389286AJDE1604_3

Attewell. (2001). The first and second digital divides. *Sociology of Education, 74*(3), 252-259.

Bejerano, A. R. (2008). The genesis and evolution of online degree programs: Who are they for and what have we lost along the way? *Communication Education, 1*(3), 408 – 414.

Burbules, N. C. (2000). Does the Internet constitute a global educational community? In N. C. Burbules & C. Torres (Eds.), *Globalization and education: Critical perspectives* (pp. 323-355). New York: Routledge.

Carnes, L. W., Awang, F., & Marlow, J. (2003). Can instructors ensure the integrity and quality of online courses? *Delta Pi Epsilon Journal, 45*, 162–172.

Cheong, P. H. (2008). The young and techless? Internet use and problem solving behaviors among young adults in Singapore. *New Media & Society, 10*(5), 771–791. doi:10.1177/1461444808094356

Connaughton, S. L., & Shuffler, M. (2007). Multinational and multicultural distributed teams: A review and future agenda. *Small Group Research, 38*, 397–412. doi:10.1177/1046496407301970

Dennen, V. P. (2005). From message posting to learning dialogues: Factors affecting learner participation in asynchronous discussion. *Distance Education, 26*(1), 127–148. doi:10.1080/01587910500081376

Doo, M. Y. (2006). A problem in online interpersonal skills training: do learners practice skills? *Open Learning, 21*(3), 263–272. doi:10.1080/02680510600953252

Dutton, W. H., Cheong, P. H., & Park, N. (2004a). An ecology of constraints on e-learning in higher education: The case of a virtual learning environment. *Prometheus, 22*(2), 131–149. doi:10.1080/0810902042000218337

Dutton, W. H., & Cheong, P. H. & Park, N. (2004b). The social shaping of a virtual learning environment: The case of a university-wide course management system. *Electronic Journal of e-Learning, 2*(1), 69-80.

Ess, C., & Sudweeks, F. (2005). Culture and computer-mediated communication: Toward new understandings. *Journal of Computer-Mediated Communication, 11*(1), 179–191. doi:10.1111/j.1083-6101.2006.tb00309.x

Hall, E. T. (1959). *The silent language*. Garden City, NY: Anchor Press.

Hall, E. T. (1966). *The hidden dimension*. Garden City, NY: Anchor Press.

Hargittai, E. (2002). Second-Level Digital Divide: Differences in People's Online Skills. *First Monday, 7*(4), Retrieved Sep.2002 from http://firstmonday.org/issues/issue7_4/hargittai/index.html

Hewling, A. (2005). Culture in the online class: Using message analysis to look beyond nationality-based frames of reference. *Journal of Computer-Mediated Communication, 11*(1), 337–356. doi:10.1111/j.1083-6101.2006.tb00316.x

Hofstede, G. (1980). *Culture's consequences: International differences in work-related values.* Beverly Hill, CA: Sage.

Hofstede, G. (1997). *Cultures and organizations: Software of the mind* (Rev. Ed.). New York: McGraw-Hill.

Horrigan, J., & Jones, S. (2008). *When technology fails*. Retrieved Nov. 15, 2008 from: http://www.pewinternet.org/pdfs/PIP_Tech_Failure.pdf

Kanata, T., & Martin, J. N. (2007). Facilitating dialogues on race and ethnicity with technology: Challenging "Otherness" and promoting a dialogic way of knowing. *Journal of Literacy and Technology, 8*(2), 1–40.

Kelly, H. F., Ponton, M. K., & Rovai, A. P. (2007). A comparison on student evaluations of teaching between online and face-to-face courses. *The Internet and Higher Education, 10*(2), 189–101. doi:10.1016/j.iheduc.2007.02.001

Kerckhoff, A. C. (1995). Institutional Arrangements and Stratification Processes in Industrial Societies. *Annual Review of Sociology, 21*, 323–347. doi:10.1146/annurev.so.21.080195.001543

Kim, K., & Bonk, C. J. (2002). Cross-cultural comparisons of online collaboration. *Journal of Computer-Mediated Communication, 8*(1).

Koss, F. A. (2001). Children falling into the digital divide. *Journal of International Affairs, 55*(1), 75–90.

Lazarus, W., & Mora, F. (2000). *Online content for low-income and underserved Americans: The digital divide's new frontier.* CA: The Children's Partnership.

Mandernach, B. J., Gonzales, R. M., & Garrett, A. L. (2006). *Journal of Online Learning and Teaching, 2*(4). Retrieved February 23, 2009 from http://jolt.merlot.org/vol4no1/Vol2_No4.htm

Martin, J. N., & Davis, O. I. (2001). Conceptual foundations for teaching about Whiteness in Intercultural Communication courses. *Communication Quarterly, 50*, 298–313.

Martin, J. N., & Nakayama, T. K. (1999). Thinking dialectically about culture and communication. *Communication Theory, 9*, 1–25. doi:10.1111/j.1468-2885.1999.tb00160.x

Merryfield, M. (2003). Like a veil: Cross-cultural experiential learning online. *Contemporary Issues in Technology & Teacher Education, 3*(2), 146–171.

Merryfield, M. M. (2001). The paradoxes of teaching a multicultural education course online. *Journal of Teacher Education, 52*(4), 283–299. doi:10.1177/0022487101052004003

Mestre, L. (2006). Accommodating diverse learning styles in an online environment, (Guest Editorial). *Reference and User Services Quarterly, 46*(2), 27–32.

Mitchell, W. J. (1999). Equitable access to the online world. In D. A. Schon, B. Sanyal, & W. J. Mitchell (Eds.), *High technology and low-income communities: Prospects for the positive use of advanced information technology* (pp. 151-162). Cambridge, MA: MIT Press.

Natriello, G. (2001). Bridging the second digital divide: What can sociologists of education contribute? *Sociology of Education*, *74*(3), 260–265. doi:10.2307/2673278

Olaniran, B. A. (2001) The effects of computer-mediated communication on transculturalism. In V. H. Milhouse, M. K. Asante, & P. O. Nwosu (Eds.) *Transcultural realities: Interdisciplinary perspectives on cross-cultural relations* (pp.83-105). Thousand Oaks, CA: Sage.

Olaniran, B. A., & Agnello, M. F. (2008). Globalization, educational hegemony, and higher education. *Multicultural Education & Technology Journal*, *2*(2), 68–86. doi:10.1108/17504970810883351

Santovec, M. L. (2003). A model for evaluating online courses. *Distance Education Report*, *7*(8), 7–8.

Schwartzman, R. (2007). Refining the question: how can online instruction maximize opportunities for all students? *Communication Education*, *56*(1), 113–117. doi:10.1080/03634520601009728

Selwyn, N. (2004). Reconsidering Political and Popular Understandings of the *Digital Divide. New Media & Society*, *6*(3), 341–362. doi:10.1177/1461444804042519

Sieber, J. E. (2005). Misconceptions of and realities about teaching online. *Science and Engineering Ethics*, *11*, 329–340. doi:10.1007/s11948-005-0002-7

Stanford-Bowers, E. (2008). Persistence in online classes: As study of perceptions among community college stakeholders. *Journal of Online Learning and Teaching*, *4*(1). Retrieved February 23, 2009 from http://jolt.merlot.org/vol4no1/stanford-bowers0308.htm

Stanley, L. D. (2003). Beyond Access: Psychosocial Barriers to Computer Literacy. *The Information Society*, *19*, 407–416. doi:10.1080/715720560

Thompson, L., & Ku, H.-Y. (2005). Chinese graduate students' experiences and attitudes toward online learning. *Educational Media International*, *42*(1), 33–47. doi:10.1080/09523980500116878

van Dijk, J. (2005). *The Deepening Divide.* Thousand Oaks, CA: Sage.

Warschauer, M. (2003). *Technology and Social Inclusion: Rethinking the Digital Divide.* Cambridge, MA: MIT Press

Witkin, H. A., Moore, C. A., Goodenough, D. R., & Cox, P. W. (1977). Field-dependent and field-independent cognitive styles and their educational implications. *Review of Educational Research*, *47*(1), 1–64.

Young, J. R. (2005, August 12). Professors give mixed reviews of Internet's educational impact. *Chronicle of Higher Education*. Retrieved August 21, 2005 from http://chronicle.com/weekly/v51/i49/49a03201.htm

ENDNOTE

1 Interview with Jill Schiefelbein, Director of Online Programs in the College of Liberal Arts and Sciences, November 26, 2008.

Chapter 7
Application of VoiceXML in e-Learning Systems

A. A. Azeta
Covenant University, Nigeria

C. K. Ayo
Covenant University, Nigeria

A. A. Atayero
Covenant University, Nigeria

N. A. Ikhu-Omoregbe
Covenant University, Nigeria

EXECUTIVE SUMMARY

This chapter examines the learning environment of visually impaired students in the school for the blind. The level of Information and Communication Technology (ICT) utilization and adoption is reported with specific interest in VoiceXML and its application areas. As a case study, a prototype voice-based e-Learning application for course registration and examination was developed and reported. The system was evaluated using ISO 9241-11 usability criteria. The outcome of the usability evaluation is also presented. The voice-based e-Learning technology described in this chapter will improve accessibility to education, including distance learning for learners who are visually impaired in the school for the blind.

BACKGROUND

The use of the Internet and web based instructional aids is now viewed as an integral part of the learning environment. As a result, students now have real-time online access to e-Learning contents and opportunities, and most tertiary institutions now offer courses through distance learning. Although some people would argue against the merits of e-Learning, it is clear that with the pace of e-Learning implementation, students such as those with visual impairments have been left behind due to the lack of an accessible content delivery system to ameliorate their disabilities.

The various options available for most learning environments are face to face, telephone, electronic mail, chat room, instant messaging, etc. However,

DOI: 10.4018/978-1-60566-942-7.ch007

this becomes a more difficult task for those with disabilities. A blind person cannot see or communicate through mail or electronic means that require ability to see the screen. Lack of provision for voice in the existing learning methods has excluded support for people with limited capabilities such as the visually impaired that affect either data entry, or ability to read (and therefore check) what they have entered, since these applications are visual in nature and require sight to see the blackboard or computer screen and manipulate the computer keyboard.

Several e-Learning design methodologies have been proposed in literature. However, not too many works were dedicated to the design and implementation of e-Learning for the disabled (Sirithumgul *et al.*, 2007, p. 1). The blind and vision impaired students, who are particularly affected by the technological change, face a range of difficulties from the act of typing a letter to the use of computers in educational institutions. The increasingly widening gap between the people who are technologically able and those who are not gives cause for great concern.

This is the case of a particular school for the blind, a privately owned educational institution located in Lagos, Nigeria, that provides a learning environment for the blind and partially sighted children at primary and secondary school levels. The school also admits people who became blind in the course of their life, for rehabilitation at higher education level (at university level). The school is headed by a principal assisted by a vice principal. There are thirty five teachers in the school and they all report directly to the school administrator while the school administrator reports to the principal.

The school's foundation was laid by the Catholic Church on the 16th of June, 1960 and it was officially opened in 1962. The total number of pupils in 1962 was four, two boys and two girls in the primary school category. The population later increased to accommodate secondary school students. Thereafter, the federal government took over all schools in Nigeria but later handed over the ownership and management of the school back to Catholic Church missionaries, the original proprietor in the year 1970. The school provides the traditional form of learning, where the teachers meet physically with the students in class.

Presently, the school is managed by the Catholic Church of Nigeria and funded by charitable individuals and organizations. It operates the same primary and secondary school curriculum as other private and public institutions within its category in Nigeria. The school spends an average of six million, six hundred thousand naira (N6,600,000) annually on capital and recurrent expenditure while its annual income is an average of seven million, two hundred and fifty naira (N7,000,250). The total number of students is one hundred and six, and they are all accommodated in the school premises.

The report presented in this chapter examines the learning environment of vision impaired students in the school used as case study. The resulting information was used to provide an assistive voice-based e-Learning platform to support learning in the school. A number of challenges were identified after the implementation of the project. However, suggestions and recommendations were made on how to overcome them. The educational institution used as case study in this project is referred to as 'the school' in the subsequent sections of this chapter.

SETTING THE STAGE

This section examines the technology utilization of the school prior to initiation of the project. It also describes the application areas of voice technology that was used to provide a solution to the case studied.

The ICT Infrastructure of the School

The school has a total of ten of Personal Computers (PCs) running on Microsoft Windows XP operating system. There are ten Uninterrupted Power Supply (UPS) units attached to the PCs. There are two DeskJet printers available in the school. The school has an Information and Communication Technology (ICT) department. A computer technician normally comes around to carry out maintenance work on the computers and printers. The school neither currently uses any e-Learning application nor provides any e-Learning service to the students. The word processing software used is Microsoft Word. The school has one land line telephone and ten personal mobile phones owned by the members of staff. The phones are mainly used for communicating with the parents of the students, amongst others. Internet service is available for the teachers and students but was reported by the management of the school to be unstable most times.

The following ICT products and services are not available for use in the school: development/programming languages, computer network, Extranet and Intranet. However, the school's management considers the importance of ICT as very high, particularly for the visually impaired. Their wish is to develop the ICT infrastructure further in the future if they have access to enough funds. For instance they would want to provide computer networks, replace all the PCs with new ones and computerize their examination processes.

The conventional learning methods for teaching the visually impaired students in the school include interaction between the teachers and the students which requires the physical presence of the teacher in the class. The equipment used for learning in the school are Slate and Stylus, Mathematics board and figures, Braille, Typewriters, Abacus, etc. The challenges with these resources are as follows: 1) they are very expensive to provide per child; and 2) they are imported into the country and cannot be sourced locally. The cost of maintaining the equipment is high and the technicians responsible for maintaining the equipment are very scarce.

Two major problems are associated with the utilization of the aforementioned equipment for teaching, learning and examination. First, the coordination of visually impaired students during course registration period at the beginning of the term or semester is cumbersome. As a result of the sight challenge of the students, the teachers in the school are most times not sufficient to guide the students for course registration that will lead to minimal errors by the students. Consequently, too much time is spent on the course registration exercise at the expense of pursuing other school activities. Second, the teachers are extremely busy during examination either invigilating in the classes or coordinating the logistics affecting examination. Sometimes, any of the equipment can fail during usage for lectures or examination.

More so, the hostel where the students are accommodated and the lecture halls are some distance apart. It normally takes the visually impaired students a lot of effort to navigate the foot path leading to the class rooms. They sometimes cause obstructions on the way to themselves and the teachers in attempt to locate their class rooms.

Voice learning means the use of mobile phones or landline telephone to access learning contents in the Internet or Intranet anytime and anywhere by dialing a telephone number. In determining the perception of voice learning provision to the management of the school, the school strongly agrees that: 1) a voice-based e-Learning application will complement existing supportive technologies to meet the needs of students with a range of disabilities such as visual impairment, etc, that make reading and writing difficult, and 2) voice-based e-Learning will be available on multiple platforms to all users as well as boosting access to education for the physically challenged, particularly the sight impaired in the developing countries of the world.

The management of the school was asked to rate the degree of their institution's concern (as Least Concern, Concern or Most Concern) when considering providing telephone-based learning using mobile and land lines in terms of Reliability, Usability and Cost. These were their responses: Least Concern for Reliability, Most Concern for Usability and Concern for Cost. This shows that usability on the part of the students who are mainly from poor backgrounds may constitute a hindrance to having a 100% acceptance of the application. They also believe that the visually impaired student in their school will rely heavily on the application once it is fully deployed for access by the students and the necessary infrastructure provided by the school.

Voice Technology and Areas of Application

A VoiceXML (known as voice extensible markup language) platform is the foundation for developing and operating voice-based applications (Rouillard, 2007, p. 27). The VoiceXML platform also provides the speech processing capabilities (speech recognition, speech synthesis, voice authentication, etc). During the human-computer interaction, it executes the commands and logic specified by applications written in VoiceXML.

Voice-enabled e-Learning systems allow users to access information on the Internet or Intranet through a telephone interface. It uses technologies such as speech recognition and text to speech (TTS) conversion to create a user interface that enables users to navigate through a dialogue system using telephone and voice commands (Gallivan *et al.*, 2002, p. 1).

A typical telephone web-based e-Learning application provides e-Learning materials that can be accessed via the web as well as via the telephone. Some students have used speech recognition systems successfully for their studies and for exams, and the use of this technology has helped them to overcome their difficulties and go on to higher education (Paul, 2003, p. 1).

Voice-based e-Learning system is a system of learning that can take place anytime, anywhere with the help of a mobile or land phone by dialing a telephone number that connect users to an application that is resident in a web server. Voice-based learning is a type of "assistive technology", used by the physically challenged. The World Wide Web Consortium (W3C) defines assistive technology as software or hardware that has been specifically designed to assist people with disabilities in carrying out their daily activities (Adaptive, 2005, p. 1). These technologies aid the learning process for learners with disabilities. People with partially sighted vision have difficulty accessing e-Learning systems due to small print or the inability to sufficiently see the position of text blocks on the screen. For blind people, e-Learning systems are often inaccessible due to the nature of the process requiring sighted information.

In addition to the provision of alternative platform for normal users, voice-enabled e-Learning systems can be helpful for people with physical access difficulties (e.g. arthritis, high spinal injury) that make writing difficult (Donegan, 2000, p. 4). It can also be effective for students with reading, writing or spelling difficulties (e.g. dyslexia) and for those with visual impairment (Nisbet & Wilson, 2002, p. 1).

Development of voice applications using VoiceXML for higher institutions of learning has remained an open area of research all over the world. For instance, Gallivan *et al.,* (2002, p. 4) presented a VoiceXML absentee system that enables students to report of their class absence through a telephone call. The Absentee System was developed basically for Pace University students to report class absences and have them stored in the university database. The VoiceXML absentee system was designed to include record keeping of absentee calls from students, faculty and university staff. Voice-driven interfaces will also be of great benefit to people who are unable to leave their home due to disability, providing them with a learning portal using a telephone handset.

Chin *et al.*, (2006) recommended that one can actually make use of VoiceXML technology to build speech applications that can serve educational purposes or in other words, build an online learning system that provides better accessibility to users. One of the e-Learning applications that can be provided using speech technology is one that delivers basic teaching by simply listening. For example, students can check their scores or other information by calling a particular telephone number and getting the information they want.

Voice-based applications have also been developed in several other areas such as in banking transactions (Azeta *et al.*, 2008, pp. 59-72) and a lot more, to assist the visually impaired and provide an alternative platform for normal users.

CASE DESCRIPTION

A sample case of user interface description and implementation process of the application is described in this section.

Sample Case (Call Flow) of the Application

Below is a sample case (call flow) for the VUI of voice-based e-Learning system (see Figure 1) for the school.

Implementation Process

As a follow-up to providing a solution for the problems experienced involving use of the conventional learning equipment in the school, two application modules were identified. 1) Course registration, and 2) Examination. Although VoiceXML is easy to learn, building a successful VoiceXML application requires not only software development skills, but other skills like understanding human factors for the telephone interface, linguistics, speech recognition and audio production.

The implementation of a voice-based e-Learning system should contribute to the success of education for the visual impaired students in the school. The institution wants a solution based on technology that allows a student to learn independently. In order to meet up with this requirement, a voice-based application was proposed that allows access using a telephone. The project was accomplished using VoiceXML application development life cycle (VoiceXML, 2007, p. 1).

The VoiceXML application development life cycle is one of the software development models used for developing voice applications. It is similar to that of a web application development process but includes voice user interface (VUI) design and speech recognition system. The development cycle consists of five phases: They include: Problem Definition; Systems Design; Systems Development; Systems Testing; Pilot and Deployment. The life cycle was engaged in the implementation of a telephone-based e-Learning application for course registration and examination modules as follows:

Problem Definition

The existing learning methods in the school does not allow students to learn independently (i.e. on their own) irrespective of location. Students have to be physically present in the class room. The process of guiding/directing visually impaired students from their hostel to the class room is cumbersome.

It is the opinion of management of the school that a system that allows the students to learn on their own would minimize the problem of scarcity of teachers in the area of teaching in a classroom setting, among others.

Systems Design

A Unified Modeling Language (UML) class diagram was engaged to represent the data flow for the course registration and examination module. The

Figure 1. A sample case (call flow) for the voice-based e-Learning application

```
IVR:    Welcome to the School for the Blind e-Learning application.
IVR:    What is your user name.
Caller: admin
IVR:    What is your user password.
Caller: admin
        IF NOINPUT then
IVR:    I did not hear any user name.  Please try again.
        IF NOMATCH then
IVR:    The user name you entered is not recognized.
        Please try again.
        IF NOMATCH then
        IVR: The user name you entered is not recognized.
               Please try again.
        ENDIF
        IF MATCH then
               IVR: You have successfully login.
        ENDIF
IVR:    Select any of the following options.
               say one for course registration.
               say two for examination.
               say three to exit.
        IF one is selected then
               IVR: for the following, listen and say select or deselect
               at the hearing of each subject.
               Mathematics, Social studies, Government.
        ENDIF
```

UML is a visual language that provides a means to visualize, construct and document the artefacts of software systems (Simeon *et al.*, 2005).

Figure 2 contains the class diagram for the course registration and examination module. The Class diagram has five classes – Student, Course, CourseRegistration, ApproveRegdCourses and Examination. Each class has three compartments, the top compartment contains the class name, the second contains the attribute names and format, and the third compartment contains the operations to be carried out on the attributes.

The lines labelled with a directed arrow connecting two classes show associations between the classes as follows: (i) "* register 8", means all students must register for 8 courses (ii) "* take part in 8", means all students must take part in examination for 8 courses, (iii) "8 approve 1", means 8 registered courses by students must get an approval from class teacher. (iv) "* must undergo 1", means all registered courses by students must undergo an examination.

The Architectural Framework

Software Architecture

Figure 3 gives the logical overview of the architecture of e-Learning application. The software architecture shows the location of each of the modules in the application. It consists of the presentation tier, business logic tier and data tier. The database is separated from the client through the middleware, here referred to as the business logic tier.

Figure 2. Class diagram for Course Registration and Examination module

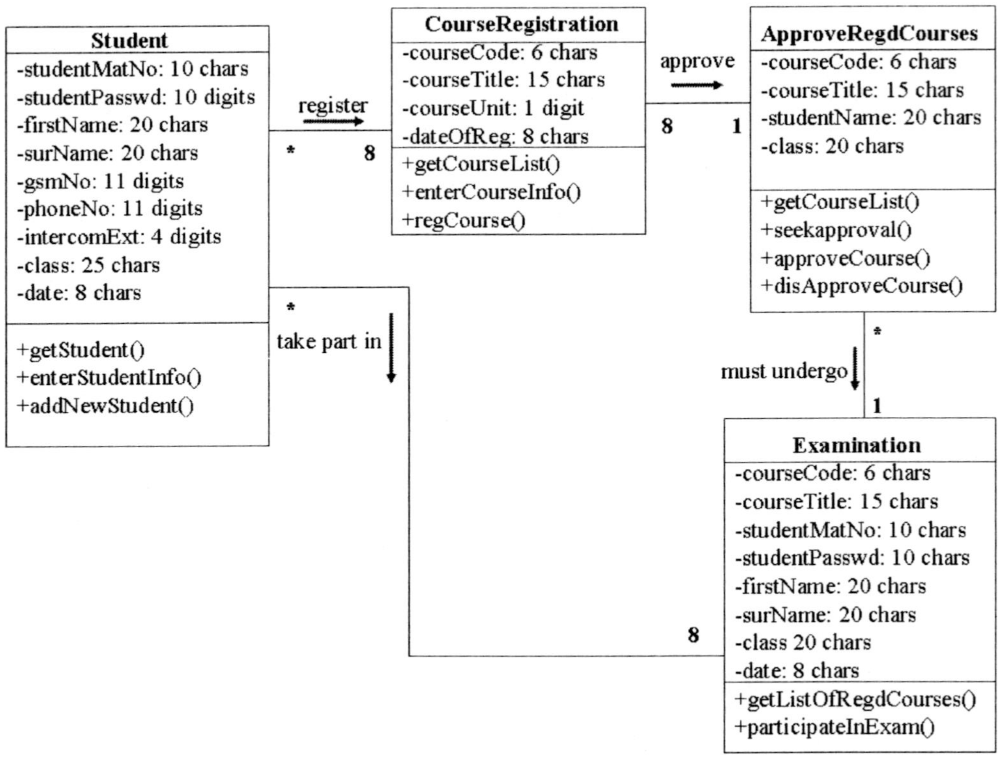

The Presentation Tier

The presentation tier provides clients access to the e-Learning application through the middleware. The components of the clients' interface are i) Course Registration, and ii) Examination. These components do not store or process any form of data. They only provide an interface for the middle tier and the data tier. Data or files or voice browsers are not stored on the mobile phones due to resource constraints associated with hand-held devices. The application is developed to use telephone and allows voice browsers (running in the voice gateway) to be used as the interface. The information from the database is presented in a compatible form to the client using voice. The voice browser simply receives any call into the application and submits them to the voice gateway for further processing.

The Business Logic Tier

The presentation tier communicates with the voice gateway component of the middle-tier through the voice browser. The middle-tier contains the voice gateway and the application/business logic. Users access the application from various mobile telephone devices and land line telephone, anywhere, anytime. Once a user has been authenticated, the user's query is translated by the automated speech recognition (ASR) to text and passed to the database server for execution. The text-to-speech (TTS) does the reverse of translating text to speech. A user can only access the module for which he or she is authorized. The client application interfaces with the business logic tier using the voice gateway.

Figure 3. Overview of a three-tier Telephone-based e-Learning Architecture

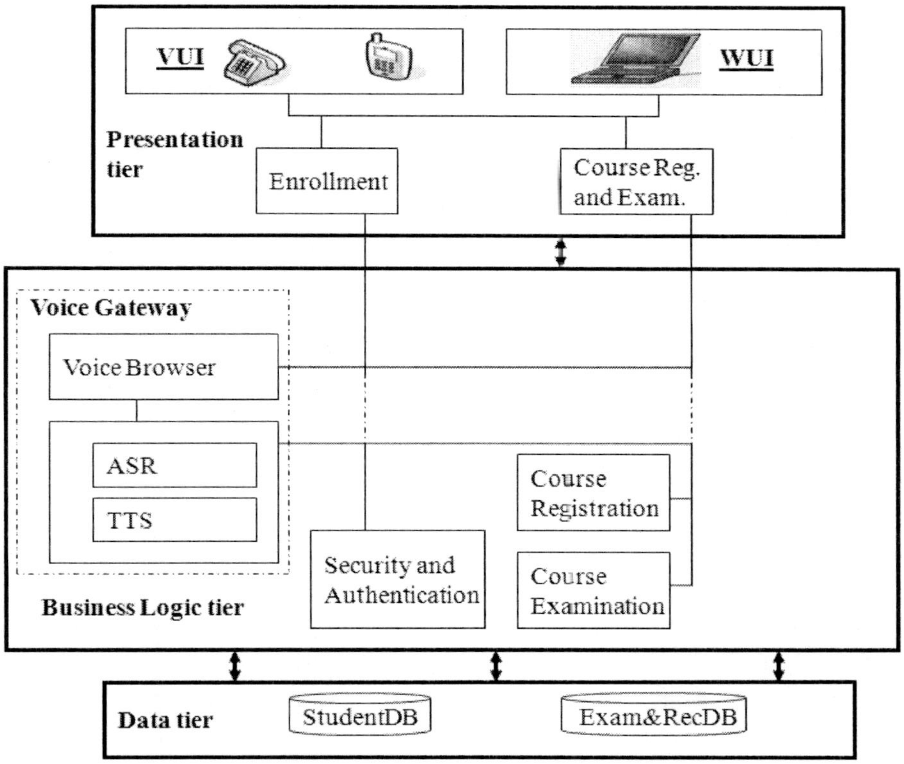

Data Tier

The data tier contains the application database. It provides data services and data base management system function. The data tier is responsible for changing, adding, or deleting information in the database within the system. We have used MySQL database for the implementation of the data tier.

The Hardware Architecture

The hardware architecture consists of client devices; servers and database (see Figure 4). The client devices include the web and hand-held devices such as mobile phones and personal digital assistants and land telephones. In a situation where students are not allowed to carry mobile phones or where cost is an issue, an alternative for them is to use the PC phone such as Skype through Voice over Internet Protocol (VoIP). The servers contain the voice server and application server. The database contains MySQL database.

Systems Development

Every authenticated user of the application will pass through some questions and answers sections, which will be matched against the content of the grammar, and the result received by the user through voice. Figure 5 describes a pseudocode for course registration and examination module.

The prototype client application for the telephone was developed using VoiceXML for the VUI. PHP and Apache constituted the middle-ware and MySQL database as the back-end component. VoiceXML was chosen because it is a foundation platform for developing and operating voice automation applications (Rouillard, 2007). PHP, Apache and MySQL database were selected because of their benefit as free and open source software (Siemens, 2003, p. 4).

Figure 4. A hardware architecture of the voice-based e-Learning application

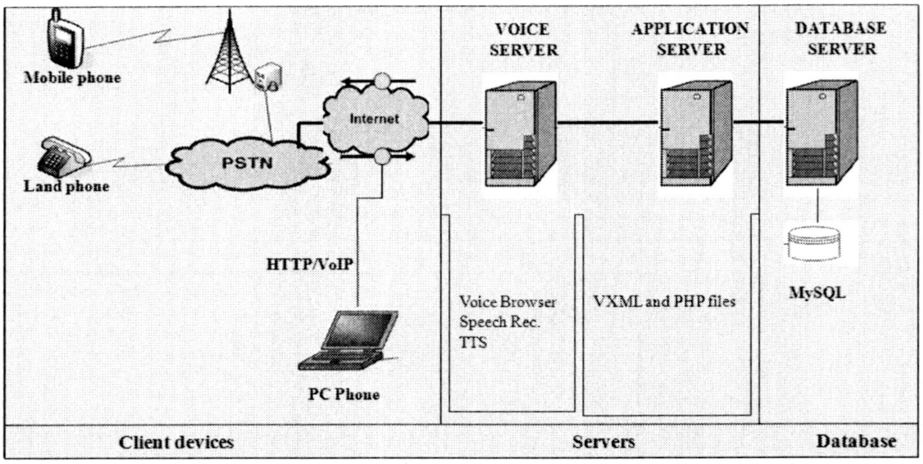

Systems Testing

Testing is a vital stage in the development of any application. The prototype VoiceXML-based application was deployed and tested using sample student data. The user logs onto the application using a "username" and "password" specific to their registration profile.

Pilot and Deployment

There are two methods for accessing the application: either on a local computer or the Internet. First, for a *local computer*, Voxeo Prophecy was installed to run on a local computer before connection and subsequent voice interaction could commence. A headset was connected to the local computer for the caller to get voice response and also be able to supply voice input. Clicking

Figure 5. Pseudocode for course registration and examination module

```
BEGIN
        SYSTEM PROMPT "Welcome to the School for the Blind e-Learning application
        //Caller supply a username and password
        SYSTEM AUTHENTICATES A CALLER;
        SYSTEM PROMPT menu selection;
        SYSTEM REPORT Student information from database;
        WHILE NOT EOF DO
                //Caller select a menu option;
                IF menu option is REGISTRATION THEN
                        SYSTEM process and report registration information;
                ELSE
                IF menu option is EXAMINATION THEN
                        SYSTEM process and report examination information/result;
                ELSE
                IF MENU option is EXIT THEN
                        SYSTEM report good bye message and exit;
                ENDIF
        ENDDO
END
```

the Dial button from the Voxeo Prophecy SIP Softphone keypad (www.voxeo.com/prophecy) provided connection to the application. The application was developed and tested using a local computer and latter deployed on the Internet when it was confirmed to be functioning without errors for access using a public telephone.

Second, on the *Internet,* Voxeo voice server (Voxeo, 2003) provided a free hosting service to deploy the prototype VoiceXML application, which can be accessed from any telephone using the format: <source country international dial-out number><destination country code><destination area code><generated voice network 7-digit number>. Dialing: 009-1-202-6849430 from any mobile or land phone from Nigeria (009) will connect and execute the application. The default username and password is "*admin*".

Once connected, the application prompts with a welcome message and goes ahead to authenticate the user name and password before any transaction can take place. The application will ask for the services demanded by a student and goes ahead to process the request, either course registration or examination. A sample list of registered courses as stored in the database is depicted in Figure 6.

At the end of deploying the application, the teachers and students were given an oral guideline on how to connect to the application and use it for course registration and examination. One of the key points mentioned during the presentation of the application to the school was that the examination module only handles multiple choice examination ("objective") questions only.

CURRENT CHALLENGES/ PROBLEMS FACING THE ORGANIZATION

After developing and deploying the system, the application was evaluated for usability to determine the level of effectiveness, efficiency and users' satisfaction. A set of questions was designed and administered through a questionnaire to the teachers and students mostly from secondary

Figure 6. A sample list of registered courses

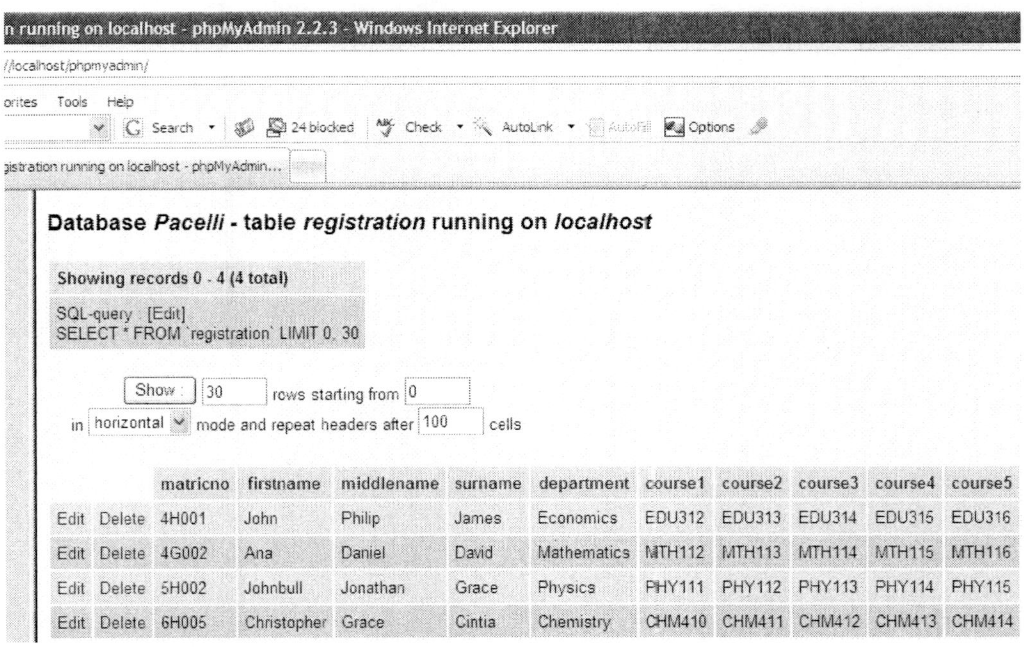

		matricno	firstname	middlename	surname	department	course1	course2	course3	course4	course5
Edit	Delete	4H001	John	Philip	James	Economics	EDU312	EDU313	EDU314	EDU315	EDU316
Edit	Delete	4G002	Ana	Daniel	David	Mathematics	MTH112	MTH113	MTH114	MTH115	MTH116
Edit	Delete	5H002	Johnbull	Jonathan	Grace	Physics	PHY111	PHY112	PHY113	PHY114	PHY115
Edit	Delete	6H005	Christopher	Grace	Cintia	Chemistry	CHM410	CHM411	CHM412	CHM413	CHM414

school and higher education level (at university level). The questionnaire contains of five sections names: background information, user experience with mobile phone and the system, effectiveness of the system, efficiency of the system and user satisfaction with the system. The questionnaire aims at eliciting information from the school in order to measure the usability of the voice-based e-Learning application provided.

The system evaluation questionnaire was designed using the information acquired from 1) the analysis of requirement elicitation questionnaire, and 2) personal oral interview conducted during several visits made to the school. Some of the current challenges facing the school after deploying the application were also derived from the evaluation result. A sample of the questions from each section in the questionnaire is presented as follows:

Question one
Gender: Male [ ] Female [ ]

Question two
Would you be able to afford a mobile phone to call the e-Learning application?
[Yes] [No]

Question three
I was able to complete my task successfully and correctly using the application 1 2 3 4 5
Strongly Disagree Strongly Agree

Question four
I was able to complete my task on time 1 2 3 4 5
Strongly Disagree Strongly Agree

Question five
I am satisfied with the performance of the system in accomplishing my tasks 1 2 3 4 5
Strongly Disagree Strongly Agree

System Evaluation

The evaluation of a product is a fundamental requirement in determining the practical usability of a product (Ikhu-Omoregbe, 2007, p. 14). The usability of the e-Learning application was measured to specify the features and attributes required to make the product usable using ISO's standard of usability (ISO 9241-11, 1998) as consisting of three distinct aspects:

Effectiveness, which is the accuracy and completeness with which users achieve certain goals. Indicators of effectiveness include quality of solution and error rates.

Efficiency, which is the relation between 1) the accuracy and completeness with which users achieve certain goals; and 2) the resources expended in achieving them.

Satisfaction, which is the users' comfort with and positive attitudes towards the use of the system.

Data Analysis

For all the learners, an overall score was computed for each of the usability dimension by averaging all the ratings on the questionnaire that was used. Microsoft Excel was used to generate the frequency distribution and mean and all the relevant charts for the ratings.

Discussions

With the assistance of some of the teachers, the respondents were taken through a short training on how to dial a telephone number from a mobile phone that will connect the learners to the application and how to navigation within the application. The ratings for the usability attributes as collected are presented below:

Table 1. Descriptive statistical analysis of questionnaire data

Summary of study variable				
Usability Measures	*Teachers*	*Students*	*Total # of respondents*	*Total Mean Rating*
Effectiveness	11	42	53	3.38
Efficiency	11	43	54	3.33
Satisfaction	10	42	52	3.35

Effectiveness

The "effectiveness" was evaluated for each of the tasks performed by each learner. The mean rating for "Effectiveness" is 3.38. This is shown in Figure 7.

Efficiency

The rating for "Efficiency" indicates the time it takes to achieve a task, most of the learners were able to realize their task on time as indicated in Figure 8 with a mean rating of 3.33.

Figure 7. Effectiveness analysis

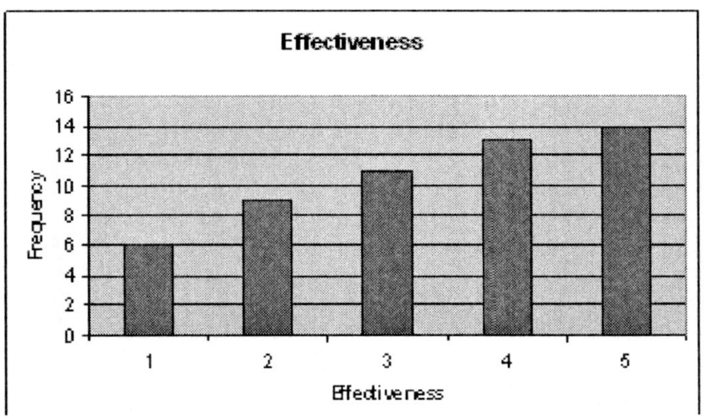

Figure 8. Efficiency analysis

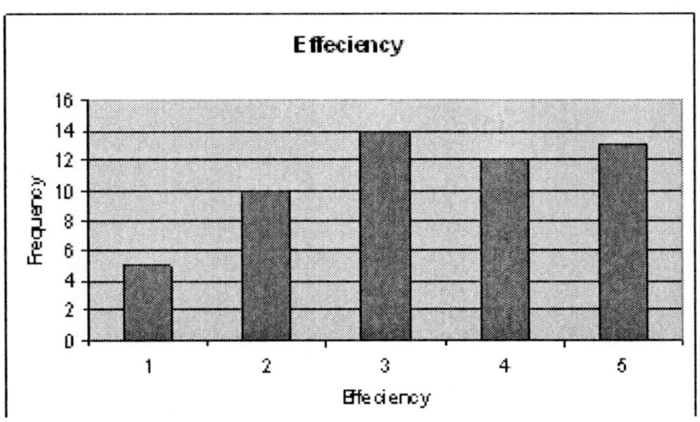

Figure 9. Satisfaction analysis

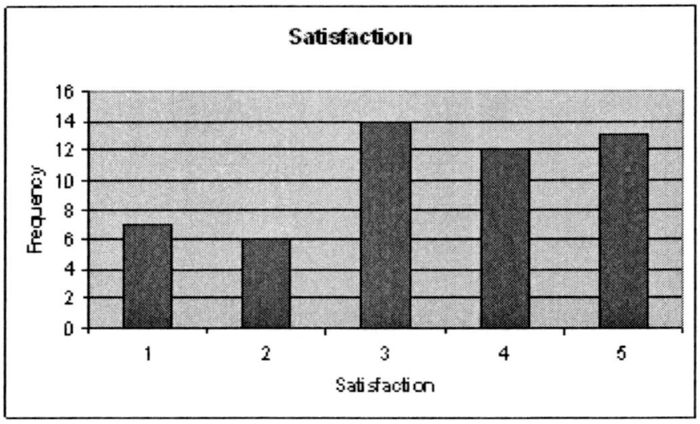

Figure 10. Usability Attributes Analysis

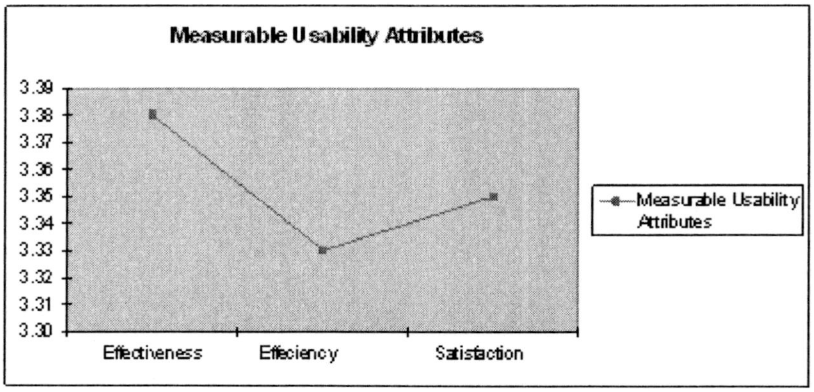

Satisfaction

The system has a mean rating of 3.35 for "Satisfaction". Figure 9 shows the values for "Satisfaction" attribute.

Several studies on usability suggest the system with "Good Usability" should have a mean rating of 4 on a 1-5 scale and 5.6 on a 1-7 scale (Sauro, *et al.*, 2005). Therefore, we can conclude that the prototype application developed for the school has an "Average Usability" based on the following mean ratings of the given usability scale of Table 2 and usability attributes of Table 3.

The voice-based e-Learning application provided to the school for the blind used as case study is a first trial which led to a bit of resistance from users. For this reason, on a total usability scale of 5 anything above 3 is considered successful, while less than 3 is a failure. Therefore, the case study presented in this chapter is successful.

The ratings for the three usability attributes are depicted in Figure 10.

Navigation Analysis

The mean overall ratings of "Navigation" is 3.62 as shown in Figure 11. This is expected since an IVR content (the words that make up the voice input and response) is required to be moderate at a particular time of call transaction.

Table 2. Usability scale

Scale	Meaning
1	Very Bad Usability
2	Bad Usability
3	Average Usability
4	Good Usability
5	Excellent Usability

The under listed challenges were faced by the school after completing the project and while the students were using the application. These challenges brought about the bit of resistance recorded from users and usability rating of approximately 3 out of 5 scale "Average Usability" depicted in Table 2 and Table 3.

- Most of the students in the school are from poor background and were not able to afford a mobile phone and the subsequent cost of calling the e-Learning voice application.
- Some of the students are under rehabilitation to be integrated into the society after their education. This set of students did not show much enthusiasm towards using the application.
- Poor ICT infrastructure and in particular slow Internet access that deter students from making use of free web-based PC phone such as Skype using VoIP for those who cannot afford to buy or maintain a mobile phone.
- Low financial and material donations from individuals, government, private and public organizations to cater for the general needs of the institution. This was also the case prior to the implementation of the project.

Table 3. Usability attribute ratings

Usability Attributes	Mean Rating
Effectiveness	3.38
Efficiency	3.33
Satisfaction	3.35

Summary and Recommendation

In this chapter, we have explored the learning experiences of students to bring to light the problems encountered by visually impaired learners for effective learning support. We have also developed a prototype voice-based e-Learning application using the VoiceXML application development life cycle to proffer a solution that will complement the existing learning methods in the school used as case study.

The voice-based e-Learning application provided for the blind has many implications. Learning can be realized faster and more efficiently than existing learning methods. The application can achieve just-in-time learning with greater reach irrespective of location (whether on the move, at home or work), speed of response and consistency of message. While claims that voice learning is an expensive form of education can be misleading, voice learning can reduce the traditional face-to-face related expenses such as lecture halls and other learning delivery facilities associated with physical presence between the students and teachers. It can also provide economies of scale at higher levels, as cost of each additional learner is negligible once the lecture materials have been developed and hosted in a central server.

The voice-based e-Learning technology will improve accessibility to education, including distance learning for learners who are visually impaired in the school for the blind. By doing so, the target group will not be completely neglected in the scheme of promoting ICT in education and learning. Loss of sight is one of the most difficult disabilities to come to terms with. The assistive

Figure 11. Navigation Analysis

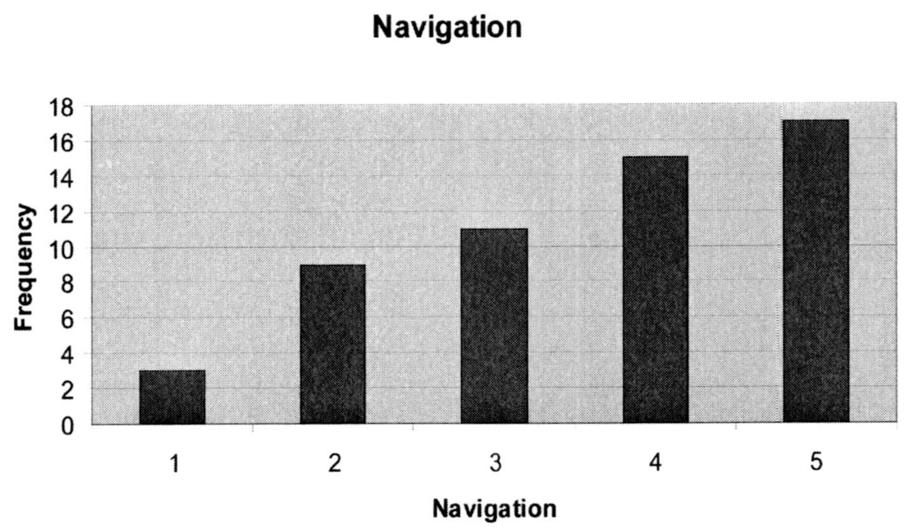

technology reported in this article has the ability to fundamentally change the way teaching and training is delivered to the students of the school for the blind used in this case study and other schools alike.

ICT driven revolutionary change in the education sector has created ever-changing knowledge and skills requirements, and traditional approaches to learning are struggling to keep up. The face-to-face method of learning has support from school teachers in the area of teaching. Meanwhile, there are growing staff shortage in the school while the demands for teachers continued to increase. Voice learning as a component of e-Learning has been deployed in the school for the blind as a means of fuelling the expansion in the school within resource constraints. The voice learning was able to provide the visually impaired learners with a more participatory educational experience.

This study has shown that there is a very low level of ICT development in the school for the blind used as case study as it is with so many other schools for the visually impaired. However, the school will appreciate any financial support towards improving their level of ICT to enable them fully embrace the new technology supported learning known as voice learning. This study also makes a contribution in the field of ubiquitous learning. Any researcher wishing to provide an assistive technology that is based on speech technology to complement existing learning methods in schools for the blind will have something to take from this article.

The voice-based e-Learning application provided to the school for the blind in this case study is the first trial of voice technology. It is also the first major technological revolution experienced after the installation of Internet services in the school. Hence, the bit of resistance recorded in the case study. Consequently, on usability scale of 5, between 3 and 5 is considered a success while anything less than 3 is a failure. Therefore, this case study is a successful e-Learning practice since the usability result recorded is approximately 3 out of a total of 5 scale.

Other e-Learning application developers trying to develop a similar voice application for the blind will need to consider several factors. To deliver and access (electronic-enabled) e-enabled voice-based learning materials will require a new level of competence and awareness with ICT on the part of the users. There are the complexities of some users having to develop the basic ICT skills before using voice learning. There should be blended model with the face-to-face sessions at the initial deployment stage to prepare users for voice learning, a practice often referred to as parallel changing over from the existing learning methods to the new voice-based learning. This to some extent will minimize resistance from users. Resistance from students and teachers arise because they are used to attending classroom sessions physically. Such cultural expectations will change over time.

Another level of resistance may come from the perception of restrictions of mobile devices in terms of output and input capabilities. Mobile devices often have limited screen display sizes and limited capacity to support audio and video data. The remedy to this issue is the fact that voice application only requires dialing a telephone number to connect to an application, which does not require any additional resource overhead on a mobile phone. However, some visually impaired learners may still prefer to hear the teacher in a class room setting. Generally, the problem of resistance may be more severe in developing countries where human resources and capacity development may be less robust and the economy less unstable. One further step to addressing the resistance issue is to understand why some learners and teachers resist. There may be a variety of reasons including 1) fear on the part of the teachers that the technology will make them obsolete and may lose their jobs, and 2) unfamiliarity with technology and fear that they will look stupid in front of others if they do not use it correctly.

While voice learning can appear to be the best option for specific learning requirements such as the visually impaired, cost of development can be prohibitive if provided through a service provider. The cost of developing the smallest voice-based e-Learning application is enormous, an amount that most schools for the blind may not be able to afford. A cost effective alternative would be to develop materials and content in-house and employ the services of an ICT administrator to manage the learning content that will be accessed by calling a telephone number. This will serve as a remedy to the issues of cost and skill since most schools for the blind in developing countries such as Nigeria does not have the required funds, skills and knowledge to make it work effectively.

There is need to increase training and development capacity for teachers in the school for the blind. It is unlikely that classroom-based delivery alone can provide sufficient capacity to bridge the gap between the developed and developing countries in the area of technology supported learning for the blind. Digital divide challenges facing developing countries in terms of ICT deployment and in particular the neglected populace such as the school for the blind are numerous. These include digital illiteracy, lack of adequate infrastructure such as electricity, etc, and lack of suitable ICT legal framework to support the visually impaired learners. In the developed and developing countries, several methodologies exist for implementing e-Learning applications. Meanwhile, not too many of them considers the plight of the visually impaired during the analysis and design stages of these applications.

Some of the challenges currently faced by the school for the blind used as case study can be overcome if individuals, Non Governmental Organizations (NGOs) and other charity organizations come to the aid of the school and render financial and material support to boost their income and asset level. This type of support will enable the management of the school to invest more on ICT including provision of mobile phones, computers and reliable Internet facility for the voice learning application that is developed and reported in this study.

The government of Nigeria should endeavour to provide the necessary ICT infrastructure in the country to support the deployment of the prototype VoiceXML-based application. Government should also formulate policies that will reduce the tariff paid on importation of ICT products and other equipment used for learning by visually impaired students.

REFERENCES

Azeta, A. A., Ikhu-Omoregbe, N. A., Ayo, C. K., & Atayero, A. A. (2008). Development and Deployment of VoiceXML-Based Banking Applications. *Journal of Nigeria Computer Society, 15*(1), 59–72.

Chin, C. C., Hock, G. T., & Veerappan, C. M. (2006). *VoiceXML as Solution for Improving Web accessibility and Manipulation for e-Education.* School of Computing and IT, INTI College Malasia. Retrieved July 2008 from http://intisj.edu.my/INTISJ/InfoFor/StaffResearch/10.pdf

Donegan, M. (2000). *Voice Recognition Technology in Education: Factors for Success.* Oxford, UK: Ace Centre.

Gallivan, P., Hong, Q., Jordan, L., Li, E., Mathew, G., Mulyani, Y., et al. (2002). VoiceXML Absentee System. In *Proceedings of MASPLAS'02.* The Mid-Atlantic Student Workshop on Programming Languages and Systems, Pace University, USA.

Ikhu-Omoregbe, N. A. (2007). *A Formal Framework for usable Operations Support Systems in E-Health.* Unpublished doctoral dissertation, Covenant University, Nigeria.

ISO 9241-11. (1998). Ergonomic requirements for office work with visual display terminals (VDTs) - Part 11: Guidance on usability.

Nisbet, P. D., & Wilson, A. (2002). *Introducing Speech Recognition in Schools: using IBM ViaVoice.* Pub. CALL Centre, University of Edinburgh, Scotland.

Paul, D. (2003). Speech recognition for students with disabilities. *Nisbet, CALL Centre, University of Edinburgh,* (pp. 1 – 6). Retrieved January 2009 from http://callcentre.education.ed.ac.uk/downloads/speech_recognition/hitmesse2003.pdf

Rouillard, J. (2007). Web services and speech-based applications around VoiceXML. *Journal of Networks, 2*(1), 27–35.

Sauro, J., & Kindlund, E. (2005). A Method to Standardize Usability Metrics into a Single Score. *ACM, CHI,* April 2-7, Portland, OR.

Siemens, G. Free and Open movement (2003). Part 1 – History and Philosophies. *Elearn space, everything elearning* (pp. 1-4). Retrieved February 2009, from http://www.elearnspace.org/Articles/open_source_part_1.htm

Simeon, B., Skelton, J., & Lunn, K. (2005). *Schaum's Outlines UML,* (2nd Ed.). London: McGraw-Hill International, UK

Sirithumgul, P., Suchato, A., & Punyabukkana, P. (2007). *E-Learning Transformation For the Visually Impaired.* Assistive Technology Laboratory. Retrieved 3rd March 2009 from http://cu-ocw.eng.chula.ac.th/cu/eng/cp/e23e30e40e1ae35e22e1ae27e34e18e35e27e34e08e31e22e17e32e07e27e34e28e27e01e23e23e21e04e2de21e1ee34e27e40e15e2de23e4c-research-methods-in-computer-engineering/Assistive%20Paper%20I.pdf

Voice, X. M. L. (2007), *VoiceXML application development life cycle.* Palo Alto, CA: Hewlett Packard Development Company.

Voxeo (2003). *Voice server.* Retrieved October, 2008 from http://community.voxeo.com

Chapter 8
Technophobe to Technophile: Entering the Internet Culture

Pamela L. Anderson-Mejías
The University of Texas—Pan American, USA

EXECUTIVE SUMMARY

This chapter describes a successful means of introducing returning, older students to online education in a university setting. After presenting basic background from the literature on retention within online classes, the case is presented in detail as to how 16 fearful learners became confident and successful through the instructor's taking time for preparation, establishing a sense of achievement using the technology, creating interconnections with peers, and demonstrating the usefulness of the virtual class over the face-to-face class. The author hopes that by describing in detail the case and the principles found, future educators can prepare their traditional students for the culture of virtual learning environments, thus expanding options for their programs while addressing university administrative concerns about student retention.

BACKGROUND

The goal of this case is to present a concise means to engage traditional, returning learners in online education delivered via internet exclusively. These "older" learners, who are returning to university after one career or who are required to update their teaching (or other types of) credentials while still working often have less experience with technology than the typical undergraduate student. They may be terrified when confronted with newer course delivery systems ranging from web-enhanced learning to completely online courses or programs. This case fits within the topic of "bridging the e-learning divide" although it does not fit the usual meaning of those who have access vs. those without access to internet resources. These students served in this case, who are sometimes called digital immigrants (Prensky, 2001) or, as some younger people today say "t'phobes," live in a society which values rapid information exchange, has easy access to computers, online technology and electricity but who, because

DOI: 10.4018/978-1-60566-942-7.ch008

of their experiences both in education and life, are not computer savvy. They need to become accustomed to the culture of online learning.

The case is located at the University of Texas—Pan American (UTPA), on the Mexican-U.S. border in deep south Texas. UTPA serves approximately 15,000 undergraduate and 2,000 graduate students in six colleges. Today the university has 56 Bachelor's level, 57 Master's level and 3 Doctoral level programs of study. UTPA's Carnegie classification is Master's Colleges & Universities (Larger Programs). The student population is 87.9% Hispanic for undergraduates and 75.1% Hispanic for graduate students (UTPA, 2008). Among strategic goals of the university is the incorporation of technology in teaching and there is a long-standing desire to see programs, as well as individual courses, taught with a significant online component, if not with reduced seat time or entirely online.

SETTING THE STAGE

The case described below is based on a number of courses which required practicing teachers to use the internet for updating their credentials as part of a master's program in second language teaching. Age of learners ranged from 23 – 61, with the average being 45 years, 8 months. These teachers were the program population from 2002 through 2006. Many of these learners were what have been called digital immigrants (Prensky 2001); they were new to the territory! Not to take the analogy too far (and not necessarily to validate all of Prensky's conclusions), I do find that age often correlates with experience in the online environment and with fear of using computers in general. To describe a couple of individual's comments on hearing that a given course would contain significant online requirements (all names have been changed to maintain anonymity) I present the following introductory information.

Student one, male, 56 years young, a grandfather who had served overseas for many years and decided to get a Master's degree for personal satisfaction, had the following comments. "I just can't get that computer machine to work right. Every time I need to do something, it just quits. I don't think I can possibly really learn by using it."

Student two, female, 48 years young, an "empty-nester" whose spouse is a busy professional decided to get the Master's degree also for personal satisfaction. She commented that she did not want to learn online because she liked to come to class and talk with everyone; she was sure she would miss the interaction and fun of the in-class environment. She was quite technologically savvy and connected to her children by the latest cellular phone, but did not use a Blackberry or other internet connection by phone because "those darned letters are too hard to hit."

The above two students were part of the program during a transition phase where the majority of courses included web-enhanced instruction using a variety of formats. All courses used e-mail and discussion boards in the WebCT platform. Some courses had additional requirements which permitted students to meet together online in lieu of one 3-hour class meeting per month. One course, which I personally taught, met online in three-week blocks and face-to-face only four times during the semester. Other instructors conducted parts of their courses entirely asynchronously. The goal was to familiarize our students with virtual learning so that eventually any or all of the program could be online. As a program this has been successful; in 2007-08 an instructor taught all of her graduate and undergraduate courses online while she was in Indonesia. The case described below took place just as the program was expanding to allow for such events.

Suddenly in 2005, the program included students who were truly distance learners. One was accessing from China, another from the interior of Mexico, and then there were local students. Because of the inclusion of these international students from abroad, the university Center for Distance Learning and Excellence in Education

(CDL) was considering use of Wimba from Horizon (http://www.horizonwimba.com/). The Wimba collaboration suite included a number of tools specifically geared for educational settings including higher education. "Classroom" instruction could be carried out synchronously where students and instructor hear/speak and see slides, write on these during the presentation, communicate using a written chat section, etc. The CDL director asked that I take the online trainings and use this technology for the courses described within the following case, which I did.

The program selected for this case is a Master's in English as a Second Language (ESL). This is an applied linguistics program training both local and non-local native speakers of English, local bilingual speakers of English and Spanish, and non-local non-native speakers of English to teach English as a second language (ESL) or English as a foreign language (EFL) in a variety of possible settings. Many of these MA students are already teaching in schools from early childhood through grade 12 in the U.S., as well as similar schools in Mexico. Others teach in private institutes—some associated with international businesses, some associated with higher education in the U.S., some associated with higher education abroad, and some private schools K – 12. Others have never taught English but have studied or taught other languages and are changing fields.

One key issue for the course, the program and the university overall was that of "student retention." Since some online courses in other departments had previously lost students, reportedly up to half of the official census date numbers, there was concern among our program faculty about the issue. The "culture" of online courses was seen as contributing to the lack of persistence among the learners. Some faculty feared that those local learners who were accustomed to face-to-face (f2f) settings and did not need to access online would resent having to accommodate others who were from outside the local access area. As instructor of record and coordinator for the program, I was concerned with ensuring minimal shrinkage and yet still reaching out to students who were not traditional and/or local.

PROBLEM STATEMENT

The major problem which this case describes is the enthusiasm of university officials, particularly those from the CDL, for testing the technology in a course filled with mostly traditional, older learners some of whom had no experience with the online environment at all and who were thus fearful and ready to drop out. While the university is committed to use of online instruction, it also must retain students. The faculty in the program must consider both issues as well as the program needs to expand our services beyond the local, physical area since many potential students cannot leave their homelands and families in order to take two (or more) years to study in the U.S. The dilemma then is helping local, older, more traditional students bridge the e-learning age/experience divide in order to interact with their peers in other countries and help the university explore uses of online technology.

A key challenge in meeting the needs of all our potential students was that of retention of these learners in the course and program. Naturally this intersects with persistence at the university itself. Issues of retention and persistence among university students built upon research among students at previous levels. A key factor found early in the literature considering youth drop out has been student involvement in school and school-related activities (see for example Finn, 1989). The importance of retention for Latin@s in particular has been discussed with suggestions for policy including the need for inclusion of these students (Vélez & Saenz, 2001). As our university data indicate, 87.7% of the student body in 2005 were Latino/a (UTPA, 2005). This percentage rose to 88.0% for 2006 (UTPA, 2006).

In working with our center for distance learning (CDL), I had previously been successful at teaching an entirely online version of our practical teaching course (Anderson-Mejías, 2005) but with students who had become accustomed to the online environment gradually throughout our program. We had seen none of the higher attrition rates reported elsewhere (e.g., Carr, 2000) and likewise had not found less satisfaction with the course. However, with students who had not had previous experience in the online environment, I believed a blended or hybrid course would help as described in Martyn (2003). This would increase the sense of collaboration and community which Rovai & Jordan (2004) found greater in the blended graduate course than either traditional or entirely online courses. Yet, this was impossible due to the true distance students—or so it seemed.

The two students who had been working in the program off-and-on and becoming accustomed to the online environment, were nearly ready to graduate. As research has noted (Dupin-Bryant, 2004) number of previous courses completed with online components would indicate no significant challenge in keeping these students in the course and program. In addition, however, there were two or three brand new students who had not been gradually introduced to online education. Among those were the following two gentlemen.

Student three, male, 46 years young was in the process of changing professions. His previous experience had been in another discipline and he was now moving into language teaching. He had begun the program earlier then stopped for awhile. He was now finishing up his final course and it was to be taught entirely online using Horizon Wimba for synchronous meetings. It is probably not an exaggeration to state that he was terrified! At first, he wanted to wait for another year or two until the required final course would be offered face-to-face but his earlier coursework would then be forfeited because of the time lapse, so he opted to try—kicking and screaming all the way.

Student four, male 35, and returning to update his credentials for teaching was just beginning the program and simply opted to wait the year or two for the next offering of the course. He stated, he had tried one online course and it was just "too much." When asked, he elaborated that there was too much reading, too much writing, and just not enough "teacher teaching" for his taste.

There were other students among those in the program courses—many were younger, in the 22 – 30 range, who readily accepted the online course requirements as a challenge—or just the norm for them. There were others who were somewhat older, in the 31 – 45 range, who were nervous, but due to their heavy schedules teaching or working, taking care of a family and often extended family members as well, welcomed the idea of accessing their university courses from home and/or partly at a time they selected rather than needing to drive to campus, find parking, miss seeing their children after school, etc. Then there were the 46 – 61 year old group. These age steps have been used by Hagedorn (2005) who calls the first "young adults," the second "prime timers," and the third "last chancers." I have described these learners in order to give the reader an idea of the variety among older adults returning to university who suddenly are faced with a new means of learning. Thus, the group of older learners discussed in this chapter is quite varied. Nearly all, however, benefited from the process described in this case and, in the final course evaluation, all were happy to varying degrees with their learning not only of course material, but about the virtual course as well.

THE CASE

The course chosen for this group was the teaching practicum which is required of all MA students in the program. It usually meets once per week for a 2 ½ hour block of time from 4:30 to 7:00 p.m. over the course of a 16 week semester. Since many of the students in the program are teaching full time in an EC – 12 school setting, this may require them to leave their campuses early, drive between 5 and 60 minutes to the university, park and then attend the class (often without eating dinner until after the course).

This particular student group of 16 members is at the graduate level. Many have had little or no instruction using the online medium and none had experience with an entirely online virtual class. Key issues from the literature include the importance of attitude toward school life where the sense of achievement influences student leaving (Glass, 1996), importance of building a sense of community (Rovai & Jordan, 2004), importance of creating meaningful application of technology which enhances quality of life for older learners (McNeely, 1991), and finally the educational preparation which will engender student readiness prior to the distance course (Diaz, 2002). It was important to all concerned that the computer environment be perceived as a means to the end of students thinking in meaningful ways and, as VanSlyke (2003:5) noted "…it is the teacher's responsibility to structure and support the students' learning experience. The computer is a medium, whereas the learner and the teacher are the mediators."

Therefore, it is the responsibility of the teacher to 1) build communities, 2) create meaningful use of the online medium, 3) prepare the students prior to an entirely virtual course, and 4) engage learners in activities where they believe they are achieving useful results. Not a small task—especially given the fact that a hybrid course was out of the question. Following are the steps I used to achieve these goals.

Preparation

Because there were true distance learners who would be accessing from differing time zones—China 13 hours ahead, Mexico 1 hour behind the local time—this course could not be held with some f2f in-class meetings. As noted above, the technology staff suggested using Horizon's Wimba as a platform for the course. It was believed by the director for instruction that this technology was user-friendly and could be implemented without any need for student preparation.

As instructor and advisor for most of the students, however, I disagreed. I believed that a few weeks of introduction to the online environment of the virtual course would help alleviate fears and better prepare the 31 – 61 year old prime timers and last chancers for a successful online experience—which in turn would promote retention in the course and program.

Prior to beginning the semester, the technology team and instructor worked together to practice using Wimba online. The instructor took the Wimba introductory courses, and had previously taken (and taught) the university Teaching Online six-week course required of all faculty who plan to teach entirely online. The technology team reserved a computer classroom for the first three class meeting evenings in the semester and arranged for the instructor to have a laptop set up within the control room behind the classroom and not visible to the students. The instructor e-mailed all students to meet in the specially prepared room as well as contacting the true distance learners regarding time for the synchronous virtual Wimba course meeting in their respective time zones. The syllabus was disseminated online through e-mail; a key requirement was the headset with microphone in addition to the usual textbooks.

Three local students replied that they preferred to "meet" online using the technology rather than drive to the campus for the special introductory meetings. The technology staff and instructor prepared a handout, sent electronically to these students, which showed how to access the course through WebCT which included the Wimba location on the homepage. All was prepared.

First Steps

The first class meeting occurred in the computer lab with nearly all local students present (two of those who thought they would access from off campus came to this first meeting anyway). There were also three staff members from the CDL technology support center, the usual computer lab personnel, the instructor and the Director of the Center for Distance Learning. The instructor welcomed everyone, introduced the staff, and turned the class over to the instructional designer who walked everyone through the procedures for entering the course, opening the Wimba connection, and getting started.

Meanwhile, the instructor disappeared behind a one-way mirror in the observation room and accessed the course with the laptop. Both true distance learners were connected and ready to go. Through the "chat window" in Wimba, the instructor was chatting with them and telling them of the steps going on in the computer lab. Predictably, there were two local students whose e-mail + password combinations were not working, so the staff members made special arrangements for their immediate access and gave them instructions as to how to remedy this challenge during usual working hours. The hands-on help from the technology center staff was invaluable in making student fears of the machine disappear.

Class began as the local students joined those already in the session. The instructor elicited participation using the "hands up" feature to a number of fairly simple yes/no questions. Key issues, like remembering to take the hand down after answering were explored via experiential learning! Another important issue that came to light was the need to open the microphone switch in order to talk to the one another. This included experiential learning by the instructor! In the adjoining computer lab room the local students could hear me talking, but they were not getting this input through their computers. This nearly caused mass confusion! The students thought they had "done something wrong" when, in fact, it was I who had not pressed the "talk" button!

The first class proceeded rather slowly from the perspective of usual instruction. However, this initial period of gaining comfort with the technology was worth the time invested—students believed that they were going to be able to access and participate because of this time spent working together with assistants in the practice session. Thus, by the end of the first class, all students felt they had achieved the ability to access the course and use the basic features of the technology. After the online course ended and the distance learners had logged out, I stopped several of the students in the hallway to ask how they were feeling about being online the next week. Of the five I talked with, four believed they would be fine but were planning to come to the campus computer lab again while one thought she would try to access the course from her own school's lab. John, student three mentioned above, was still unsure he would continue in the course. I convinced him to come back the next week for the class and assured him I would be there and that there would be a CDL staff member present as well.

After briefly talking with the students, the CDL director and staff talked with the instructor. All were very pleased with the results of the class. It was decided that only one staff member would return the following week and the instructor would be available before and after the class to debrief the students who accessed on campus and try to persuade them to venture off campus.

The Second Week

The purpose of the second week was to accomplish two things. First, for the course itself, it was critical to set up collaborative learning groups which would discuss different topics about a model lesson together during the online class. The instructor would not be available to these groups as they were composed of three students each who were to role-play different parts based on the prompts.

At the end of the fifteen minute discussion sections wherein the groups would meet using chat but not their microphone/headset, the whole group would reconvene online and each collaborative sub-group would present its key points discovered from the role play about the given classroom situations. The second key purpose was to solidify the students' belief in their abilities to access the course and use the computer for learning.

Eight students came to campus to access the course in the computer lab. The challenges with username and passwords had been fixed for all of them. The online students from China and Mexico were already in the course when the group in the computer lab entered, and there was individual and group chatting while waiting for the other local students to access from their various locations. Among the local virtual students, two had no difficulties but two others could not enter the course. One drove from her school where she had been trying to enter the course to campus—about 35 minutes. The other had gone home and was having difficulty because her home computer was dial-up based. The phone number and code had been restricted somehow and she never was able to get into the course that evening. She did, however, call the instructor's cellular phone to let everyone know she could not get in this particular evening. The CDL director worked with her to get the phone codes in order and she was able to access later in the semester. Wimba's archive feature saved all of the course information for this student, but unfortunately, the collaboration work for her did not get a practice session.

During this second course meeting, the class itself began to take off. Information was easily presented using the power-point slides prepared in advance by the instructor and which formed the background on which to write during discussions. One note made during this session was to insert blank slides for students to use to write their comments and notes on during the class discussion since those with prepared materials seemed to inhibit students from writing on them. A second note from this early session was that color and power-point backgrounds could reinforce differing areas of content being taught rather than preparing the same style slides for the whole week's class. This was a small step which reaped great rewards in helping orient students both during the class sessions and when reviewing through the archives.

There was one unanticipated result of working in collaborative groups online while still in the same computer lab with others. Although two groups were required to be working online through dialog in the chat area since not all of the students were in the lab, one group was all present in the computer lab and found it much easier to talk with each other rather than type into the chat. The fourth group (a pair), missing the dial-up class member, also resorted to the easier face-to-face chat rather than typing online. The CDL staff member, not really an instructor, felt that she could not intervene; and the instructor, hidden from the class behind the one-way mirror, could not hear that there was oral discussion taking place. In later class meetings where everyone was accessing virtually, this actually came back to haunt the pair and group of three who did not practice online small group work! And, for the final class together, the instructor decided it was important to leave the door open between the lab and the monitor room.

At the end of this second class, students had formed rudimentary collaborative groups which would be used periodically during the course. This step began to build community among the class members. In addition, among the eight students who had come to the computer lab, five had gained sufficient skills and self-esteem using the computer, WebCT and Wimba, that they planned to access from off campus the next week. The instructor asked students whether they would be comfortable meeting without her behind the mirror the next week, but the three remaining students did not feel ready to "go it alone" quite yet. So in order to scaffold these learners with a bit more support, the CDL staff member and instructor reserved the computer lab for the third (and final) week of being online in the same room!

The final online-together class. The third week of the course had three major goals. First, this was to be the final week of contact where the instructor would be available face-to-face before, during, and after the class. The next week the instructor would access from off campus and there would be no CDL staff member available to help students who were in this course. In order to prepare for this event, after one hour the CDL staff member was planning to leave the computer lab. Students were to become self-sufficient by the end of this class session.

The second key goal of this class was two-fold. Based on course information to be presented, the instructor planned to require students to conduct an immediate search online using Google Scholar for key research about an issue presented via another "pushed out" online source which was part of the students' textbook appendices available only online from the publishing house. The goal was to address students' recognition that using the online format was the most meaningful method for accomplishing the learning task and to engage these learners in activities where they knew they were achieving useful results that could not be better achieved in any other (non-online) manner.

The third goal of this class was to cement community through use of pair-work online leading to individual student's being given the "pen" to lead discussion orally and write their own notes on the blank slides of the presentation. By the end of this final f2f class, the instructor wanted students to realize they could rely on one another as much as, or even more than, on the instructor; to solidify the collaborative groups and the community; to ensure that most, if not all, students recognized their own achievement using the medium of instruction; and to cut the cord with campus.

Life, unfortunately, does not always comply with the best prepared plans. The class session got started with four students in the computer lab; the three who were not quite ready from the previous week and the one whose dial-up issues had not yet been resolved. All of the others accessed this session from off campus. Just as the CDL staff member was preparing to leave, there was an electrical brown-out of the whole building! This resulted in all of the computers losing power except the laptop used by the instructor. From the wireless connection, I was able to continue talking with our student from China, 13 hours and thousands of miles away, as well as some of the others who had accessed from off campus. Unfortunately, I could also hear the confusion in the computer lab room! As the power came back up and computers rebooted, the students in the lab spent nearly 15 minutes regaining access to the course. However, that access was imperfect and it was hilarious (and irritating) to hear my own voice giving instructions or talking with one of the distance students, then hear the echo of my voice coming over the computer lab system in the next room, plus hear through my ears the students in the lab responding to me or one of their peers online then a few seconds later hear the same through the headset. The lag between China and campus was about 20 seconds; the lag between the computer lab and the monitor room was about 90 seconds!

Needless to say, the CDL staff member stayed with the group and I closed the lab door! As the class session proceeded, it was clear that these four students really did not need further hand-holding. Although the brown-out intervened in the planned course activities, it showed that these four students were able to handle an unanticipated system failure without panicking that THEY had somehow done something which hurt the computers. All but one of the students in the lab had simply rebooted the machine and started the procedures for entering WebCT and then our course and the Wimba site. Student three above had been slightly taken aback and did not immediately know how to reboot the computer system. After that was taken care of by the lab personnel, he proceeded to enter the course although he claimed that it was his "bad electric aura" which had blown the system in the first place.

By the end of this third f2f meeting, the students were confident in their abilities to handle the online learning situation. The goal of cementing the community of learners had been accomplished. The goal of recognizing the usefulness of the online environment for learning was partially accomplished. The push-out of the appendix material worked fine for the students in the computer lab; however, not all of the online students were able to receive the material through the Wimba. In order to remedy this situation, the CDL staff member suggested that the instructor give the original web location of this appendix material through the chat box to all of the students having difficulty. This worked wonderfully and I would not have thought of it had not my trusty CDL staff realized I was at a loss even more so than the students! Nonetheless using the Google search feature was a great success and did emphasize that there are some things which the internet provides far better and more immediately than other means can.

One final lesson from this third session can be drawn. Observing the students presenting to their peers while talking through the headset and trying to write on the presentation area of Wimba showed that this duality was confusing both to the presenter and the students watching and hearing that presentation. In future virtual sessions, whenever a group presented its results of discussion to the whole, one member would take the pen to write notes while a different one talked through their points. This separation allowed everyone to participate AND those watching and listening to understand without becoming confused by time lags or unclear notes. It alleviated undue stress as well.

Ready or not, the cord was cut. After this final class together, everyone was to access independently. Some students, who did not have computers at home, had requested use of a smaller computer room so they could talk using their headphones and microphone but not disturb the rest of the students studying or using the labs. The next 12 weeks included all of these students interacting, uploading their teaching demonstrations, commenting on one another's teaching, and studying various aspects of good classroom management, techniques of addressing student variation, and evaluating one another and themselves as leaders of online groups (Anderson-Mejías, 2006).

The learning objectives of the online course were met as exemplified by end of course grades where all passed with a grade of B or above as well as the end of program portfolio which included the teaching demonstrations and paper from this course. Each and every student exceeded program learning goals for theory, application of theory to a specific environment, and innovative, student-centered teaching as demonstrated by the work from this course.

After the final meeting of this semester-long course, all local students were invited to meet with the CDL director and instructor at a local restaurant close to campus for coffee/dinner/snacks. Those attending (8 of the 12 local students) were requested by the CDL director to anonymously complete a survey evaluating the course and their own attitudes toward the content as well as the online delivery of the virtual course. This consisted of four general areas: effectiveness of the course (6 items), learning effectiveness (5 items), teaching effectiveness (3 items) and preference for virtual learning (2 items). Each item was evaluated on a five point Likert-type scale. Among these students, who conceivably could have taken the course face-to-face, the overall satisfaction with the virtual course was 80%. Teaching effectiveness was rated 100% on all three items. 25% of these students believed that they contributed more to discussion in the virtual class than they otherwise would have in a f2f one, and 37.5% felt they paid more attention. 75% said they found learning as effective in the virtual class as the f2f traditional format and 75% felt they were connected to their classmates more in the virtual class than a traditional one. In terms of motivation to keep abreast of the readings and prepare for the class, 50% stated that they were more motivated and 37.5% said there was no difference in their motivation levels to prepare for the two types of learning situations. Only one learner answered strongly disagree to any questions; these were four key questions—namely, 'I found learning in the virtual classroom as effective as in a f2f class,' 'I felt connected to other classmates in the virtual classroom,' 'I prefer to learn in the virtual class rather than in a f2f class,' and 'I would take another course in the virtual classroom format.' Despite these results, all students were retained in the course; in addition, all completed the program and graduated with the MAESL from the University.

Bridging the Divide from Technophobe to Technophile

This case fits within the topic of "bridging the e-learning divide" although not with the typical meaning. Usually we consider the divide to be between those who have access to electronic media on a regular, daily basis (primarily in the developed, first-world) vs. those who have little access to electronic media or even educational facilities such as books in places like Burkina Faso or Xinjiang, China. However, this case presents those who reside in a society which values and has easy access to computers, online technology and electricity but who, because of their experiences both in education and life, are not "computer savvy" or are, as some younger people today say "t'phobes." These are the immigrants rather than the digital natives.

Socio-psychologically these learners, generally prime timers or last chancers, need to overcome their fear of the computer itself, their inexperience with online delivery and lack of comfort in the virtual environment, as well as the isolation resulting from online courses where students are physically separated from one another.

The first step must be to prepare the newcomers for their online experiences. In this case study, this step was addressed by demonstrating the technology during three sessions of online meetings progressively removing outside guidance while training the students to rely on one another. In the first meeting, students followed a staff member demonstrating how to access the course and could rely on two additional staff members and the course instructor who were present. In the second meeting, those students still wanting support, could rely on one staff member circulating as a problem solver. Although the instructor was present in a separated room viewing through the one-way mirror, she was not "reachable" except by interaction in the online course. For those requiring further scaffolding into the virtual environment, the third class session began as the second one had ended, but then the staff member left and the course was finished entirely without physical contact. After each class session, there was a short debrief of students on the campus; additionally the CDL staff and instructor met to discuss strengths and challenges.

The second, closely related step to bridge the transition from technophobe to technophile is to create a community of learners. This case study began that process during the second class session by creating peer collaborative groups, using the chat feature to discuss among the group members without the whole class observing, then using the features for student presentations in Wimba to elicit discussion orally from the entire class about the issues experienced or discussed among the smaller groups. Because peer grading of participation and leadership among these graduate students was embedded as part of the course grade, all students recognized the importance of each other as colleagues rather than subordinates of the instructor. The same collaborative groups were used throughout the course, although other groupings were also used from time to time in order to vary the interactions.

The third step in helping students make the transition is creating a meaningful, necessary atmosphere for using the internet and virtual environment effectively as part of the course. In the present case, this was begun during the third and final transition session by asking for short searches to support or dispute a prompt delivered online during the class session. For the adult learner who needs to see a reason for the pedagogy, this was a vital stage. Because the students were enrolled in a course to improve their own teaching skills, unmotivated choices in delivery of content would be counterproductive.

SUMMARY OF THE CASE

Psychologically adult, non-traditional, learners must feel success in their ability to cope with new environments or cultures. They must overcome the fear inherent in new situations or they will simply drop out of the uncomfortable setting. Using a computer, interacting actively rather than passively in a classroom setting, typing as well as speaking, finding the "right" button or icon or area of the monitor—all are new and for some uncomfortable. Scaffolding the expected and critical behaviors through directed observation (as in the first class session), followed by facilitated problem solving as needed (in the second session), and finally independently with help available (as in the third), will help alleviate stress and engender a sense of achievement. This building of self-esteem by structuring individual student success encouraged our students to stay with the course, the program and in the university.

For the students psychologically the preparation to enter a new culture—the virtual course—was worth the additional time and effort required. Those students who were already comfortable with online delivery actively supported the halting steps of the new arrivals, much as happens when second language learners make mistakes among native speakers who repeatedly engage them until communication is accomplished. In both cases, the learner stays within the new culture because she or he has received the support of others.

Socially, people want to stay in groups which have meaning for their lives. Many adult learners, beyond the traditional ages of university students, have numerous outside influences which can cause them to simply disappear from a course, a program, and the university. By creating a community, this drop out rate is reduced. Leaving friends is much more difficult than leaving a group of unknown people.

Finally, when not only the course content but the method of course delivery is utilized to engage students due to its very nature, then a key human question, namely, why must I attend this course taught in THIS manner, has been answered. Using the critical features of the online world—immediate access to vast amounts of information, quick reproduction of stimulating material in audio, video and kinesthetic forms simultaneously in multiple locations, multiple modalities available simultaneously (like the various sections of the screen used by Wimba)—all clearly are available in the online teaching of courses which are not available in traditional face-to-face teaching. By building these into the course from the earliest point, the instructor or instructional designers have met the need for adult learners to gain meaning from the presentation format as well as the content. These factors together encourage student retention.

CHALLENGES FOR THE FUTURE

Universities throughout the U.S. will undoubtedly continue to feel the crunch of financial need, requiring student retention efforts and expansion into new media of course delivery. Balancing these systemic needs with the learning objectives and individual needs of students who must learn using these new technologies will continue to require thoughtful preparation of instruction. This case has presented one means of moving older, returning learners from the traditional to the virtual classroom. Without some bridge for these students, they will be lost to the system—they will not be retained because they can choose not to attend university programs which use virtual classrooms and online education systems. Recognition of and providing time and space for psychological factors of fear and leaving, as well as social factors that the virtual classroom is, in fact, a new and different culture for these individuals suggests the critical need for systematic preparation as presented in this case.

REFERENCES

Anderson-Mejías, P. L. (2005). Online training for English as a second/foreign language teachers. In M.O. Thirunarayanan & A. Perez-Prado (Eds.), *Integrating technology in higher education* (pp. 143 - 154). Lanham, MD: University Press of America.

Anderson-Mejías, P. L. (2006). A case study in peer evaluation for second language teachers in training. In T. Roberts (Ed). *Self, peer, and group assessment in e-learning* (pp. 17-63). Hershey, PA: IGI Global.

Carr, S. (2000). As distance education comes of age, the challenge is keeping the students. *The Chronicle of Higher Education, 46*(A), 39-41.

Diaz, D. P. (2002). Online drop rates revisited. *The Technology Source*. Retrieved December 4, 2008, from http://technologysource.org/article/online_drop_rates_revisited/

Dupin-Bryant, P. A. (2004, December). Pre-entry variables related to retention in online distance education. *American Journal of Distance Education, 18*(4), 199–206. doi:10.1207/s15389286ajde1804_2

Finn, J. D. (1989). Withdrawing from school. *Review of Educational Research, 59*(2), 117–142.

Glass, J. C. Jr. (1996). Factors affecting learning in order adults. *Educational Gerontology, 22*(4), 359–372. doi:10.1080/0360127960220405

Hagedorn, L. S. (2005, January) Square pegs: adult students and their "fit" in postsecondary institutions. *Change*. Retrieved December 15, 2008, from http://www.coe.ufl.edu/Leadership/ihe/TRUCCS/Files/Square_pegs_for_Change.pdf

Martyn, M. (2003). The hybrid online model: good practice. *Educause Quarterly, 1,* 18 – 23. McNeely, E. (n.d.). Computer-assisted instruction and the older-adult learner. *Educational Gerontology, 17*(3), 229–237.

Prensky, M. (2001). Digital natives, digital immigrants. *On the Horizon, 9*(5). Retrieved November 2, 2008, from http://www.marcprensky.com/writing/Prensky%20-%20Digital%Natives,%20Digital%20Immigrants%20-%20Part1.pdf

Rovai, A. P., & Jordan, H. M. (2004). Blended Learning and sense of community: A comparative analysis with traditional and fully online graduate courses. *The international review of research in open and distance learning, 5*(2), 1-13. Retrieved November 5, 2008, from http://www.irrodl.org/index.php/irrodl/article/view/192/795

UTPA. (2005). *Fact book.* Retrieved December 16, 2008, from http://oire.utpa.edu/publications/minifactbook2005.pdf

UTPA. (2006). *Fact book.* Retrieved December 16, 2008, from http://oire.utpa.edu/publications/minifactbook2006.pdf

UTPA. (2008). *Stats at a Glance.* Retrieved February 12, 2009, from http://oire.utpa.edu/publications/minifactbook2008.pdf

VanSlyke, T. (2003). Digital natives, digital immigrants: some thoughts from the generation gap. *The Technology Source*. Retrieved December 16, 2008, from http://technologysource.org/article/digital_natives_digital_immigrants/

Vélez, W., & Saenz, R. (2001, December). Toward a comprehensive model of the school leaving process among Latinos. *School Psychology Quarterly*, *16*(4), 445–467. doi:10.1521/scpq.16.4.445.19897

Wimba, Inc. (2009). Retrieved from http://www.wimba.com/products/wimba_collaboration_suite/

Chapter 9
An e-Training Support Program for Regional and Local Development

Vassilis Syrris
Aristotle University of Thessaloniki, Greece

Fenia Tsobanopoulou
Aristotle University of Thessaloniki, Greece

EXECUTIVE SUMMARY

New forms of learning such as distance training and consulting constitute a significant field that presents considerable advantages compared to the traditional educational practices. Computer and communication technologies like World Wide Web/Internet and broadband networks enrich the knowledge environments and grant new perspective to learning mechanisms. In this case study we analyze the technological, cultural and social issues involved in an online distance training program implemented to address in particular farmers, animal-breeders, unemployed and low-salary workers. Distance consulting focuses on subjects concerning entrepreneurial skills and personal training. The project scope includes decentralization, local intervention for employment purposes and bridging of geographical and technological distances.

INTRODUCTION

The dynamics of the new global economy form an environment characterized by uncertainty, insecurity and risky decision-making. This established status urges countries to implement and incorporate a digital policy in all financial and social sectors in order to achieve solid, sustainable development. Contemporary politicians, businessmen and citizens are compelled to think in digital terms; to see future economic viability and social growth under a digital umbrella.

The prospect of a digitized world, that is, an organizational configuration where computers, network communications and almost every device equipped with electronic circuits and processors will control and administer procedures and actions, presents in parallel both challenges and threats. One major issue that arises, known as digital divide, is the possible marginalization of those human groups that will be unable to adjust or be incorporated in the new setting. For example, observers noted that people

DOI: 10.4018/978-1-60566-942-7.ch009

with higher Internet access had greater access to education, income and other resources that help people get ahead (Bucy, 2000; Strover, 1999).

European Union policies highlight the importance of a balanced approach to the development of information society including the construction of broadband infrastructure in both central locations as well as in remote, rural ones. Specific projects reinforce the capacity of regional authorities to plan, manage and implement Information & Communication Technologies (ICTs) strategies, supporting thus, EU policy objectives within the digital divide context. In this direction, the role of learning and education is vital: better educated rural citizens of all ages and backgrounds, with ample life-long learning opportunities and access to information society and knowledge economy assets, can protect the natural resources of their countryside regions, resist urbanization tendencies, understand the new challenges and respond with flexibility, initiative and new proposals.

Various international forums and agreements on information access and technological capacity have acknowledged the importance of technology sharing, especially throughout under-developed or developing areas. Accordingly, this study focuses on digitally isolated rural areas, under the consideration that it is relatively more difficult for rural people to keep up with the revolution taking place in technology and management, and that they need better access to information, capital and dexterities in order to implement innovations and improve their professional and social status.

DIGITAL DIVIDE

The term digital divide or digital gap is the distance, in terms of socio-economic factors, among individuals, families/households, enterprises and geographical regions with regard to their opportunities for accessing Information Technology and Communications (ICTs) and to their use of the Internet for a wide variety of activities (OECD, 2001)–translating into their difficulty in entering a wide spectrum of activities (Gurstein, 2003). The majority of studies define three general categories:

- The world divide between developed and developing world (James, 2005; Wade, 2004): technologically and industrially developed nations versus countries with lack of elementary infrastructure.
- The divide between countries: strategic advantage acquisition in sectors such as economy, army, energy etc. The result is the creation of specific strong political or economic lobbies in decision-making and policies implementation.
- The inter-social divide within a state -it is also referred to as the 'technological divide' or 'the lack of digital inclusion' (Rice, 2001): differences between big urban centers and rural regions (Rao, 2005); or, between educated and not educated people, upper or lower class, financially independent or dependent, men and women etc. Age, sex, location, culture, social position and personal or political beliefs are, in general, parameters that can cause an individual to abstain from or not, to be given an option for or not, in technological evolution.

The digital divide affects the growth of individuals, social groups and countries disproportionately (see Table 1; Africa, for instance, accounts for about 14% of the world's population, but only 5.6% of its population can connect to the Internet). However, digital gap is a multifaceted issue and someone cannot attempt to confront it univocally. Van Dijk (2006) claims that the digital divide cannot be understood without addressing issues such as attitudes towards technology, the channels used in new media diffusion, educational views of digital skills, and cultural analyses of lifestyles and daily usage patterns. We pinpoint that even

Table 1.

World Internet Usage and Population Statistics			
World Regions	Population (2008 Est.)	Population % in World	Penetration (% Population)
Africa	975,330,899	14.5	5.6
Asia	3,780,819,792	56.3	17.4
Europe	803,903,540	12	48.9
Middle East	196,767,614	2.9	23.3
North America	337,572,949	5.1	74.4
Latin America/Caribbean	581,249,892	8.4	29.9
Oceania / Australia	34,384,384	0.5	60.4
WORLD TOTAL	6,710,029,070		23.8
Data Copyright © 2001 - 2009, Miniwatts Marketing Group. All rights reserved worldwide. (WIUS, 2001-9)			

sophisticated users such as businessmen, students, teachers and practitioners experience the digital divide impact when their companies, schools, educational institutions do not advance quickly enough to provide new content and to implement suitable qualification programs.

Factors Leading to Digital Divide

It is well-understood that the digital divide should not be viewed as a negative technology tendency; instead, a broader interpretation of the conceptual context is necessary (Joseph, 2001; Luyt, 2004). A policy maker needs first to detect the origins of the phenomenon under consideration. Some common factors are:

- The existence of poverty in both developed and developing countries; when food, health and security are unreachable goods, the occupation with technological stuff seems an unattainable and unrealistic scenario. Alcántara (2001) argues that people in low-income countries are limited not only by their lack of access to modern means of communication and sources of information, but also by a complex network of constraints ranging from unresolved problems of poverty and injustice in their own societies (*economic causes*).
- The phenomenon of analphabetism; lack of basic educational services (especially in developing countries) and the need for survival that appear from the early years of a child (observed also in developed countries in certain social groups like the gypsies, people living in distant rural areas etc) limit dramatically the chances for their digital inclusion (*socio-economic/cultural causes*).
- Technologically advanced countries handle their superiority as a competitive advantage. The ability to create, acquire and adapt new technologies is a critical requirement for competing successfully in the global marketplace. Although many IT products and services have now been transferred, are constructed in and provided by less advanced countries (especially in Asia) still companies coming from these countries are rarely capable of competing with the big brand names. Furthermore, even when the big brands have branches in less advanced countries, still, the grand technological advancements concerning army, space research, energy, security,

health, and other high-technology projects are kept as top secrets, concealed from local governments or even from the actual companies they develop them. Therefore, easy and unimpeded technological growth is not a privilege allowed for low income countries; they are rather preferred as consumers or inexpensive workforce. (*economic/political causes*).

- Political decisions based on interest, class priorities, social or racial discriminations might lead to planned negligence towards the advance or development of certain areas, social groups, second- class citizens, useless or dangerous -for the status quo- groups of people (*socio-political causes*).
- Governmental control of the information flow in authoritarian regimes (*political causes*).
- A non revised curriculum or suitably adapted educational policy unwilling or unable to incorporate the digital dimension in all learning levels, resulting in inadequate technology users or ignorance or indifference at a personal level (*educational/political causes*).
- Inefficient training strategy. Lack of specialized scientists or trained employees who could better contribute to productivity and could suggest new directions in the products market and the services provision along with ultimately improving working conditions and advancing the socio-economic status of individuals and local areas as a whole (*educational/political causes*).
- Some other shortages resulting from inappropriate or slow national policy, such as lack of creation of high-quality digital content in the national language (the bigger sources of information is available only in English), the controlled flow of information, lack of formulation and boosting of a technological culture etc (*economic/political/technological causes*).
- Lack of basic infrastructure such as communication/electric power networks, roads, railways etc. When one cannot find technology nearby then one is not in the position to utilize or apply it in any way (*economic/political/technological causes*).
- Cost of equipment requirements-hardware, software, networks access, high taxation, and both the time and money required for the development of computer skills are parameters connected mainly to personal wealth in contemporary capitalism. For instance, Tukiainen (2004) found that income is one of the significant factors affecting ICT adoption in Finland, Ireland, Netherland and Sweden (*political/economic/technological causes*).
- Although efforts have been made to make technology user-friendly, some types of disability do not permit or impede the user's access and familiarization both with hardware and software (*health/technological causes*).

A more analytical or elaborate discussion of the aforementioned factors is beside the scope of this study. However, other than cases where abstinence from the digital world is due to a personal life attitude or due to apathy or disinterest or fear (e.g. that increased digitization may increase government control, which in turn will diminish the opportunities for informal labor, building and agriculture), a crucial point to consider is that digital divide is a deeply political issue which can be resolved only through courses of actions and measures applied on wide geographical areas and on a big population scale.

In this case study, the focus of interest is upon the inter-social digital gap and more particularly, on the problematic access to knowledge and information as observed in rural regions in the developed world.

DISTANT CONSULTANCY

In general terms, a consultant according to Wighton (1993) is someone with experience in a specific field, who gives advice and shares his/her experiences with others. The consultee (client) is the one who receives advice and direction.

Consulting, we believe, constitutes a 'mild' form of mentoring. A mentor holds a wide and deep knowledge in several cognitive areas in order to be able to address issues requested by the consultee (mentee in this case). The mentor views his/her interlocutor as a complete entity, with personal, social and professional needs located in the same frame or context. In addition, in most cases, mentor and mentee develop a long-term relationship. The time mentors and mentees have for the tele-mentoring relationship affects their choice in engaging or not in these relationships (Lynch, 2003). In a concrete, consultancy framework now, a consultant implements problem-solving tactics during the procedure of his/her own expertise and experience transfer. Proposition of ideas and initiatives, aid in decision-making process, promotion of creativity and novelty, provision of justified and structured analysis on topics of interest and custom-tailored solutions are some main priorities applied by an experienced consultant. Critical factor in consulting is not so much the relationship duration, the depth of the effect or the consultee's change of attitude, but the adequacy and effectiveness of the proposed solutions. Consultants offer know-how, promote professionalism, help develop self-confidence, activate the consultee and his/her potential, occasionally even, help develop leadership dexterities.

Consulting addresses both individualized and group needs. We distinguish four feasible scenarios:

1. One-to-one consulting: it appears more demanding in terms of cost and time; however, it is considered more effective due to the content customization upon the consultee's requests. It takes place either on-line or live (face-to-face session, consulting through phone, e-chat, video/audio-conference) or off-line (by post, e-mail, e-fora).
2. One-to-many: this type of session is realized in real-time and resembles a typical learning approach met at academic rooms. It takes the form of teaching or training in case the presented material is defined by the consultant and addresses some general needs of the audience. Otherwise, the audience poses subjects for discussion (specific needs) and either the consultant and/or members of the audience (peer-learning) attempt to resolve them.
3. Many-to-one: This scenario presupposes the existence of more than one experts (consultants) who provide advice to the same person. This type of session can take place either in real-time or in off-line mode. Except for the case of free internet-based services like newsgroups and e-fora, this method is in general more expensive than others but more accurate and fast in terms of offering multiple replies and specialized information.
4. Many-to-many: It constitutes a combination of 2 and 3. It is a process usually accommodated by the Internet (e-mail, chat, newsgroups, e-fora, mailing lists, e-encyclopedias, video-lectures etc).

Distance consulting is an extension of the traditional face-to-face approach, attempting to bridge geographical distances when the participants are located at different places. The case where a computer operates as an interface between consultant and consultee is called computer-mediated distance consulting or electronic consulting. It can be realized in a synchronous mode (the participants discuss at the same time –online session such as video-conference, telephone, chat channels) and an asynchronous mode (the participants communicate off-line, via e-mail, post, newsgroups, e-fora and digital interactive content released on web-sites, cd/dvds, etc). When do we need to apply an electronic distance consulting approach? With the new technology advances, this question has multiple answers:

- Feasibility of consulting session and cost reduction. There is no need for commuting when the advisory session participants are not in the same geographical location.
- Time saving; even when both consultant and consultee are in the same city, it is more convenient and quick to communicate without leaving their specific location.
- Some people prefer this type of contact; electronic mail, newsgroups and chatting in particular, where there is absence of direct speech, help them express their thoughts and desires more open and precisely.
- Very often, learning procedure relies on the positive image of mainly the instructor. Distance in communication delays demystification.
- Computer-mediated learning enriches the constructive interaction by using specialized software and additional digital material. Also, the meetings can be easily recorded; in such a way, a consultee history is generated and can facilitate the consultant's work or be employed as an exemplar case for new sessions.

On the contrary, distance consultancy approach is not suggested in cases that require emotional support, trust building and high security-when sensitive (secret or very personal) data is exchanged.

CASE STUDY

It is common fact in both rural and urban areas that inadequately skilled working force is a major obstacle for the application of new production technologies and management practices, while lack of specialized knowledge is the principal obstacle in making use of ICTs. For the European Union both access to broadband infrastructure and services, and the strategic use of ICTs constitute one of the main drivers for economic growth and social inclusion.The European strategy is "to ensure that the benefits of the information society can be enjoyed by everyone, including people who are disadvantaged due to limited resources or education, age, gender, ethnicity, etc., [and by] people with disabilities as well as those living in less favored areas" (i2010 European Strategic Plan, 2005). Life quality improvement and solid balance between urban and rural areas are also priority axes, reflected in the cohesion strategic guidelines for 2007-2013. However, the viability of rural areas needs to place rural activities in a broader context, taking into account the increasing competitiveness of the agricultural and food sector, having the protection of rural income as the final goal.

This case study describes a nine-month action that took place in border regions of the Greek periphery aiming at dexterities enhancement of financially weak groups, such as farmers, animal-breeders, low-salary workers and unemployed ones (Greeks and immigrants). The project goal was the efficiency testing of feasible distance training approaches customized to the special requirements and the cultural and social environment of the target groups.

Background

Education in Greece is obligatory for all citizens until the age of 15.The Greek educational system is structured in three stages: 1) Elementary education: primary school (ages: 6-12), 2) Secondary education: high school (ages: 12-18) which is divided in general and technical education (ages: 15-18) and 3) Academic education: universities (adults). Besides, recognizing the need for life-long education, the educational system is sup-

Figure 1.

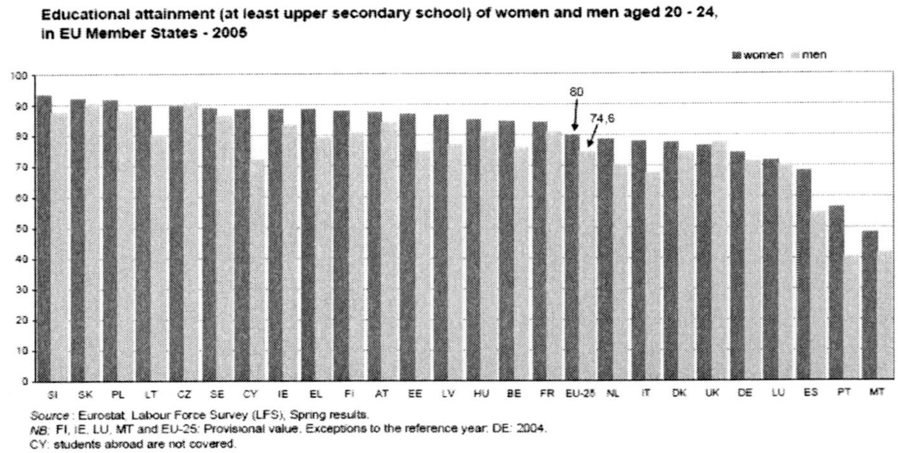

ported by institutions that provide qualification training for adults (Figure 1 & Figure 2). Common feature of all these levels is attendance of a common educational program structured so as to address groups of people, while tuition takes place in specific location (lecture halls, classrooms). In some cases the latter has been omitted, with the introduction and application of distance learning tools (e-learning web sites/vortals, electronic/traditional mail etc). As for adults, rarely does the instructor approach the student *online* or trains him/her at his/her own place of residence or work.

In late 2005, a group of professionals specializing in career orientation training, organization, management and development of rural areas founded a non-profit association aiming at researching issues of personal development and dexterities enhancement leading to proper professional orientation. Its members had gained experience and specialization working in the private sector and through their participation in similar European Programs, for instance 2003/055–435. The action of the group was funded mainly by its members and partly externally sponsored.

The project described in this paper aimed at researching the digital divide between urban and rural areas while experimenting with new training methods combining extended ICTs application. The particularity of these methods was their easy application with efficiency and low-cost on behalf of the state authorities (such as local administration and life-long learning education foundations) as well as of private companies, in cases of workforce distance training. The participant consultants worked on a voluntary basis.

Target Domain

Our group was situated in Thessaloniki, the second biggest city in Greece, recently developed as a major economic, scientific and cultural center of the country and the Balkans in general. The Serres area where our action was realized is a border area (the Bulgarian and Former Yugoslavian Republic of Macedonia-F.Y.R.O.M. border lie in the north) and its urban center is 80km away from Thessaloniki. Both Thessaloniki and Serres lie in Macedonia, the northern part of Greece. Here follows a brief description of certain geographical and economic facts of the area:

Employment structure of the economically active population of the urban area of Serres shows important concentration of activities in the sectors of trade, transport, communications, banking, safety, services etc. The industry sector

Figure 2.

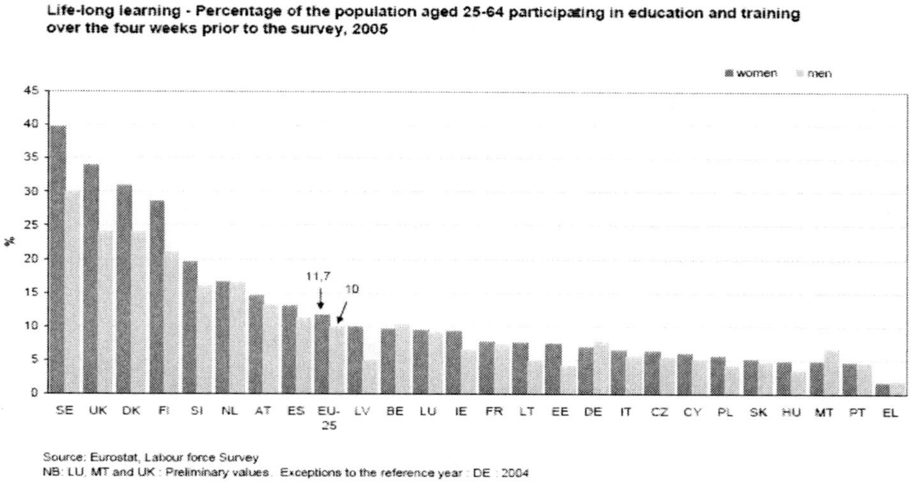

is characterized as of low-medium capacity (milk, tomato, sugar, cotton, wood, pastries and sweets). The sector of rural products trade is connected immediately with the local rural production. Local economy is based mainly on agriculture (fruits and vegetables) and livestock-farming (cow, pork and chicken); however, in the last years, these sectors have suffered continuous shrinking. Population records at 2001: 200,916; distribution of population: rural: 106,599 and urban: 94,317.

Consultees & Consultants

The best distance education practices depend on creative, well-informed instructors (Kruse, 2002). Selecting the specialists to act as consultants is of crucial importance for the development and the success of the program. Basic features and qualities of a consultant are:

- Specialized theoretical knowledge of the subject
- Practical experience
- Analytical thinking
- Respect for and acceptance of the interlocutor/Patient listening
- Communication ability/Getting across the gist without tiring the interlocutor/Clarity and precision in thought and speech
- Flexibility and adaptability
- Ability to trace, analyze and process information

During the program 4 experts were used, in the following cognitive areas: one economist, one psychologist, one European issues expert (funding, project types, policies and future tendencies) and one agricultural development expert.

The target group comprises people (residing or being professionally active in the application border area) with ages ranging from 28-55, either belonging to the active productive sector (engaging with agriculture and animal-breeding) or are unemployed.

Additionally we thought that a distant consulting program should also take into account the importance of internal migration movement, that is, to explore the factors that force citizens to move to other areas (from the country area to large city centers and vice-versa). Several issues needed to be addressed: labor availability, finding affordable accommodation, integration of internal migrants into the community and the new social structures, decline of agricultural income, existence or not of facilities and opportunities to upgrade their skills etc.

Besides internal migration issues, the phenomenon of economic immigration, i.e. foreign citizens coming to Greece seeking permanent or temporary employment, was also explored. During the last 17 years there has been an immigration flux in Greece from countries such as Albania, Bulgaria and Rumania. For the latter ones, immigration was expected to rise from 2007 onwards when these two countries would join the European Union and legal access to Greece would be facilitated.

What we realized was that in many areas of northern Greece temporary economic migration was the norm i.e. permanent migrants: 1-3% in Serres, Drama, Komotini and Alexandroupoli, even Thessaloniki has only 7% (Baldwin-Edwards & Kiriakou, 2004)). Another important feature is that the Greeks residing in border areas are usually of old age (Tsobanopoulou, 2008) while immigrants range from 15-64 years old. Foreign economic immigrants were included in the program in order for them to achieve better information on post availability and job opportunities in Greece and subsequently pass this type of specialized knowledge to their home countries.

Due to the cultural particularities of the target group (cognitive level, experiences, habits, mentality) a social anthropologist was employed to offer experts advice on cultural parameters and communication issues. Interpreters were employed when necessary for the Albanian and the Bulgarian language. Two technicians were also employed for the application and proper usage of the technological equipment used as platform for the e-training program.

E-Consulting Objective

Traditional methods of education and training often prove inadequate - the developed countries included- to cover the rapidly growing and constantly changing framework of dexterities enhancement. In the past, industrial growth simply required the improvement in quantity and quality of the basic technical education and encouragement of within-firm training. The new competitive settings entail greater emphasis on high-level and specialized training; there is close interaction between education and economic organizations in order to assess and communicate evolving needs. Moreover, another important issue is the development of cognitive skills relevant to information technology (Bresnahan et al., 2002).

The confluence of technology, demographics and work/family requirements make life-long learning imperative (Berge, 1998). The structure of distance learning gives adults the greatest possible control over the time, place, and pace of their education. Their combination, that is, distance learning applied for life-long educational needs was the basic axis of our project. Originally, distance learning was proposed as a means to bring educational institutes to students, regardless of whether the students were in a farmhouse, in an urban center, or in the suburbs; it was that is, to reach those that could not have access to the traditional classroom. But now, the demographic changes have resulted in the growth of non-traditional learners who could benefit from distance learning. Adults over 40 years of age are the fastest growing segment of the new student population (Levy, 1998).

The project objectives were based on a systematic analysis of the real economic potential and needs in each region, recognizing the role of closer rural-urban linkages for stimulating rural economies. We discussed with the local Labor Syndicates, the representatives of municipalities, local councils and production organizations, we recorded their demands and needs and designed the educational material accordingly. The objectives encompassed by the program were:

a) Entrepreneurial and management dexterities: Cultivation and promotion of business spirit and leadership skills; stimulation of consultees for creative expression; acceptance and handling of uncertainty; active participation in different social groups (interpersonal skills, socializing); development of critical thinking; diagnosis and cultivation of personal/social skills (development of initiative, self-confidence, cooperation); comprehension of some elementary concepts: needs and satisfaction, pricing, relationship between state-banks-companies (in global economy), business/marketing plan, budget-review, scheduling, risk, social contribution/responsibility of an individual/company, international trade, products and services, production means; communication: foreign languages, technical terminology, stress handling.

b) Demonstration of ICTs capacities: internet, web services, electronic commerce, e-mail, information searching (search engines, newsgroups, portals/vortals), video-conference etc. After physical access to ICTs, the development of computer hardware/software literacy skills for those excluded from the information society is considered to be of a high priority for tackling the digital divide.

c) Information on labor market issues: permanent or seasonal labor needs, essential features and dexterities required for each post, experience acquisition programs, funding possibilities (European or national programs, bank products, collaborations etc), market development (tendencies) and rural areas pre-scheduled development design by the European Union. Beyond the e-consulting session, information on the above topics took place in many ways, following Figure 3. To aid the adopted distance training approach we constructed a web site, newsgroup and e-forum (using free license applications such as Apache server, MySQL database and PHP script language) where all up-to-date information was announced.

d) Specialized issues such as: product modeling, alternative agriculture, organic farming and products, new channels of product distribution -how the producer reaches the consumer directly without any intermediaries, production organization, exports/demand for special products, clusters forming, alternative energy sources.

Precondition for the consultee's function in the consulting session is for him/her to feel the consultant as an ally. The consultant should always be concerned with "how to contribute to the consultee's viability" and help provide or sustain an income for the consultee. Additionally, the latter

Figure 3. Information dissemination model

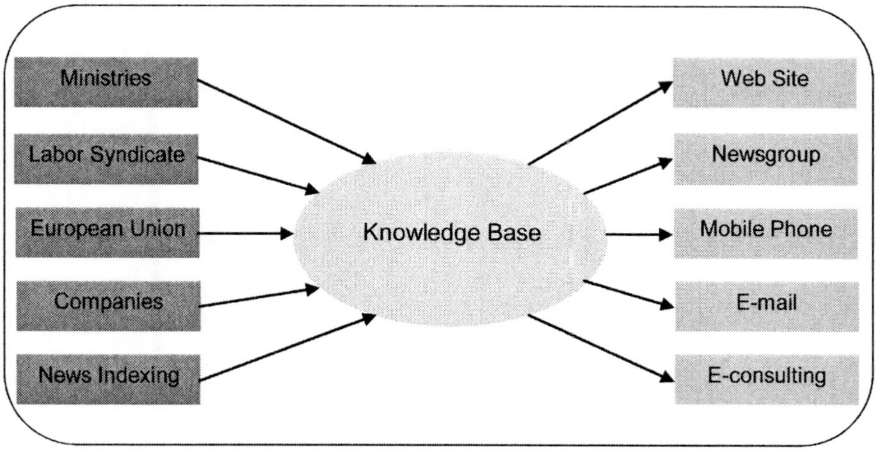

must be persuaded into changing of perspective and constant adjustment to new conditions; his/her self-confidence must be enhanced and the need for programming and self-organization must be stressed; all the above are critical in a successful decision-making procedure.

In order for a trusting atmosphere to be created, a particular consultant was chosen to deal individually with each consultee. In practice, though, multiple demands on behalf of the consultees led to the employment of more than one consultant for each consultee, a fact that delayed the creation of the appropriate cooperation climate. The general scheme followed during the sessions is displayed in Figure 4.

Technological Tools

Technology can highly contribute to human progress, help poverty reduction, living conditions improvement, social inclusion and participation. The utilization of open broadband standards and the implementation of advanced ICT architectures can lead to several benefits:

- Overcome the challenges of terrain, infrastructure, and cost to increase access.
- Extend the benefits of digital education to previously unreachable populations.
- Expand access to educational resources and broad information availability.
- Improve lifestyle for citizens regardless of socio-economic status.
- Create unprecedented opportunities to exercise entrepreneurial skills regardless of user background.
- Promote communication and cooperation.

More than one billion people can connect to the Internet and benefit from its broad array of information-rich experiences. The remaining 5.5 billion people, however, do not have access to computers or the Internet (WIUS, 2001-9).

Video-Conference

Today, the Internet and compressed video have taken distance learning in new directions, allowing distance learning to occur in real time (Valentine, 2002). Video-conference is a network-based technology that permits visual contact and oral communication in a synchronous mode. The communication becomes real through picture (video), sound (audio) and other types of files (texts, presentations etc.) in a bi-directional way. The signals can be transmitted in various ways, via Internet, via satellite or through telephone lines (ISDN, ADSL). The data administration at the logical-level takes place through protocols such as MBONE, TCP/IP, H.323, T.120 and software applications such as MeetingPoint, CU-SeeMe and NetMeeting.

In our case, the e-Training platform was based on video-conference technology through two-side satellite Internet. What is particular about the specific solution is that one end-point is movable, a fact that allowed covering long geographical distances without the use of extra equipment and without allowing for inexistent technological facilities in remote border areas to act as an obstacle. The technical requirements for the equipment are described below.

Connectivity Architecture

Over the last years, satellite technology has gained in popularity as an alternative, high-quality communication means. The average end-user especially, has been reached thanks to the development of small, low-cost stations known as Very Small Aperture Terminals (VSAT). Connectivity providers use satellite communications to bypass the increasingly clogged terrestrial and submarine networks to complement their backbone connectivity or to supplement them where they are not yet available. Satellite technology allows the easy and fast connection between computers and networks via satellite. This technology is divided into two categories, depending on the operation way:

Figure 4. The consulting scheme

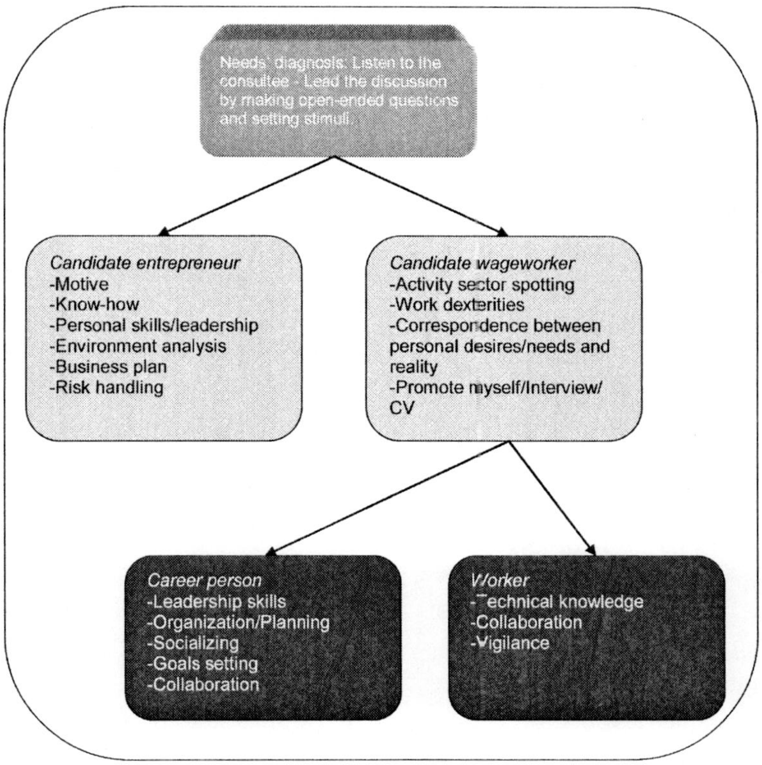

- One way satellite connection: the required equipment consists of a satellite dish (for data downloading) and a conventional dial-up PSTN, ISDN modem or even an ADSL router (for data uploading). The modem sends the users' requests to the network server which processes them and routes the requested data packets through the satellite service. The satellite dish receives the data, forwards them into the satellite card which decodes the signal. The modem and the satellite card are installed into a typical personal computer. The downloading data speed can range from 1Mbps up to 20Mbps. This type of connection is recommended in cases where local broadband supply does not exist or is not adequate or is very expensive.
- Bi-directional satellite connection: this case presupposes at least two satellite dishes (one central and one terminal) installed in different locations. The communication is based on the Digital Video Broadcasting-Return Channel via Satellite (DVB-RCS) technology. The network is constituted by the satellite, a terrestrial station (HUB) and the user interactive terminals. The signal transmission is realized via two channels: a) The forward channel from the central satellite land station towards the satellite and afterwards to the terminal. b) The return channel by the terminal station to the satellite and afterwards to the central satellite land station. A key trait of this technology is the simultaneous connection of a big number of terminals; the satellite Internet uses Internet Protocol (IP) multicasting technology, which means up to 5,000 channels of communication that can simultaneously be served by a single satellite. The

bi-directional broadband transmission services support all data types, such as audio and video, at very high transfer rates (compression reduces the size of the data and the bandwidth): current configurations can deliver data at a rate of up to 40 Mbps for data reception and at a rate up to 4 Mbps for data sending.

At this point, it should be mentioned that trees, mountains and heavy rain can affect reception of the Internet signals and the available bandwidth can change from one moment to the other depending on the number of users who are simultaneously connected to the network.

With satellite transmission, the number of potential users that can receive and decode broadcast data can range from one to many. In our project, we used two end-points connection. The interconnection model is displayed in Figure 5. At a high level, the many components of the used satellite communication network can be divided into four modules:

a) The satellite: Placed in orbit around the earth, a satellite is a specialized repeater that receives radio-frequency signals from earth stations and retransmits them to other earth stations. The satellite also amplifies the signals and switches the frequencies between the uplink and the downlink carriers.

b) The central hub: contains many components, including a large dish antenna (4 to 11 meters in diameter) as signals receiver/transmitter, a network management system which monitors and controls all components of the satellite interconnection, a baseband equipment that handles satellite access, routing between the hub and remote earth stations, dial backup, quality of service, TCP and HTTP acceleration, and some optional components such as MPEG transport coder/decoder, application servers and audio/video broadcast programming devices.

c) The terminal satellite dish that operates as receiver/transmitter unit. The dish diameter plays essential role in the data transfer speed. In order to achieve high speed the diameter must be at least 1.3-1.5m. We suggest a 2m dish; though, we keep in mind that a critical factor for the selection is the equipment purchase cost.

d) The terminal station: a personal computer or notebook with processing power of class Pentium II at least, equipped with video signal coder/decoder device, a high-resolution camera, a microphone and stereo speakers are required. In our case, we used Microsoft Windows XP Professional edition as operating system and NetMeeting as connection software. The terminal A was installed in our clubhouse in Thessaloniki, whereas terminal B was a custom-made construction installed in a van (hired for the project time period) which permitted a mobile end-point.

Using satellite technology offers several advantages:

- Gives fast access to the Internet.
- Reception is possible with a small antenna.
- Can be installed anywhere regardless of location. Connection is possible almost anywhere instantly within the footprint of the satellite, with no cabling work or delays dependent on terrestrial infrastructure.
- Can be installed quickly.
- Provides a secure, manageable connection model.
- Equipment provides high-quality services and has a relatively low cost.
- Can be integrated with other compatible services where such services exist.

Some of the main disadvantages include:

- Satellite Internet is generally more expensive than terrestrial access solutions, at least in regions where they are available.

Figure 5. The bi-directional satellite communication

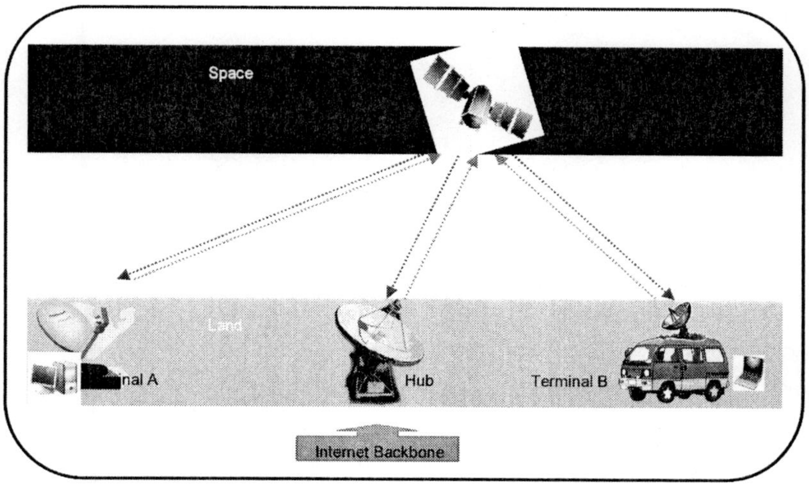

- The outdoor unit (antenna and cabling) are more vulnerable to vandalism and weather conditions.

As additional aid, we constructed an interactive internet site that functioned as a video data repository, where all session recordings were stored. In this case, agreement on behalf of the consultee was essential in order for personal data to be recorded (see Figure 6).

The programs used were based on open-source technology: Apache web server, MySQL as database and PHP script language for the web pages. The programs could communicate via ODBC (Open DataBase Connectivity) model with several extensional applications. These recordings were used as case studies in following sessions and as a diary where the consultants could study in detail certain consultees' cases. It acts as written record and history of what has been communicated and experienced. Besides, potential cross-cultural relationships may offer some unique challenges that will need to be addressed.

Upon concluding this unit, we should mention that this specific communication type allowed our intervention in areas where access to ICTs was particularly difficult. If we take into consideration the mountainous regions and the island clusters

Figure 6. The video-data repository.

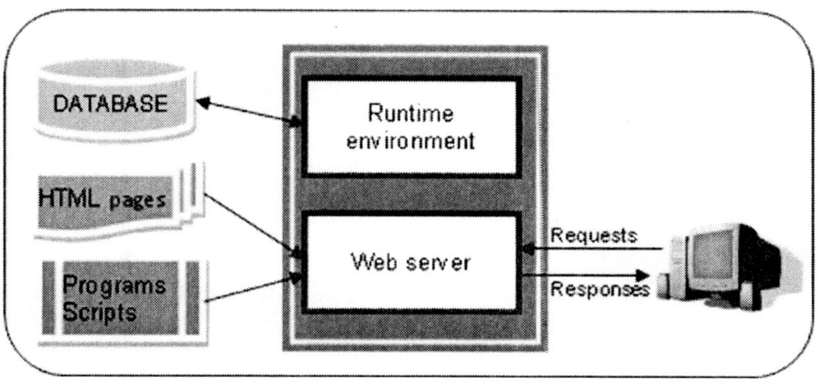

that are scattered in the Greek seas, then the satellite interconnection model sounds a promising solution for reaching the distant and isolated areas of the Greek periphery.

A Real Session

The vehicle parks at the arranged spot for the day, according to a weekly schedule, reformulated every fortnight. The consultee sits inside the van and the technician proceeds with the necessary settings for the video-conference to begin. The session lasts 30 to 45 minutes. The interlocutors mainly discuss, interact through questions and answers, exchange opinions and experiences; for instance the consultant explains how he consulted other farmers to set up a local partnership aiming at product quality standardization and packaging, or the consultee might express his problems in engaging with the national/international market or dealing with unemployment etc. The consultee expresses his/her feelings about the problems s/he faces, observes the consultant handling these feelings and these problems. Occasionally, if necessary, the consultee engages in role-playing that will enable the consultee to broaden his/her perspective and get the whole picture, to experience the feeling of the 'other', i.e. to achieve a certain level of empathy or experience the constraints of the "opposite" side, be it the merchant, the employer, the manager etc. It is not unlikely for the consultee to behave with suspicion initially; he feels insecure as he still hasn't acquired the level of trust needed in such interactive encounters; gradually, he begins to cooperate, that is be more open, more expressive, less suspicious of the consultant's honesty and ability. After some time (and depending on one's personal idiosyncrasy) he starts to draw personal satisfaction from the interaction and becomes more involved and active. As the consultant has earned the consultee's approval and recognition, he urges the consultee to contribute substantially to this instructive procedure; the consultee feels more productive and builds on his self-confidence.

Obstacles: Solutions

We encountered the existence of several barriers to information access. These barriers were physical, economic, intellectual or technological, and they often impeded the participation of rural users in the activities that contributed to the digital knowledge repository.

PROJECT EVALUATION

The evaluation focused on three areas: 1) Appropriateness and usefulness of the training and consulting content (correspondence to real needs, layout, clarity, variety etc), 2) Operation and aptness of the ICT component and 3) Effectiveness and organization of e-training program (planning, educational methodology, flexibility and support).

The program lasted for nine months. We discussed with 37 individuals and conducted 124 sessions (an average of three sessions per consultee). The consultees were farmers (16), animal-breeders (5), foreign workers (4) and 12 were unemployed. Unfortunately, all the consultees were men; there was strong unwillingness observed on the women's side (partly due to the fact that the sessions were taking place in a van with the sole presence of one male technician; female technicians or a different spot for sessions might enhance the feeling of safety and encourage female participation in future projects.)

The strengths of the project were:

- The participants benefitted from better educational services and their access to high-quality information sources, thus increasing their opportunities for social and economic mobility and achieving greater involvement in their local communities. The already active professionals had the chance to increase in productivity through

Table 2.

Problems	Confrontation
Lack of adequate hardware and the subsequent cost barrier of obtaining equipment could place undue hardship on some remote consultees (Berge, 1998). Differences in urban and rural living conditions and infrastructures are still very big.	We did not wait for the consultee to reach us; instead, using new technology, we reached the places where they live and engage in activities.
Consultee's frustration in learning and training and potentially prohibitive start up costs (Bresnahan, et al., 2002; Galusha, 1997).	We used a 'cookbook' approach: we provided solutions step by step, in a clear and simple manner. As for the cost, the consultees did not need to buy anything; we ourselves, selected moderate cost technological solutions.The duration of each video-conference did not exceed 30-45 min due to the increased telecommunications cost.
Mis-use of technology, and the attitudes of instructors, students, and administrators have an effect on the overall quality of distance learning (Valentine, 2002). Many adult students are not well versed in the uses of technology such as computers and the Internet. Using electronic medium in distance learning can inadvertently exclude consultees who lack computer, typing or writing skills. If technology is involved as a tool for communication it is important that each individual not only has the same or similar technology and software to use, but each must have a clear understanding of how to use it and the dexterities to do so.	The use of the technological equipment was restricted to the beginner's level. No previous knowledge was taken for granted; besides, a technician was always present to assist in the occurrence of any problem. The consultee's attitude cannot be treated with rules and prescriptions; we tried to address each case individually. As for the consultants and the supporters, the aims and the project objectives were clear cut from the very beginning; after all, their commitment is proved by their voluntary participation.
Isolation, lack of effective advice, costs and motivators, feedback and teacher contact, student support and services, lack of experience (Rahm & Reed, 1998). Ambiguous instructions on the web site as well as through e-mail (Bresnahan, et al., 2002). Learning disembodied nature restricts feedback leaving learners feeling abandoned (Luyt, 2004). Low use of written information in daily life and high reliance on oral communication for knowledge transfer.	The project consultants had long experience in the topics elaborated with the consultees. The video-conference approach was selected in order to achieve immediacy, oral conversation and quick response.
Understanding of basic cultural differences may be necessary in order for distance training to be effective.	The consultants were directly guided by a social anthropologist in order for them to adjust their advice and manner of speech to the cultural needs of the consultee.
It is very difficult for the screen to replace the natural presence of the instructor in the room. Lack of familiarization with the video-conference means creates embarrassment and makes the relations more "formal". Consultees find it difficult (especially at the beginning of a distance training session) to confide their problems and difficulties in their consultants.	The consultants aimed at creating an atmosphere suitable for cooperation, warm and comfortable for the consultees.
Consultees are unsure of their objectives; resentment on behalf of the individuals who do not participate; unreasonable expectations from each other.	The detection of the consultee's needs was a timely procedure. Multiple stimuli were offered to ensure the consultee's participation and activity. Realistic objectives were set; the expectations, commitment and agreement were often repeated in the sessions. The consultant stated his/her own limitations and learning benefits from the interactive procedure.

the expansion of selling channels and infrastructure construction. Also, the program stressed the need for individuals' collaboration and provided them with a broader base for decision-making. The importance of cooperation between city/border regions rather than competitiveness was denoted.

- Creation of equal opportunities for high-quality educational services and access to digital world. Access to the Internet and multimedia became available to remote communities, effectively fighting exclusion.
- Advanced and specialized know-how and expertise were gained.
- Having to use the technological tools helped participants become skilled and feel comfortable with these technologies,

which are inevitably going to expand in all areas of personal and professional life.

- Lack of establishment of broadband connections in many regions could be repaired through satellite connection. This solution is viable and economically acceptable. It can constitute an exceptional choice wherever the land networks are insufficient. Satellite telecommunications systems present important advantages; they can cover the needs of geographically extensive regions. New services can be offered through this facility independently of place and time, lifting the barriers of traditional communications (time/schedule limitations and actual distances).
- Finally, the consultees benefitted from ICTs in the following ways:
    - Searched reliable and up-to-date information from web sites such as those of Ministries, European Union, Labor Syndicates and Employment Agencies, related to project funding, adoption of new production and organization methods, information about job announcements and new marketing/financial policies etc.
    - Gained knowledge on personal promotion and CV compilation techniques.
    - Achieved quick and inexpensive communication with other parties / formed synergies / implemented projects using alternative selling channels / shared ideas-thoughts-problems.

Some weaknesses or lacks were the complete absence of women's participation, the project financial sustainability (it started based mainly on volunteer support) as services were provided for free, sometimes the lack of personal contact, the difficulty in obtaining permission from the consultees to record and release their personal information and problems in the consultees expressing clearly their ideas and feelings.

The benefits of the program extended to all the participants. Consultants comment that they enjoyed the opportunity to think about and discuss important professional queries; they appreciated the chance to share ideas and support others in their problems. They also managed to get acquainted with individuals coming from unfamiliar cultures and with foreign customs; the innovative character of ICTs appears to fight strict culture which imposed models of behavior in most educational environments. Consultees mentioned that they had the opportunity to reach their personal, social and career potential; they could discuss with someone who knew how it all operated, and hence, feel less anxiety regarding their problems. All participants report positive experience in using Internet and computer-mediated conference as a communication tool. The ICTs provides the rural people with support from experts otherwise difficult to reach, connects them to a rich collection of resources, and provides for them opportunities to learn, contribute actively in their personal development and improve their professional status.

Indicative of the above are the following extracts, representative of the feedback we got from the consultees:

C1: "It was unexpected for me at first. I had to sit in front of a computer facing a face unknown to me, who would supposedly help me. How? By distance? I wasn't sure it could work. But later, step by step, you realize that they know what they're doing, they're professionals. They tell you things about others in your position and they suggest solutions. You feel less alone in the end. There are others like you, in your situation, all over the country and you feel like you're progressing with them. You're fighting with them. It feels nice…".

C2: "I found it challenging from the beginning. I am a young farmer with aspirations to apply organic farming and make a difference. The consultant showed me the way I could do it, where to get the information, the support. I am grateful for this opportunity; it helped me realize my dream".

FUTURE WORK AND CONCLUSION

The positive project results made us reflect on and plan future activities in order to expand the benefits produced. Some future work might include:

- Designing territorial development policies is difficult due to the complexity of inter-linkages. In general, economically interlinked areas are more influenced by common resource bases and geographical features than by administrative boundaries. Studying the literature, someone finds out that there is limited empirical base on the links between agriculture and urban development. Hence, care should be taken not to generalize based on a few case studies; a systematic and constant research must be run, able of covering big geographical areas, even groups of neighboring countries. Such type of research cannot be carried out by the private sector but it must be included in the general strategy of states or unions. Our interest remains in people getting information and our intention is to be part of such policy implementation.
- A very challenging work is to retain the relationship between the consultant and the consultee after the latter has entered the production cycle. Especially, in the case of an entrepreneur who establishes a new company, there are many issues an expert could occupy with and offer support.
- Application of peer-education. As we have seen in previous sections, the video-conference system could be easily expanded to include more end-points. In such case, two

Figure 7. Internet penetration to EU

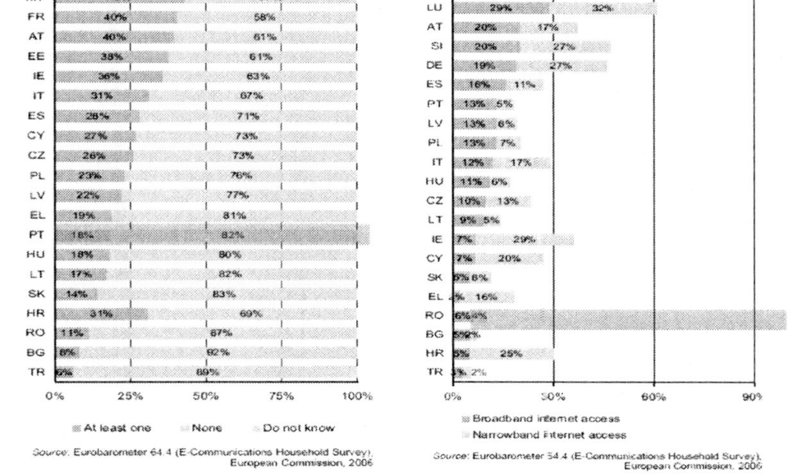

or more individuals sharing the same interests, problems and needs could creatively discuss and exchange their expertise.

In summarizing, the present case study discusses a private-sector empirical intervention in the crucial issue of technology usage and information access by populations (potentially) digitally excluded. Our political position behind this intervention was that knowledge appearing in communication networks (like the Internet) constitutes the vehicle for democratic procedures, for critical reasoning and social equality. The consequences of the digital inequality imply different life chances and opportunities for those who are not technologically enabled (DiMaggio & Hargittai, 2001; Servon, 2002). Expanding the technological benefits to previously undeserving populations generates improvements in qualification and income while providing a well-founded solution to the digital divide; innovations can aid producers and entrepreneurs respond to changing markets, cut down on costs, and improve products' and services' quality. More particularly, in rural areas this effort should focus on enhancing the content and prospects of life and, eventually, render rural areas more attractive for young farmers, prospective professionals and new residents.

As a final note we mention that lack of qualification and familiarization with ICTs has led the country to one of the last places in EU (see Figure 7), while at national level, intraregional inequalities are even bigger. Unfortunately, the investment in the education field pays in the long run; when strong, immediate needs exist, educational objectives such as the ICTs import in education are faced as second class priority objectives. The state needs to take more energetic initiatives concerning the expansion of broadband services, the operational cost reduction and the funding of special training programs in new technologies addressing mainly the active population, especially in neglected, under-developed or distant regions.

REFERENCES

Alcántara, C. H. (2001). *The Development Divide in a Digital Age An Issues Paper*. United Nations Research Institute for Social Development.

Baldwin-Edwards, M. & Kiriakou, I. (2004). *Statistical Data for the Immigrants in Greece: Analytic Study on the Available Elements and Proposals for the Conformity with Standards of the European Union*. Mediterranean Observatory of Immigration, Panteios University, Revised publication.

Berge, Z. L. (1998). Technology and Changing Roles in Education. In Z.L. Berge & M. Collins (Eds.), *Wired together: Computer-mediated communication in K-12, (1): Perspective and instructional design,* (pp. 1-13). Cresskill, NJ: Hampton Press.

Bresnahan, T. F., Brynjolfsson, E., & Hitt, L. M. (2002). Information Technology, Workplace Organization and the Demand for Skilled Labor: Firm-level Evidence. *The Quarterly Journal of Economics, 117*(1), 339–376. doi:10.1162/003355302753399526

Bucy, E. P. (2000). Social Access to the Internet. *Press/Politics, 5*, 50-61.

DiMaggio, P., & Hargittai, E. (2001). *From the 'Digital Divide' to 'Digital Inequality': Studying Internet Use as Penetration Increases*. Princeton, NJ: Woodrow Wilson School, Princeton University, Center for the Arts and Cultural Policy Studies.

Eurostat. (2009). Retrieved from http://epp.eurostat.ec.europa.eu/portal/page/portal/eurostat/home

Galusha, J. M. (1997). Barriers to Learning in Distance Education. *Interpersonal Computing and Technology: An Electronic Journal for the 21st Century, 5*(3-4), 6-14. Retrieved April 28, 2007, from http://www.emoderators.com/ipct-j/1997/n4/galusha.html

Gurstein, M. (2003). Effective Use: A Community Informatics Strategy beyond the Digital Divide. *First Monday, 8*(12). i2010 European Strategic Plan. (2005). *i2010 Communication*. Retrieved on October 11, 2007 from http://ec.europa.eu/information_society/eeurope/i2010/inclusion/index_en.htm

James, J. (2005). The Global Digital Divide in the Internet: Developed Countries Constructs and Third World Realities. *Journal of Information Science, 31*(2), 114–123. doi:10.1177/0165551505050788

Joseph, R. (2001). Understanding the Digital Divide. *Prometheus, 19*(4), 333–336. doi:10.1080/08109020110091396

Kruse, K. (2002). *The Benefits and Drawbacks of E-Learning*. Retrieved April 11, 2005, from http://www.e-learningguru.com/articles/art1_3.htm

Levy, J. (1998). *Some Observations on Distance Learning*. Ithaca, NY: Office of Distance Learning, Cornell University.

Luyt, B. (2004). Who Benefits from the Digital Divide? *First Monday, 9*(8). Retrieved from http://www.firstmonday.org/issues/issue9_8/

Lynch, M. (2003). Technology Education Through Online Mentoring: The Impact of an Online Course on Teachers' Beliefs and Attitudes about Being Mentors. *Dissertation Abstracts International, 63*(12), 4283A, (UMI No. 3075282).

OECD-Organisation for Economic Co-operation and Development. (2001). *Understanding the Digital Divide*. Retrieved January 17, 2005 from http://www.oecd.org/dataoecd/38/57/1888451.pdf

Rahm, D., & Reed, B. J. (1998). Tangled Webs in Public Administration: Organizational Issues in Distance Learning. *Public Administration and Management Interactive Journal, 3*(1). Retrieved May 13, 2005 from www.pamij.com/rahm/html

Rao, S. S. (2005). Bridging Digital Divide: Efforts in India . *Telematics and Informatics, 22*(4), 361–375. doi:10.1016/j.tele.2005.01.007

Rice, M. F. (2001). The Digital Divide and Race in the United States. *Politics . Administrative Change, 36*, 20–31.

Servon, L. J. (2002). *Bridging the Digital Divide: Technology, Community, and Public Policy*. Oxford, UK: Blackwell Publishers.

Strover, S. (1999). *Rural Internet Connectivity*. Columbia, MO: Rural Policy Research Institute.

Tsobanopoulou, F. (2008). Acrobats on Borderlines: The Konitsa Border Guards. *Balkan Border Crossings Network*. Lit-Verlag.

Tukiainen, J. (2004). *Access to Computer, Internet and Mobile Phone at Home in Finland, Ireland, Netherlands and Sweden*. Vatt Discussion Papers 324, Government Institute for Economic Research.

Valentine, D. (2002). Distance Learning: Promises, Problems, and Possibilities. *Online Journal of Distance Learning Administration, 6*(3).

Van Dijk, J. (2006). Digital Divide Research, Achievements and Shortcomings. *Poetics, 34*(4-5), 221–235. doi:10.1016/j.poetic.2006.05.004

Wade, R. W. (2004). Bridging the Digital Divide: New Route to Development or New Form of Dependency? In Avgerou et al. (Eds.), *The social study of Information and Communication Technology – Innovation, Actors, Contexts.* Oxford, UK: Oxford University Press.

Wighton, D. J. (1993). *Telementoring: An Examination of the Potential for an Educational Network*. Education Technology Centre of British Columbia.

WIUS (World Internet Usage Statistics—News and Population Stats). (n.d.). *Miniwatts Marketing Group. 2001-2009*. Retrieved from www.internetworldstats.com/stats.htm

Chapter 10
The e-Learning Puzzle in Turkey: Déjà Vu?

Selçuk Özdemir
Gazi University, Turkey

EXECUTIVE SUMMARY

This chapter aims to share Turkey's ICT integration experiences from a country-wide perspective rather than a school or classroom case. Many experiences in different countries indicate that successful ICT integration requires interlocking components, such as purchasing hardware, in-service training for principals and teachers, curriculum integration, financial resources for maintenance, technical, and pedagogical support, and an adequate amount and quality of digital learning material. Lack of one of the components may cause the failure of the whole integration process. The employment of ICT in education is a complex process comprising intricate components, much like the pieces of a puzzle. Sharing the experiences gained from national initiatives is especially important for developing countries, which should make an effort to learn from the experiences of other countries because loans granted by foreign sources make up a majority of the e-learning investment.

BACKGROUND

You can import a Ferrari into any country in the world as long as you have money; however, this purchase does not guarantee the Ferrari's performance in that country. Simply importing a Ferrari is not enough unless you have drivers skilled in driving fast cars, high quality fuel, maintenance services, and smooth highways accessible across the country. Similarly, it has been demonstrated through a number of practices around the world that e-learning, or information and communication technologies (ICT) in education, is not just about purchasing the required hardware or supplying internet connections to schools.

Technology does not deliver educational success on its own. It only becomes of value in education if learners and teachers can adopt it to a useful end. According to an OECD report (2001), in spite of the USD 16 billion spent in 1999 in

DOI: 10.4018/978-1-60566-942-7.ch010

OECD countries, the expectations for ICT use in education were mostly left unfulfilled, and ICT investments made by the public sector did not result in improved performance, quality, or access to a better education. The lessons learnt from these attempts in OECD member countries led researchers to focus on the concerns over the return on ICT investment. Many experiences in different countries indicate that successful ICT integration requires interlocking components, such as purchasing hardware, in-service training for principals and teachers, curriculum integration, financial resources for maintenance, technical, and pedagogical support, and an adequate amount and quality of digital learning material. Lack of one of the components may cause the failure of the whole integration process. The employment of ICT in education is a complex process comprising intricate components, much like the pieces of a puzzle. Each piece should fit the others well in order to form an ideal picture. Otherwise:

- Even though you buy the best hardware, if your teachers are not ready for ICT use in classroom;
- Even though you buy the best hardware and provide in-service training to teachers on how to use ICT in education, if you do not realize ICT integration in the curriculum;
- Even though you invest in the best hardware, provide in-service training to teachers on ICT in education, and integrate ICT into the curriculum, if you cannot reserve financial resources for maintenance;
- Even though you buy the best hardware, provide in-service training to teachers on how to use ICT in education, integrate ICT into the curriculum, and reserve financial resources for maintenance, if you cannot supply an adequate amount and quality of digital education material and educational software, the integration of ICT in the educational process will be incomplete and will not bring about the anticipated results.

Figure 1. The intricate components of ICT integration in education

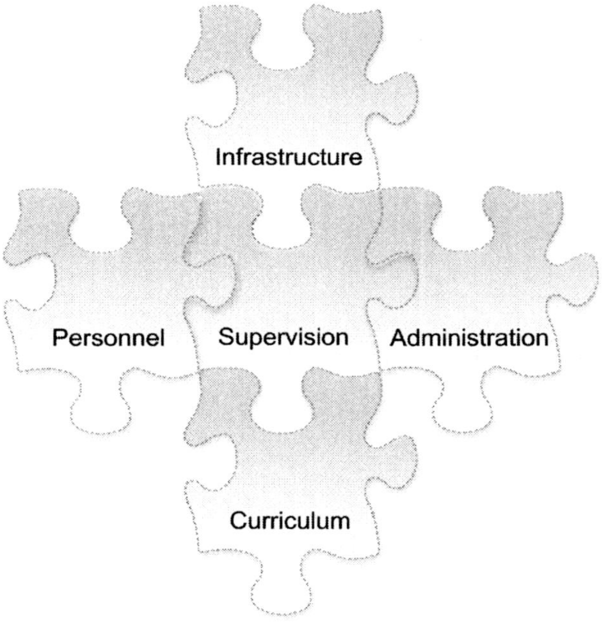

As a developing country, e-learning efforts in Turkey remind one of déjà vu, rather than a puzzle. As a careful observer of the projects designed to integrate information and communication technologies (ICT) into the Turkish educational system, I often feel as though I come across the same problems in each ICT integration project. The Ministry of National Education (MONE) starts each project with a massive hardware purchase. Then, this hardware is distributed to schools across the country. Later, researchers from Turkish universities or MONE conduct impact research studies to evaluate the effectiveness of the investment. The reports of these research projects usually indicate more or less the same findings. Unfortunately, since 1998, the research on several education projects, including the ICT integration dimension, has reported the same problems hindering the effective and efficient use of ICT in education. Experiencing the same problems in consecutive projects indicates that policy makers and project managers of MONE execute new projects without taking lessons from the prior ones and manifests that *ICT integration is not a one-off or unidirectional process*.

This chapter aims to share Turkey's ICT integration experiences from a country-wide perspective rather than a school or classroom case. Sharing the experiences gained from national initiatives is especially important for developing countries, which should make an effort to learn from the experiences of other countries because loans granted by foreign sources make up a majority of the e-learning investment.

SETTING THE STAGE

The Structure of the Ministry of National Education

Prior to the discussion of the ICT integration projects in Turkey, understanding the structure of MONE will help us see the big picture, including all of the modernization attempts in the Turkish education system.

History of the National Education System

The transformation implemented by Ataturk, the "Turkish Revolution", is the main driving factor behind the national educational system. As the founder of the Republic of Turkey, Ataturk regarded education as the key instrument to ensure that the upcoming generations are endowed with contemporary citizen consciousness and patriotism nourished with national culture. Ataturk believed that illiteracy was the foremost enemy blocking Turkish society's path to living in peace and welfare; therefore, the educational system had to try to eliminate illiteracy and raise future generations with practical and useful knowledge (Akkutay, 1996). Following the announcement of the Turkish Republic's foundation, MONE aimed to improve the quality of education and expand it to include the whole country. Especially with the adoption of the new Latin-based alphabet, the new Republic started a campaign to promote equality, democratic education, secularism, and a scientific way of thinking through formal and informal educational channels. The radical change in the philosophy of education during the establishment of the Republic required the use of contemporary educational technology. Thus, in order to secure a quality education, maps, overhead projectors, testing apparatus, etc. were imported and developed by the MONE.

To determine a road map for the new education system, the founder of the Republic invited several famous scientists, like Albert Malche and Beryl Parker, to Turkey. John Dewey was the first scientist to be invited and carried out an investigation on the Turkish education system for two months. The report he prepared pointed out the importance of practical knowledge in education, the dissemination of education to the country as a whole, good-quality in-service training for teachers, the establishment of an educational statistics unit within MONE for timely diagnosis of needs, the decentralization of the administration of the educational system in order to identify problems adequately and quickly, the development of curricula in accordance with the conditions of the geographical regions, supervisors responsible for promoting best practices throughout schools, the quality of pre-service teacher education, and the quality of contemporary educational materials (Akkutay, 1996). During the first years of the Republic, the reports prepared by Dewey and other foreign scientists proved helpful in constructing a modern education system, and consequently the literacy levels of the society rose from 7% in 1923 to 18% in 1935 and to more than 95% in 2007.

MONE with Numbers

As a democratic and secular nation and the 16th largest economy in the world, and considering the volume of its education system, the endeavors of the Turkish education system may be regarded as a lesson from which much may be learned, especially by developing countries around the world.

The schooling ratios for primary and secondary education were 96.34% and 86.64%, respectively, by 2007. There are 41,500 schools, 636,490 teachers, and 14,115,892 students at the primary and secondary education levels. The 2008 budget allocated to MONE by the government is approximately $16,000,000,000, the third largest budget of all governmental institutions in Turkey.

The Organizational Structure and Strategies of MONE

The educational system in Turkey can be analyzed in two parts: formal education and informal education. Formal education covers preschool, primary, secondary, and higher education institutions. Informal education covers all educational activities outside of formal education. At all levels, education is provided free of charge.

As a candidate country for European Union (EU) membership and in accordance with EU's Lisbon 2010 objectives, the Turkish government aims to (DPT, 2006);

- Educate its youth so as to be able to respond to the requirements of the new, global knowledge-driven economy;
- Equip the graduates of secondary schools with basic computer literacy skills;
- Ensure that at least 30% of the population benefits from the electronic services provided by governmental organizations;
- Make sure that at least 50% of citizens are internet users.

Employing ICT in education is one of the main strategies of MONE in the struggle to reach these goals. Equipping the classrooms with the necessary hardware to meet the demands of the 21st century and using remote education and computer-assisted education methods to expand the coverage and the quality of education are the fundamental strategies of MONE regarding technology use in education (MONE, 2001).

CASE DESCRIPTION

Information and Communication Technologies and The Turkish Educational System

The huge amount of ICT investments realized in the last decade in Turkey is quite striking. The Ministry of National Education (MONE) has purchased a substantial amount of ICT hardware to be installed in Turkish primary and secondary schools. As of today, Turkish primary and secondary schools have a substantial amount of ICT hardware compared to the conditions seen in 1998. This investment in Turkish schooling has been implemented within the scope of a plan *to increase the qualitative and quantitative aspects of schooling with the help of information and communications technology.*

The Last Decade in ICT Projects

The fundamental purpose of ICT use in the Turkish educational system can be understood clearly when the Basic Education Program (BEP) (1998-2007) and the Secondary Education Project (SEP) (2006-2010) are analyzed. In terms of funding and the number of pupils involved, BEP was one of the largest national educational development programs in the world. The program included three sub-projects with an approximate budget of $700,000,000 that was funded by local sources, the World Bank, and the European Union and aimed:

- To raise the basic educational level of Turkey to the standards of developed countries;
- To improve interest and quality in Basic Education;
- To turn primary schools into learning centers for the public.

To realize these objectives, hundreds of thousands of computers, projectors, printers, etc. have so far been distributed to schools. Through the Basic Education Project Phase I (1998-2003), the Basic Education Project Phase II (2004-2007), and the Education Frame Project (2002 and 2007), 98% of secondary school students and 93% of primary school students have been connected to the Internet via ADSL.

The Secondary Education Project was started with an €80,000,000 budget in 2006. It aims to restructure the general, vocational, and technical secondary schooling, to improve the related curriculum, to supply more in-service training, and to furnish the schools with contemporary educational technologies. Since the project will end in 2010, there are not yet any data to evaluate the success of the project.

The Basic Education Project (Phase I) (1998-2003)

This was the first large-scale project aiming to realize the objectives of the Basic Education Program. The Republic of Turkey received a loan of 300 million dollars from the World Bank toward the cost of the Basic Education Program in 1998. With regard to ICT integration, MONE decreed that there should be ICT classrooms in at least two pilot schools in every county nationwide, 18.000 computer coordinators, and 200,000 teachers trained in ICT to (Özdemir & Kılıç, 2007):

- improve the quality of primary education by providing the pilot schools with ICT courseware, games and reference materials;
- provide ICT facilities for all classrooms in the pilot schools;
- teach pupils how to use the technology and how to access information, learn, problem-solve, apply, evaluate, and present their learning via ICT;
- create learning environments in which, from the very first year of schooling, pupils learn actively, individually and collaboratively;
- provide in-service training for teachers in ICT use for lesson planning, materials creation, assessment, and student-centered learning;
- use ICT in the school administration;
- achieve cooperation between communities, schools, teachers, and pupils in the utilization of ICT.

In the scope of Phase I (1998 to 2003), approximately 2,800 computer laboratories were installed in primary schools, and 1,500 notebook computers were distributed to the primary education supervisors in order to increase the quality of school monitoring and evaluation. Additionally, about 20,000 schools in rural areas were supplied with educational materials, overhead projectors, and computers.

The Basic Education Project (Phase II) (2004-2007)
MONE and the World Bank signed a second loan agreement that amounted to 300 million dollars for Phase II, which ended in 2007. This second project had the same objectives as Phase I: to construct new school buildings and to provide schools with more educational materials, computers, and educational software. Moreover, an education portal was planned to support all actors working in education through the internet in the scope of Phase II. Between the years of 2004 and 2007, 3,600 primary schools were supplied with more than 4,500 computer laboratories, and 4.600 rural schools were supplied with educational materials such as TVs, DVD players, and microscopes. In addition, to improve the skills and knowledge of teachers, various in-service training sessions were organized and implemented. Even though educational software distribution to the schools was one of the main objectives of this project and about 35 million dollars of financial resources were reserved for this purpose, problems such as the low number of producer companies, the few options for educational software and disharmony among the responsible agencies in MONE inhibited the purchase of educational software.

The Education Frame Project (2002and 2007)
To realize the objectives of the Basic Education Program, the Turkish Government signed two loan agreements amounting to €100 million in 2002 and 2007 with the European Investment Bank. The aim of the project was to establish computer laboratories in the primary schools. By 2007, 1,610 schools with 1,487,967 pupils were equipped with computer laboratories. Moreover, the computer laboratories established in the scope of Phase I were renewed and updated during the course of this project.

CURRENT PROBLEMS FACING THE ORGANIZATION

The Recurring Problems

The repetitive problems in the Turkish education system discussed in this chapter are based on the related literature. Two major reports prepared for MONE and six articles focusing on the ICT integration processes in the Turkish education system are analyzed to discuss the problems below.

One of the reports, the ICT Integration Baseline Study (ICTIBS, 2007), was prepared for MONE by academicians from three universities in Turkey. This report aimed to evaluate the current situation of primary schools in terms of ICT integration level. Both qualitative and quantitative data were collected from the 308 primary schools that were equipped with computers, projectors, etc. in the scope of Basic Education Project Phases I and II. The following individuals were interviewed as part of the report:

- All principals and vice principals responsible for ICT in each school (n=516),
- Teachers from different teaching fields (n=3063),
- Students at each class level (one class each from fourth, sixth and eighth class levels) (n=2324),
- Parents of these students (n=1265).

Interviewees were selected from 308 schools in 26 different cities in Turkey as a representative sample of the 7202 schools equipped with computers in the scope of the Basic Education Project Phases I and II.

The second report, called Impact Research of ICT Classrooms (IRIC, 2004), was also prepared for MONE by academics from different universities in Turkey. This report aims to evaluate ICT classrooms in primary schools in the scope of Basic Education Project Phase I. The needed data were collected from 2,398 primary schools' administrators and computer teachers.

1. **General Problems:** In the ICTIBS (2007) report, the lack of a budget and ICT integration plan for the schools was seen one of the main problems influencing and triggering other problems. In this report, when 58 MONE provincial and county administrators were asked about a budget reserved for ICT integration activities, 54 of them (93.1%) responded that there was no such budget. According to the responses, the central government planned and provided all the needed expenses, but they had no financial resources for local needs such as purchase of hardware or software, upgrading, maintenance, and staff training. Moreover, 478 school principals and vice principals were asked whether there was an ICT integration plan for their school, and 367 of them (76.8%) confessed that they had no such plan. Although 111 of them (23.2%) said they had an ICT integration plan, they could not provide any printed document. The school administrators said they did not need an ICT plan because:
   - ICT classrooms were newly installed or not running (39.5%)
   - Computers were out of service or too out of date to run, and they had insufficient technical support (18.1%)
   - Lack of teachers qualified to use ICT in education (20.3%).

Interestingly, the reasons school administrators listed for the nonexistence of an ICT integration plan for schools contradict the policies listed in the national policy document on information and communication technology in education announced by MONE in 2004. In the document, MONE categorized its policies and practices under four headings: (1) ICT Infrastructure, (2) Educational Content, (3) Human Resources and (4) Prevention of Digital Gap. In this macro-level ICT integration plan, MONE aimed to improve and upgrade the ICT infrastructure of schools, provide digital educational content and technical support staff, and assist with the professional development of teachers and administrators on the use of ICT in education.

2. **Infrastructure-Related Problems:** The school buildings and classrooms do not constitute a suitable physical space and have a negative effect on the installment of hardware. Since most of the school buildings were constructed without taking into consideration the installation of extra hardware, an existing classroom had to be modified and turned into an ICT laboratory; however, these classrooms were neither large enough nor suitable (Akbaba-Altun, 2006; Usluel, Mumcu & Demiraslan, 2007).

In the Turkish case, the scarcity of computer access in the schools is another factor hindering the effective use of technology in education. The schools where ICT laboratories were installed within the scope of the Basic Education Project Phases I and II had a relatively small number of computers (Kılıç & Özdemir, 2006; Toprakci, 2006; Akbaba-Altun, 2006; Usluel, Mumcu & Demiraslan, 2007). In the ICTIBS report (2007), the average number of computers in 302 schools was 30. Only 24 of them were used for instructional aims. When the average number of students (700) in the same schools is taken into account, it is clear that the number of computers is quite low.

The amount of student-computer interaction was almost less than one class hour for 80.7% of the 1800 students in the ICTIBS report (2007), even if we assume that all computers were functioning properly. Such brief interaction with a computer cannot create the expected results for students' academic achievements or technological skills. The reason for the short supply is the distribution of a limited number of computers to schools across the country. Instead, the same number of computers could have been distributed to fewer schools in order to increase the duration of student-computer interaction. A bureaucrat from MONE confessed that MONE preferred to distribute the purchased computers to more schools because of political pressures from members of Parliament (Akbaba-Altun, 2006; Kılıç & Özdemir, 2006).

Computers, printers, projection devices, and notebooks are quite sensitive and may break down easily. Moreover, certain parts of these tools require frequent upgrading. In the Turkish case, delays in repairs of broken hardware or upgrading of parts negatively affected the number of accessible computers. Any broken computer has to be repaired in five days within the warranty period, but due to the geographical conditions and the limited number of authorized computer service points, the needed repairs were not always made on time (Akbaba-Altun, 2006). In addition to this, the MONE did not allocate any budget to or make contingency plans for maintenance or upgrading after the end of the warranty period (Kılıç & Özdemir, 2006, Toprakci, 2006). In the ICTIBS report (2007), 84.5% of 306 primary schools did not have any technical support staff to repair broken hardware. The school administrations were held responsible for finding the required financial aid, but due to the economic conditions in Turkey, the administrators were not able to raise the required funds from the parents and businessmen in the region. The schools' administrators were frustrated with technical problems. Providing teachers with technical skills and training might also help with maintenance of the peripherals, but the limited amount of in-service training is not enough for teachers to gain the required knowledge and skills. Employing technicians to deal with the technical matters may help maintain the ICT classrooms to the necessary extent and standards.

3. **Personnel-Related Problems:** The British Educational Communications and Technology Agency (BECTA) (2006) warns that schools should have long-term plans for training sessions for teachers so that they can practice the skills with which they have been familiarized. The teachers should be able to access short, goal-oriented, effective, and high-quality in-service training sessions whenever they need to improve their skills and increase their knowledge on ICT use. In order to gain a better understanding of the situation Turkish teachers are facing, the in-service training sessions for ICT use provided to them must be evaluated. In total, 859 of 1279 teachers (67.2%) who responded to the Impact Research of ICT Classrooms (IRIC) (2004) and 1860 of 2758 teachers (67.4%) who responded to the ICTIBS report conducted for MONE said that they had participated in at least one computer-literacy in-service training session provided by MONE. Various case studies on the Turkish educational system have pointed out that the content, number and quality of the training sessions were not sufficient to satisfy the expectations (Akbaba-Altun, 2006; Usluel, Mumcu & Demiraslan, 2007; Özdemir & Kılıç, 2007). In addition, Somyürek, Atasoy and Özdemir (in press) showed that the in-service training supplied to teachers related to ICT use can be regarded as "familiarization presentations". At the level of familiarization, teachers may appreciate the advantages of technology, but they have no opportunity to practice or use it in the classroom environment. The training programs provided by the MONE to help prepare the teachers were essentially technological rather than pedagogical. The ICTIBS report (2007) indicates that while 47.9% of schools have been using computers for administrative tasks for more than seven years, just 15.7% of them have been using computers for instructional goals. However, 34.1% of the schools have been using computers for instructional goals for one year. Toprakci (2006) demonstrated the limited training of the school staff on ICT as an obstacle in the Turkish case. Özdemir

and Kılıç (2007) found that the education actors, like teachers, administrators, and supervisors, were given little or no training on the rationale and nature of the ICT integration projects; this was due to the lack of funding available for this type of training after all the hardware and software was purchased. Even though Turkish teachers had positive attitudes toward ICT use in education (Özdemir & Kılıç, 2007; Usluel, Mumcu & Demiraslan, 2007; Akbaba-Altun, 2006), further and permanent in-service training sessions for all actors are critical in increasing their understanding of the impact of employing technology in learning.

The limited number of in-service training sessions in which the administrators participated gave way to other problems as well. MONE stressed that the administrators and supervisors would be responsible for reporting the problems that would arise as the projects progressed. However, it became clear that the responsible personnel were given little or no training as to the rationale and nature of the projects or in developing ICT knowledge and skills. Hence, the administrators performed their duties and tried to solve the problems related to ICT use in education based on their own experiences or knowledge. This led the administrators to underestimate problems and needs, believe that computer laboratories increased their workload, and often to keep the ICT classrooms under lock and key to protect against theft, damage or improper use of the ICT peripherals (Özdemir & Kılıç, 2007; Akbaba-Altun, 2006). Similarly, lack of training and criteria for supervision complicate the use of ICT and make it difficult to solve the problems that occur. Özdemir and Kılıç (2007) found that the ICT integration projects also lacked performance indicators by which processes and outcomes could be monitored and evaluated. MONE prepared a checklist for the supervisors, but this mainly focused on minor matters such as the cleanliness of the ICT classrooms and the protection of the hardware and software, rather than the effective use of these devices in learning and teaching processes. Akbaba-Altun (2006) found that supervisors are aware of their limited level of computer literacy and confess that they are not well trained to supervise computer laboratories and student work.

4. **Curriculum-Related Problems:** The related research indicates that the teachers agree on the lack of emphasis on ICT use in the existing curriculum (Özdemir & Kılıç, 2007; Akbaba-Altun, 2006; Usluel, Mumcu & Demiraslan, 2007). A requirement analysis was not conducted prior to the preparation of the new primary education curriculum in 2003; therefore, the courses do not include the required content for ICT. A MONE expert's response to the question, "Is any member of the group preparing the new Turkish curricula an educational technologist?" was quite meaningful and shed light on the ICT-related problems in the curriculum: "We do not need any educational technologists in the group because we already know how to use computers!" Scrutinizing the current Turkish primary school curricula, the existence of a gap between the expected and the implemented curricula for ICT use can be seen easily. While MONE aims to integrate use of ICT within the learning and teaching process, the current curriculum treats the information and communications technologies as an ordinary classroom tool and does not direct the teachers as to when and how to use ICT in learning and teaching. In the ICTIBS report (2007), 430 school principals and vice principals (79.8%) stated that the existing computer technologies in their schools were used to make the students computer literate.

Lack of adequate digital learning material is another factor that bears negative effects on the use of ICT in education (Özdemir & Kılıç, 2007; Akbaba-Altun, 2006; Usluel, Mumcu & Demiraslan, 2007; Toprakci, 2006). Just 23% of the school administrators stated that they had educational software in their schools (ICTIBS, 2007). Digital learning materials are important because they include the content that constitutes the teaching objectives of the curriculum. The lack of adequate digital material may cause subject-specific insufficiencies (Wall, Higgins & Smith, 2005). Digital educational material brings about the advantages of multimedia, including static and dynamic images, sounds, videos, animations, web sites, presentations, and text editors. The limited availability of digital learning material for schools may lead to resistance from teachers who already have concerns about ICT use in education (Miller & Glover, 2002). Unfortunately, digital learning material is the one component neglected by MONE as a whole. In fact, the schools were provided with a limited amount of software, but since the language of these materials was English, the teachers could not use them. In 2004, MONE announced that it had reserved 35 million dollars to purchase educational software. This purchase seemed as if it would be able to respond to all of the digital learning material needs of the schools, but the bidding was cancelled by MONE due to the low number of supplier firms and the inappropriateness of the prospective software. However, since MONE was aware of the chain reaction caused by the lack of adequate digital learning material, they offered the use of digital learning object to respond to the educational software needs of the schools in the *Information and Communication Technologies Politics Report* (MONE, 2004a). Activating the inner dynamics of MONE, the new approach will help to increase the number and quality of the digital learning objects in the Learning Object Repository running on MONE's Learning Management System (LMS). To speed up the digital learning object production, MONE established a center for design, development, testing and dissemination of learning objects, employing about fifty graphical designers, field experts, computer programmers, multimedia experts, etc. In addition to this center, MONE aims to incorporate the teachers in the process of digital learning object production. Some of the teachers in the Turkish education system have already produced digital learning materials to use in their own classrooms using different kinds of software. While there is a non-negligible amount of production going on at different schools, these materials do not adhere to any standards and are used only by a limited number of teachers. MONE's target is to create a new digital learning object production source by supporting these self-motivated teachers technically and pedagogically and by establishing standards for the production processes so that these digital learning objects can be made available for use by the entire education system in Turkey.

The Lessons to Learn From the Turkish Case

Remember the Ferrari metaphor at the beginning. If you won a lottery and received $1,000,000, you would probably not buy a Ferrari costing $1,000.000. You would use this financial resource to optimize your happiness, comfort and life quality. Similarly, if you borrow such a financial resource from a bank, you must be careful in order to obtain a successful return on the investment.

Turkey is an important case providing various lessons for the countries that intend to learn from others before investing in ICT in education. The most important lessons learned are listed in the ICTIBS report (2007). School administrators and teachers indicated five problems listed in Table 1.

It is likely that there are some other problems as well. However, the list above shows the prior problems hindering the successful use of ICT in school environments. First, if the practitioners – teachers and school administrators – are not well educated or skilled, they cannot use the invested hardware to meet the expectations or goals. Pelgrum (2001) affirms that ICT use for instructional

Table 1. Comparison of problems encountered during the integration of ICT into the teaching-learning process (ICTIBS, 2007)

Problems	School Administrators		Subject-matter Teachers	
	f	%	f	%
Lack of training	156	*57,5*	808	*63,0*
Lack of hardware	115	*42,6*	610	*49,3*
Out of date and insufficient hardware	66	*27,0*	273	*26,3*
Lack of technological infrastructure	33	*17,2*	199	*22,2*
Lack of educational software	41	*17,8*	165	*18,9*

aims usually does not progress if teachers are not provided with the skills and knowledge needed to carry them out. Training teachers is a very expensive activity that is often neglected in large-scale investments. As seen in the heading "Personnel-Related Problems" above, only supplying technical training or training at the familiarization level will not help teachers or administrators to use technology effectively or efficiently. Moreover, the British Educational Communications and Technology Agency (BECTA) warns that "training is not a one-off issue" (2006). Policy makers and project managers have to consider that the administrators and teachers are able to participate in in-service training as often as needed. These actors should be able to participate in continual professional development activities to keep pace with technological improvements.

The second point that policy makers and project managers have to bear in mind is the lack of hardware, or in other words, student-computer ratio. MONE purchased a substantial amount of hardware in the last decade. On the other hand, the number of students in Turkey far outnumbers the available computers. The limited hardware was distributed equally to the whole country because of political concerns. Thus, the student-computer ratio was inadequate to meet the expectations for academic achievement. Instead, the hardware could be distributed to specific regions so that the student-computer ratio would be higher, and this investment could create a positive effect.

The third problem to consider is maintenance and upgrading of the technology. Careful distribution of hardware also includes planning for technical support and budget needs concerning broken down and out of date technology. Since technological equipment is sensitive and is used very intensively in schools, it requires maintenance very often. When preparing technology distribution plans, managers should also plan how to supply technical support and financial resources for maintenance.

The fourth problem is curriculum integration and availability of adequate educational software. ICT use can facilitate active learning. Active learning requires the learner's easy access to information sources (Pelgrum, 2001). If a country wants to use ICT in education in order to facilitate active learning and improve the quality of education rather than to make the students computer literate, the curriculum should be redesigned to allow use of ICT in student-centered educational contexts. Teachers have to understand clearly why and when to use ICT during instruction. Moreover, an adequate amount and quality of digital content should be available. Because of their reusable, shareable, repurposable and easily-renewable structures, learning objects are quite a suitable solution for the lack of digital content.

The last issue is the need for macro- and micro-level ICT integration plans for countries and schools. Such plans will be a kind of road map for policy makers, project managers, school administrators and teachers. With the help of these plans, they can foresee the next steps to take for successful implementations and how to allocate the budget.

Conclusion

In the last decade, the Ministry of National Education of Turkey successfully developed several web portals with administrative goals. To illustrate, the *e-library portal* allows access to various online books, journals, and videos; the *e-school portal* informs parents about their children's grades, dates of examinations, the average grade of a specific exam, etc; the *ilsis portal* runs all business processes of MONE's provincial organization in 81 cities country-wide; the *mebbis portal* informs the teachers about their personnel affairs, etc; and the *e-graduate portal* traces graduates of vocational schools. In contrast to such successful projects, the problems discussed above are obstacles hindering the effective and efficient use of the disseminated hardware in schools. The purchase was realized with loans granted by foreign sources, which make up the major part of the e-learning investment.

A worldwide research study conducted in 26 different countries shows that the expectations related to use of ICT in education shift from viewing the learners as passive and mere consumers of education to learners as active knowledge producers. "It seems that the current belief is that ICT is not only the backbone of the Information Society, but also an important catalyst and tool for inducing educational reforms that change our students into productive knowledge workers" (Pelgrum, 2001). In order to utilize ICT in classroom settings, the missing and problematic components should be completed or remedied. If the return on investment in the use of ICT in education fails to be sufficient, then a common public opinion that it is nonessential to spend money on technology in education is formed as a consequence. Before such an opinion prevails in public opinion and ICT in education loses its novelty to teachers and administrators, the necessary measures should be taken by the relevant institutions.

ADDITIONAL READING

Alessi, M. S., & Trollip, S. R. (2001). Multimedia for Learning. Massachusetts. Allyn and Bacon Publishings.

Anglin, G. J. (Ed.). (1995). Instructional Technology: Past, Present and Future. Englewood, Colorado. Libraries Unlimited Inc.

Khine, S. M., & Fisher, D. (Eds.). (2003). Technology-Rich Learning Environments: A Future Perspective. London. World Scientific Publishing Co.

Mueller, J. (2008). Identifying discriminating variables between teachers who fully integrate computers and teachers with limited integration. *Computers & Education, 51*, 1523–1537. doi:10.1016/j.compedu.2008.02.003

Osin, L. (1998). Computers in education in developing countries: Why and How? *Education and Technology Series, 3*(1), 1–14.

Reiser, A. R., & Dempsey, J. V. (Eds.). (2002). Trends and Issues in Instructional Design and Technology. Upper Saddle River, New Jersey Columbus, Ohio. Merrill Prentice Hall

Wang, Q., & Woo, H. L. (2007). Systematic Planning for ICT Integration in Topic Learning. *Educational Technology & Society, 10*(1), 148–156.

REFERENCES

Akbaba-Altun, S. (2006). Complexity of integrating computer technologies in education in Turkey. *Educational Technology & Society, 9*(1), 176–187.

Akkutay, Ü. (1996). *Foreign Expert Reports in National Education*. Avni Akyol Kültür ve Eğitim Ümit Vakfı.

BECTA. (2006). *Getting the most from your interactive whiteboard: A guide for secondary schools.* Retrieved October 15, 2008, from http://foi.becta.org.uk/content_files/corporate/resources/foi/archived_publications/getting_most_whiteboard_secondary.pdf

DPT. (2006). *Knowledge Society Strategy.* Retrieved October 11, 2008, from http://www.bilgitoplumu.gov.tr/btstrateji/Strateji_Belgesi.pdf

Kılıç, E., & Özdemir, S. (2006). Evaluation of the Distribution and Keeping the Continuity of the IT Classrooms Established in the Scope of Ministry of National Education (MONE) Basic Education Project. *Journal of Turkish Educational Sciences, 4*(2), 129–140.

Miller, D., & Glover, D. (2002). The interactive whiteboard as a force for pedagogic change: The experience of five elementary schools in an English Education Authority. *Information Technology in Childhood Education Annual*, (1): 5–19.

MONE. (2001). *National Education at the Beginning of 2002.* Republic of Turkey Ministry of National Education Research, Planning and Coordination Board (RPCB). Retrieved October 10, 2008, from http://www.meb.gov.tr/Stats/apk2002ing/apage00_0.htm

MONE. (2004a). *Basic Education Program ICT Politics Report.* Ankara, Turkey: Ministry of National Education.

MONE. (2004b). *Impact Research Report of ICT Classrooms.* Ankara, Turkey: Ministry of National Education.

Odabaşı, F. (1998). Computer Assisted Instruction. In Y. Hoşcan (Ed.), *Computer.* Eskişehir. Anadolu University, Open Education Faculty Press.

OECD. (2001). *E-learning The Partnership Challenge.* Centre for Educational Research and Innovation. OECD Publishing.

Özdemir, S., & Kılıç, E. (2007). Integrating information and communication technologies in the Turkish primary school system. [BJET]. *British Journal of Educational Technology, 38*(5), 907–916. doi:10.1111/j.1467-8535.2006.00678.x

Pelgrum, W. J. (2001). Obstacles to the integration of ICT in education: results from a worldwide educational assessment. *Computers & Education, 37*(2), 163–178. doi:10.1016/S0360-1315(01)00045-8

Somyürek, S., Atasoy, B., & Özdemir, S. (in press). Board's IQ. What makes a board smart? *Computers & Education.*

Toprakci, E. (2006). Obstacles at Integration of Schools into Information and Communication Technologies by Taking into Consideration the Opinions of the Teachers and Principles of Primary and Secondary Schools in Turkey. *e-Journal of Instructional Science and Technology (e-JIST), 9*(1).

Usluel, Y. K., Mumcu, F. K., & Demiraslan, Y. (2007). ICT in the Learning Teaching Process: Teachers' Views on the Integration and Obstacles. *Journal of Hacettepe University Education Faculty, 32*, 164–178.

Wall, K., Higgins, S., & Smith, H. (2005). The visual helps me understand the complicated things': pupil views of teaching and learning with interactive whiteboards. *British Journal of Educational Technology, 36*(5), 851–867. doi:10.1111/j.1467-8535.2005.00508.x

Section 3
User Centered Focus in E-Learning

Chapter 11
Users' Satisfaction with E-Learning:
A Case Study of the University of Botswana

Adeyinka Tella
University of Botswana, Botswana

EXECUTIVE SUMMARY

This chapter examines a case study of the user's satisfaction with e-learning at the University of Botswana. The study drawn on 415 undergraduate students who are users of e-learning from across six faculties and 39 departments of the university. Data was collected through an adapted and validated questionnaire. The result reveals generally that students were satisfied with e-learning system at the University of Botswana. Overall, 87.3% were adequately satisfied, satisfied, and moderately satisfied; while on the other hand, 11.8% were less satisfied and not satisfied. Perceived usefulness, perceived ease of use, system quality, content quality teaching and learning effectiveness dimensions were indicated to have the capacity to determine users' satisfaction with e-learning. Furthermore, the results demonstrate that the entire user satisfaction dimension positively and significantly correlate with and adequately predict and determine satisfaction with e-learning. Challenges indicated facing use of e-learning system are log on problems, loss/forgotten password, network/ server failure, access, and long download time for large adobe and PPT files. Upon these findings recommendations such as increase in the number of access and bandwidth of the system to allow it to work faster than before were suggested.

BACKGROUND TO THE UNIVERSITY OF BOTSWANA (UB)

The **University of Botswana** was established in 1982. This was after the break up of the multinational and multi-campus University of Botswana, Lesotho, and Swaziland, which had been established in 1964 to serve the three Southern African countries of Botswana, Lesotho and Swaziland. The University main campus is situated in **Gaborone**, the capital city. During 2006/2007, the University had a total enrolment of 16,238 students of which 12,934 were fulltime. Approximately 51% of the students are females. Of the total enrolment,

DOI: 10.4018/978-1-60566-942-7.ch011

15,248 are pursuing undergraduate programmes (University of Botswana, 2007). The University has six faculties, namely: Business, Education, Engineering and Technology, Humanities, Science and Social Sciences. The six faculties comprise thirty nine (39) departments. The University has a School of Graduate Studies and several specialised centres and research units. The University has staff strength of 2,640 of which 994 are academic staff. The academic programmes are offered at certificate and postgraduate levels (University of Botswana, 2008).

E-Learning at the University of Botswana

The implementation of **e-learning** at the University of Botswana was motivated to fulfill the University's responsibility to among other things (UBel, 2002:16): prepare students for effective participation in the wider information society, use ICT to increase the success rates of students, provide the opportunity for the University to enhance flexible learning anytime, anywhere and at student's own pace, enable access to relevant national and international resources, handle large classes The University of Botswana has policy-guided e-learning program emphasises a blended approach to e-learning in which various modes, methods and media are integrated and organised for appropriate learning.

The University of Botswana embarked on a programme of e-learning in 2001 when it mandated EduTech unit within the Centre for Academic Development to technologically transform the education process at the University (Uys, 2003). EduTech carries out the training of faculty in the effective and appropriate use of educational technologies at the University of Botswana. The Unit also provides resources such as state-of-the-art computer laboratories known as smart rooms. These laboratories are fully equipped with wireless Local Area Networks (LANs), videoconferencing facility, digital projectors, scanners, and **Web Course Tools (WebCT)** e-learning platform. Smart Classrooms constructed for technology-based, open, active, and collaborative learning. This classroom is laid out with clusters of computers situated in such a way as to provide eye contact.

During 2006, 145 lecturers of the 827 faculty were using e-learning in the delivery of their courses and during 2007/08 academic session 258 lecturers of the 994 academic staff were using e-learning in the delivery of their courses. The number of students enrolling in e-learning course is also growing. During the 1st semester of 2006-2007, more than 1,300 students were added to online courses (University of Botswana, 2006). According to (UB WebCT Report, 2007:9), ''it is difficult to tell the exact number of students online because most students are enrolled in more than one course. A rough estimation of about 8000 + are enrolled on **WebCT**''.

At the **University of Botswana**, issues about the cultural and social aspects of e-learning were included in the implementation of all on-line courses, programme and content. The e-learning platform at the university was design to embrace social interaction which is an important aspect of online pedagogy in catering for diverse learners need bearing in mind that diversity in learning approaches, style, and cultural patterns are universal.

Setting the Stage

E-learning is becoming an integral part of the education process around the world. **E-learning** consists of several components such as: course content, course content management, course content management system, organization of learners, teachers' interaction, content design and development (Caliner, 2005).

E-learning at the University of Botswana is used to support and enhance teaching and learning and to ensure that all students are committed to life long learning. At the University of Botswana, e-learning provides an integrated environment for online teaching and learning. It enables academic staff with basic computer skills to provide and manage **web-based learning** materials and utilise a range of online communication tools in order to supplement, or provide an alternative to, face-to-face teaching. E-learning is also used at the University of Botswana as communication tool, students learning activities tool, content tool and student tool. E-learning lends itself to the formation of communities of practice over and above the assignments where optional group work occurs. University of Botswana students use the discussion tools to explore issues and to articulate their individual and eventually collective problems, forming connections that make their studying more meaningful to them. These connections are what help them to make sense of their experience and give meaning to what they observe and how they participate.

However, in the context of the University of Botswana, previous studies on e-learning have largely focused on online course content creation; proportion of students using online content; online content development; securing content; quality of content online; management of students marks and course materials, etc. None of these studies seem to attempt measuring the **users' satisfaction** with e-learning. The evaluation of e-learning satisfaction deserves special attention, and evaluators need appropriate guidelines as well as effective evaluation methodologies (Zacharias et al., 2002; Seddon (1997). Moreover, the number of studies addressing evaluation or satisfaction of e-learning systems is relatively small and inadequate (Squires and Preece, 1999; Quinn et al., 2005); and additionally, user satisfaction has been identified as one of the factors for measuring an information system of which e-learning is one. It should also be noted that, despite the increasing use of e-learning the world over for teaching and learning, little attention has been given to examining issues of users' satisfaction (Drury, 1998; Gatian, 1994) which are central to e-learning implementation. Moreover, **e-learning satisfaction** in the educational setting could be determined from various perspectives such as IT personnel, students, teachers, administrators and designers. In this study, satisfaction was determined from the perspective of the students. This is because they are the primary and great consumer of e-learning at the University. Determining satisfaction of e-learning from other stakeholders is hereby left to the future researchers.

This paper after introduction begins with a theoretical background, which includes a review of the relevant research. This is followed by the methodology section which provides details of this study's context and the instrumentation process are then described. The measurement construct developed to assess satisfaction is analyzed to investigate the underlining factors/dimension, and is then tested for reliability and validity. Discussions, conclusions and implications are outlined in the following order.

Case Description

This study looks at **e-learning satisfaction** at the University of Botswana. Observation has revealed that the University is investing a lot of resources annually on e-learning. A lot of facilities have been put on ground to facilitate the e-learning programme. In this regard,, the need has come to find out whether users are satisfied with the programme of e-learning in the light of the huge investment on the programme, resources and facilities provided. This necessitates the conduct of this study. The outcome from this study will provide:

- A framework for making informed decisions regarding investment in e-learning.
- An indication of where pedagogical changes are needed to optimise the use of e-learning.
- Best practices from which University of Botswana can use to improve on its e-learning.

- A framework for understanding how e-learning may be used to open up education opportunities for all those who desire it.

Current Challenges on E-learning at University of Botswana

Despite the advantage of the e-learning system to the University of Botswana several challenges are facing the use of the system. There have been several reports of poor access. Tracking of courses is tedious. Migration of courses from the old version to the new **WebCT** the e-learning platform use at the University started in September 1, 2007 (UB WebCT Report, 2007). During this migration period, some courses such as computer skills fundamental posed some problems. Some of the file names were changed especially the file pictures. The problem was fixed by changing the file name in HTML.

There is also the challenge of students' assignment submission not sorted by ID and surnames as was the case with old version. In the old version, students' submissions were sorted according to their students IDs. This made marking for the instructors and the teaching assistance easy (UB WebCT Report, 2007). In the new version, it was reported that teaching assistants marked the assignments assigned to them by pages and this is considered too tedious.

Downloading assignments from the old version was easier than on the new version. The instructor found this exercise on the new version not very intuitive because of the several steps they took to undertake to download the submitted assignments. Even after downloading, students' submissions, one still has to mark the assignment one by one unlike the old version.

The platform in use at the university (**WebCT**) by default usually rounded marks to the nearest number. In the UB WebCT Report (2007) example of GEC 121 courses was cited where instructors preferred to leave the computing skills and information marks not rounded. They prefer to round the final examination mark only. The Grade Book was used to solve this problem through the options button. Users appreciated the new features in the new version. In the old version, one had to copy and paste the marks into the **excel sheet**. In case of the new version, Grade Book converts the whole Grade Book into excel.

Objectives of the Study

The major objective of this study is to examine the users' satisfaction with e-learning at the University of Botswana. The sub-objectives of the study are to:

1. Determine the level of students' satisfaction with e-learning at the University.
2. Identify the e-learning users' dimension/ characteristics that determine users' satisfaction with e-learning at the University of Botswana.
3. Find out the dimension of e-learning users' characteristics that best predict e-learning satisfaction.
4. Find out the challenges of using e-learning at the University of Botswana.

Research Questions

The major research question of this study is: What is the users' satisfaction with e-learning systems at the University of Botswana? The sub-research questions of the study are:

1. What is the level of students' satisfaction with e-learning at the University of Botswana?
2. What are the dimensions of users' characteristics that determine satisfaction with e- learning at the University of Botswana?
3. What dimension of users' characteristics best predicts users' e-learning satisfaction?
4. What are the challenges of using e-learning at the University of Botswana?

LITERATURE REVIEW

The University of Botswana defines **e-learning** as the appropriate organization of information and communication technologies for advancing student-oriented, active, open, collaborative and life-long learning process (UBeL, 2002). A concept closely related to e-learning but preceding the birth of Internet is **multi-media learning**. Multi-media is the use of two or more media, such as text, graphics, animation, audio, or video, to produce engaging content that learners access via a computer.

E-Learning Satisfaction

Operationally, **e-learning satisfaction ELS** can be considered as a summation of satisfactions with various attributes or items. On the one hand, **ELS**, like traditional customer satisfaction, represents an exchange-specific affective response (Halstead and Hartman, 1994), an attitude-like post-consumption evaluative judgment varying along the hedonic continuum (Wesbrook and Oliver, 1991). In the information system literature, common factors or dimensions used to determine users' satisfaction have been usefulness, perceived quality of the system, perceived usefulness, perceived service quality, perceived ease of use, user acceptance, user participation (Delone and Mclean, 1992, 2003; Ives et al., 1983; Doll and Torkzadeh, 1989). Since this study was conducted in the educational setting and e-learning system whose satisfaction is focused on this study was implemented to enhance teaching and learning; therefore, factors focused include perceived quality of e-learning content, use of e-learning, usefulness of e-learning, ease of using e-learning, perceived system quality, and perceived e-learning service quality.

Domain of E-Learning Satisfaction

Perceived System Quality

This refer to the perception of the judgment of the degree to which the technical components (including hardware, software, help screens and user manuals) of delivered e-learning provide the quality of content and service as required by users/ stakeholders. This concerned with whether or not there are "bugs" in the system, the consistency of the user interface, ease of use, response rates in interactive systems, documentation, and sometimes, quality and maintainability of the program code.

Perceived Content Quality

A perceived judgment of the degree to which e-learning users/stakeholders are provided with content of excellent quality, with regard to their defined needs excluding user manuals and help screens (features of System Quality). This concerned with such issues as the timeliness, accuracy, relevance, and format of content generated by the e-learning system.

Perceived Usefulness

This is the degree to which users believes that using e-learning system would enhance his or her learning and performance. Information system use means using the system. It is expected that resources such as human effort will be consumed as the system is used. **E-learning** system use might be measured in hands-on hours; hours spent analyzing reports, frequency of use, number of users, or simply as a binary variable, namely use and non-use (Seddon, 1997).

E-Learning Impact on Individual

This is important effects resulting from the use of e-learning system. This may include improvements in learning performance, effectiveness, domain or knowledge, and decision making.

Perceived Service Quality

Marketing literature has generally treated perceived service quality and customer satisfaction as related but distinct (Bolton and Drew, 1991). While recent research appears to indicate that perceived service quality is an antecedent of customer satisfaction, debate on the causal direction between these two constructs continues (Oliver, 1993). A literature review identified a consensus on the fundamental distinction between perceived service quality and customer satisfaction constructs: namely, perceived service quality is a long-term attitude, whereas customer satisfaction is a transaction-specific judgment (Bitner, 1990). With perceived service quality and customer satisfaction now being two distinct constructs, they should be measured using different instruments. Researchers also suggest that the directionality of the relationships between perceived service quality and customer satisfaction should continue to be examined. Service quality in this study refers to the entire quality of support provided when using e-learning system.

Intention to use is a measure of the likelihood a person will employ the application. It is a predictive variable for system use. However, only when system use is difficult to assess, measuring intention to use can be worthwhile (Lederer, Maupin, Sena, Zhuang, 2000).

Perceived Teaching Effectiveness

Students' evaluation of teaching effectiveness SETE is a primary method for defining and measuring teaching quality/effectiveness, and many established instruments exist in educational psychology (Wang, 2003). Overall, the students' evaluation of educational quality SEEQ presents a comprehensive definition and measurement of teaching quality and has eight factors. Curiously, the quality satisfaction relationship is seldomly examined explicitly. One reason is that both concepts are often used synonymously (Abrami and Cohen, 1990). Thus, conceptual ambiguity between quality and satisfaction that marketing is currently exploring also appears in educational psychology literature.

This **teaching effectiveness** in this study is concerned with improved quality of teaching and learning as a result of using e-learning as perceived by the students. This is because the core business of e-learning system is teaching and learning. Therefore, in this study teaching effectiveness will mean the improvement in the way courses are delivered, through e-learning platform and the effectiveness of tutors' interaction with the students on the platform. This will be measured by the perception of users (students) in terms of learning and teaching improvement on a five points scale.

Empirical Studies on User Satisfaction with E-Learning

Gibbons and Fearweather (2000), Clark (2002) in their different studies carried out in US on students' use of e-learning found consistently that students are very satisfied with e-learning. Learners' satisfaction rate increased with e-learning compared to traditional learning, along with perceived ease of use, access, navigation, interactivity, and user-friendly interface design. DeLone and McLean's (1992) comprehensive review of different information system success measures conclude with a model of interrelationships between their six IS Success constructs (system quality, information quality, usage, user satisfaction, individual and organizational impact). Based on this model, Seddon and Kiew (1994) in a study at the University of Melbourne in Australia investigated an Accounting System (DAS) among those who were relatively senior clerical officers in each department or faculty office that uses the system for about 4-5 hours per week for maintaining the department's accounting records. The researchers examined critically the meaning of four of these constructs and the evidence of relationships between them. Tests were conducted using both conventional ordinary least squares regression path analysis and structural equation modeling. The results provided substantial support for the DeLone and McLean's model. System Quality, Information Quality, and Usefulness, were found to explain 75% of the variance in the overall User Satisfaction measure.

Clerkin (2004) in a study conducted at Berkeley College in UK on comprehensive plan for preparing online students found that the majority of online students were satisfied with the course. It was further revealed that 42.9% and 52.4% felt that online degree was very helpful and helpful. Again, the majority of Berkeley's online degree students were satisfied with the orientation. Clerkin's study and its result are relevant to this study, in the sense that this study is aimed at determining user satisfaction with the e-learning system at the University of Botswana.

The United Arab Emirate UAE Laptop Project (2004) conducts a questionnaire survey on its web-based learning programmes by the end of the winter semester of the academic year 2002/2003. The survey was initially aimed at evaluating learning quality, students' satisfaction, technical reliability, and ease of use. The questionnaire was emailed to 5,740 students who had at least one **Blackboard** course. The number of respondents was 1,435 students (25 per cent) with completed and usable questionnaires. The survey analysis revealed that satisfaction and improvement of learning and teaching process were found to be 78.4 per cent and 70.5 per cent, respectively. The implication of this finding to the present study is based on the fact that satisfaction with e-learning will be assumed if the users' satisfaction level on this study is reported to be as high as 78-80%.

Kim, Liu, and Bonk (2005) examine the challenges perceived by the students enrolled in an MBA program at Indiana State University, America. These researchers completed a survey with 100 second-year online students and conducted in-person interviews with 22 students. In that particular study, 30 percent of the participants viewed their online learning experiences in a negative manner. In contrast, delayed feedback, difficulty in communicating with team members in different time zones, and a lack of emotional connection were perceived as challenges.

Several weaknesses related to **online learning** were also described in the literature. Delay in responses is one reported weakness. In a study by Petrides (2002) at the University of California, Los Angelis, some participants reported they felt a lack of immediacy in responses in the online context in comparison to what could typically occur in a structured face-to-face class discussion. This appears to be especially obvious in asynchronous online discussions when students have to wait for others to read and respond back to their bulletin board postings or e-mail messages.

Methodology

The study seeks to provide rational explanation of the phenomenon under study (users' satisfaction with e-learning) using response to a questionnaire instrument to explain its findings. The study adopts a survey research design. Survey design has been successfully used in research similar to this one (e.g. Hussein et al, 2007; Palm et al, 2006; Garity et al, 2005; Xiao and Dasgupta, 2002). Survey strategy is chosen because it's mostly associated with the philosophical paradigm of positivist (Oates, 2007) which underpins this study. A cross sectional form of survey design was embarked upon in this study. This is to collect data at one point in time from sample (students) selected to represent the larger population of (students) at the University of Botswana. The target population of this study are the students at the University of Botswana using e-learning. The profile of the study population indicates:

- There are 16, 238 number of students;
- Students are distributed in 7 faculties and 39 departments;
- 8,000 students are using e-learning across all faculties (UB WebCT Report, 2007; UB Fact and Figures, 2006/07).

Generally, the sample size was determined from the total of students (8,000) who are users of e-learning system at the University of Botswana. The selection of sample size on this study followed (Israel, 2003) model. This means that from the 8,000 total population of students' users of e-learning, ±5% was taken; this gave a total of 381 students which represent the sample in this study. The sampling frame for this study was the UB WebCT Report 2007 which contained data/information on students and academic staff who are using e-learning at the University. Questionnaire was used to gather data from the respondent (students) in this study. The items in the questionnaire were adapted from various previous IS measures and modified to suit the purpose in this study. The questionnaire consists of 20 items organized in five domains/constructs which are intended to capture separate dimensions of **e-learning satisfaction**.

The following domain constructs were included in the questionnaire: ***Perceived Usefulness***- adapted from (Lund, 2001) usability with use questionnaire with r = 0.92 Cronbach alpha. *Perceived Ease of Use*- adapted from Doll and Torzadeh (1988) end user computer satisfaction questionnaire with r = 0.90 Cronbach alpha. *Perceived Service Quality*- adapted from (Wang et al., 2007), e-learning system success scale with r = 0.89 through Cronbach alpha. *Perceived Content Quality*- adapted from (Wang, 2003) e-learner satisfaction questionnaire with r = 0.95 Cronbach alpha. ***Perceived teaching Effectiveness***- adapted from (Feldman, 1976) students' evaluation of teaching and learning, SETE with r = 0.91 Cronbach alpha. The overall reliability co-efficient of the questionnaire returned r = 0.88. This is in line with the minimum standard of .80 suggested for basic research and .90 suggested for use in applied setting where important decisions were usually made with respect to specific test scores (Nunally, 1978). The likert response format was adopted for all the items in each of the domain. This range from strongly agree to strongly disagree.

Faculties were censured. Within each faculty, a core course was chosen at each year of study. This means 4 core courses was chosen from faculty which has courses offered through the e-learning platform in every year and level of study. The total number of students taken each of the core courses was identified and added to give a total for each faculty. From this total, ±10% was taken for precision. This represent sample for each faculty. Data collected was coded using Statistical Package for Social Science (SPSS) version 16.0 for windows. Inferential statistical analysis such as ANOVA, Stepwise Multiple Regression and Beta was also undertaken to indicate the determinant/predictive capability of each of the satisfaction dimension. Items was rated on a 4 points likert scale with end points of '4' Strongly Agree, and 1 Strongly Disagree; the two midpoints will be '3' Agree and '2' Disagree. Neutral responses such as Not Sure/Indifferent was excluded from the response choice to overcome the neutral and don't know responses (Hussein et al., 2007) which is typical about African respondents; and moreover to avoid distortion of results.

Results

A total of 415 students responded to the survey. The response rate was 99.9%. This was achieved based on the fact the **e-learning platform** at the university embraces social interaction which caters for diverse learners need and diversity in learning approaches, style, and cultural patterns. This helps the respondents in the interpretation of the survey items. Similarly, the assistance of some lecturers in many of the on-line lessons where data was collected in making sure that questionnaire was returned immediately also contributed to this high return rate. Most respondents were females (55%), 27% were aged between 25-28 years, 23.1% 21-24 years and 21.9% were aged between 29-32 years. The years and levels of respondents reveal that 38.3% were chosen from year 1, 33.3% from year 2, 16.1% from year 3 and 12% from year 4. Furthermore, the demographics reveal that 27.7% respondents were from faculty of business, 20.5% from science, 20% from education, 19% social science, 8.2% from engineering technology and 4.6% from humanities.

Question 1: What is the level of students' satisfaction with e-learning?

To answer this question, students (respondents) were asked to show their level of satisfaction on five points ratings. Table 2 presents the result.

Table 1. Respondents' Bio-data Information

Demographics	Sample	Percentage
Gender		
Male	187	45
Female	228	55
	415	100
Age		
1 17-20 years	75	18.1
21-24 years	96	23.1
25-28 years	112	27.0
29-32 years	91	21.9
33-37 years	31	7.5
38-41 years	10	2.4
	415	100
Year/Levels of Study		
Year 1	159	38.3
Year 2	138	33.3
Year 3	67	16.1
Year 4	50	12.0
	415	100
Faculties		
Business	115	27.7
Education	83	20
Engineering Technology	34	8.2
Humanities	19	4.6
Science	85	20.5
Social Sciences	79	19
	415	100

Table 2 shows the level of respondents' satisfaction with e-learning. The result reveals generally that students were satisfied with e-learning system at the University of Botswana. Overall, 87.3% were adequately satisfied, satisfied, and moderately satisfied; while on the other hand, 11.8% were not satisfied. This hereby answers the first research question. The data reveal further what students found most satisfying as the provision of useful quality content, provision of exact content needed by the students in this order. On the other hand, what the students found least satisfying was also revealed. This has to do with the system failure to provide sufficient content.

Question 2: What dimension determines users' satisfaction with e-learning?

Responses to all the dimensions provided were obtained to answer this question. The results confirm all the dimensions to be good indicators of users' satisfaction. Table 3 contains the detail.

The results in table 3 above reveal that the entire five dimensions included in the table are actually good determinant of users' satisfaction.

Table 2. Overall level of e-learning satisfaction

Level of satisfaction	Number of Responses	Percentage
Adequately Satisfied	207	50.0
Satisfied	101	23.3
Moderately Satisfied	58	14.0
Less Satisfied	35	8.4
Dissatisfied	14	3.4
	415	100

Table 3. Dimension of User satisfaction

Dimensions	Number of respondent	Mean
Perceived Usefulness	415	5.89
Perceived Ease of Use	415	5.99
Perceived Service Quality	415	5.22
Perceived Content Quality	415	7.11
Perceived Teaching Effectiveness	415	6.81

The table shows that all the dimensions have significant mean value. This indicates they all have the potentials and capacity to determine users' satisfaction with e-learning. As results above indicated, students are generally satisfied with the overall quality of content of the system except for the system failure to provide sufficient content they needed. The results hereby answer the research question by revealing **perceived content quality** as the dimension that determine satisfaction with e-learning followed by **teaching effectiveness**.

> Question 3: What users' dimension best predict /determine users' satisfaction with e-learning? To answer this question, a multiple correlations of responses to the e-learning users' satisfaction questionnaire were run. This is followed by determining the extent to which the entire dimensions when put together adequately determine/predict e-learning system satisfaction. Table 4, Table 5, and Table 6 provide the results.

The inter-correlation between the independent variables (Dimensions of satisfaction) shows that **Perceived content quality** of the e-learning system has the highest and positive correlation with e-learning satisfaction (.711). This is followed by Perceived teaching effectiveness (.556), perceived ease of use (.454), perceived usefulness (. 523). Perceived service quality shows the least positive correlation with e-learning system with (.454). The results generally suggest that the entire **e-learning users' satisfaction** dimension significantly correlate with e-learning satisfaction. The results also reveal the mean and standard deviation for each of the dependent variables. Perceived content quality again has the highest (Mean = 6.21, SD = 1.79), followed by perceived **teaching effectiveness** (Mean 6.01, SD = 1.15). Others followed in this order: perceived ease of use (Mean 5.99, SD = 1.21); perceived usefulness (Mean 5.99, SD =1.21) and perceived service quality (Mean 5.89, SD = 1.07). The gravity of the mean value and standard deviation of content quality reveal it to be the dimension that exert the most positive significant correlation with users' satisfaction with e-learning. This thereby provides answer to the research question 3.

A stepwise multiple regression analysis on the data obtained on independent variables (dimension of e-learning satisfaction) and dependent variables (e-learning satisfaction) were run. Table 5 shows that the entire satisfaction dimensions made 67%

Table 5. Multiple regression on dimension of e-learning satisfaction and users' satisfaction with e-learning system

Multiple R .4681 R. Square .6758 Adjusted R. Square .3149 Standard Error .1140 Analysis of Variance				
	Df	**Sums of Square**	**Mean Square**	**F**
Regression **Residual** **Total**	5 410 415	7.409 49.718	37.045 20384.38	.002

F. Observed = 5.0711

Table 4. Multiple correlation of the dimensions of satisfaction with e-learning system

Dimension	Mean	Std Dev	ELS	PU	PEOU	PSQ	PCQ	PTE
E-learning Satisfaction	6.80	2.13	1.000	1.000	1.000	1.000	1.000	1.000
Perceived Usefulness	5.89	1.07	.523	.409	.640	.780	.602	
Perceived Ease of use	5.99	1.21	.544	.567	.449	.556		
Perceived Service Quality	5.22	1.03	.454	.596	.435			
Perceived Content Quality	6.21	1.79	.711	.586				
Perceived Teaching Effectiveness	6.01	1.15	.556					

prediction of users' satisfaction with e-learning. From the analysis of variance performed on multiple regression, it is seen that the observed F value = 5.0711, P < .05 when the five dimensions were regressed with the users' satisfaction. This indicates that the entire dimensions have no difference with the **users' satisfaction with e-learning system**. These results further strengthen the prediction in the table 6.

Table 6 provides the co-efficient on the extent of the determinant/prediction. The essence of this is to know which of the dimension produce the most significant effect on users' satisfaction with e-learning system. The results show that perceived content quality had the greater effect (Beta = .17372, t = 4.113, P< .05). Perceived teaching effectiveness followed with (Beta = .15518, t = 3.235, P<.05). Other dimension followed in this order: Perceived ease of use (Beta = 14641, t = 3.211, P < .05), Perceived service quality (Beta = 14968, t = 3.007, P <.05) and perceived usefulness (Beta = 14863, t =2.185, P< .05) respectively. The results generally show that the entire satisfaction dimension have the capability to determine satisfaction with e-learning system except that the gravity of and the extent of the effects differs.

Question 4: What are the challenges facing users when using e-learning at the University of Botswana?

To answer this research question, respondents were asked to indicate how often they faced some identifiable challenges of e-learning when using the system at the University of Botswana. Table presents the result.

Table 7 above reveals that respondents faced all the entire problems identified in the table when using the e-learning system. This was evidence with 93% respondents indicating experience of log on problem most often and often, 96% indicating loss/forgotten password most often and often, 94% indicating network/server failure most often and often, 96% indicating problem of access most often and often and 99% indicating long download time for adobe and PPT files most often and often. This indicates the challenges and problems listed in table 9 are faced by users of e-learning at the University of Botswana. This provides the basis for answering research question 4.

Discussion

The study has explored the **users' satisfaction with e-learning** at the University of Botswana.

Table 6. Co-efficient of the prediction

Dimensions/variables	B	SE.B	Beta	T.	Sig. T
PU	.11895	.05114	.14863	2.185	S**
PEOU	.01107	.05439	.14641	3.211	S**
PSQ	.03972	.05909	.14968	3.007	S**
PCQ	.04789	.06617	.17374	4.113	S**
PTE	.01469	.06453	.15518	3.235	S**
Constant (E-learning satisfaction	31.20589	3.18019		5.342	000

Table 7. Challenges of e-learning system

Challenges of e-learning system		Most Often	Often	Rarely	Total
1.	Log on Problem	250 (60%)	137 (33%)	28 (7%)	415
2.	Loss/Forgotten Password	300 (72%)	100 (24%)	15 (4%)	415
3.	Network/ Server Failure	367 (88%)	33 (8%)	15 (4%)	415
4.	Access Problem	215 (51%)	190 (45%)	10 (2)	415
5.	Long Download Time for Large Adobe and PPT files	351 (84%)	60(15%)	4 (1)	415

The result reveals generally that students were satisfied with e-learning system at the University of Botswana. Overall, 87.3% were adequately satisfied, satisfied, and moderately satisfied; while on the other hand, 11.8% were less satisfied and not satisfied. Perceived usefulness, perceived ease of use and other dimensions were indicated to have the capacity to determine users' satisfaction with e-learning. Furthermore, the results demonstrate that the entire user satisfaction dimension positively and significantly correlate with and adequately predict and determine satisfaction with e-learning. Challenges indicated facing use of e-learning system are log on problem, loss/forgotten password, network/server failure, access, and long download time for large adobe and PPT files.

In sums, this article contribution to knowledge includes the fact it serve as bridge to the paucity of research on e-learning satisfaction particularly from the African context. The study has provide the basis for showcasing whether or not users are satisfied with e-learning system at the University of Botswana despite the huge investment on the programme, resources and facilities provided. The findings of the study could be helpful to other African universities who are developing their e-learning portfolios to improve their **e-learning** delivery.

The first research question on this study indicates generally that students were satisfied with e-learning. This was not by chance considering the fact that earlier report of related studies has demonstrates correlation between e-learning and users satisfactions. For instance Gibbons and Fearweather (2000), Clark (2002) in separate studies at the University of Leed in UK has found consistently that students are very satisfied with e-learning. Learners' satisfaction rate increased with e-learning compared to traditional learning, along with perceived ease of use, access, navigation, interactivity, and user-friendly interface design. Similarly, West et al (2007) in their study discusses the benefits and challenges of e-learning reported by students and faculty. It was acknowledged that the tool has become critical to many participants' practices. The author stated emphatically that when Blackboard e-learning system worked without technical difficulties, most students and instructors reported being satisfied with the tool because it was convenient for them to use, easy to learn, and helpful in their studies. Sixty-six percent of student survey respondents in the winter semester (67% in fall semester) preferred that their instructors use the tool, and 73% (winter) and 75% (fall) of students said it was easy to use Blackboard e-learning system. The features of Blackboard with which students were most satisfied were the announcements, course documents, gradebook, and sometimes email features. Moreover, Clerkin (2004) report that 56.8% of the students felt that e-learning covered everything they needed to know about taking an online course; and that 28.5% felt it covered most of what they needed to know about taking an online course. Therefore, the majority of Berkeley's online students were satisfied with the course. It was further revealed that 42.9% and 52.4% felt that online degree was very helpful and helpful. Again, the majority of Berkeley's online degree students were satisfied with the orientation. These results in part corroborate the finding in this study by confirming the fact that students were satisfied with the e-learning system.

Seddon and Kiew (1994) in a study at the University of Melbourne in Australia investigated an Accounting System (DAS) among those who were relatively senior clerical officers in each department or faculty office that uses the system for about 4-5 hours per week for maintaining the department's accounting records. The results provided substantial support for the DeLone and McLean's model. System Quality, Information Quality, and Usefulness, were found to explain 75% of the variance in the overall User Satisfaction measure. This in part supports the finding on this study because perceived usefulness, content quality and service quality were reported to have significant capacity to determine satisfaction with e-learning.

This study also report that **perceived usefulness**, ease of use, **service quality, content quality** and **teaching effectiveness** adequately and significantly determine users satisfaction with e-learning. It should be noted that in the information system literature, common factors or dimensions earlier used to determine **users' satisfaction** have been usefulness, perceived quality of the system, perceived usefulness, perceived service quality, perceived ease of use, user acceptance, user participation (Fisher, 2001; Delone and Mclean, 1992, 2003; Ives et al., 1983; Doll and Torkzadeh, 1989). This basically lends a good support to the present finding on this study by confirming those dimensions identified as good surrogate to e-learning satisfaction.

There has never been a good information system without one perceived shortcoming. **E-learning** is an example of what is being said here. Though advantageous to teaching and learning, yet research such as this one and others have come up with challenges associated with its use. The challenges confirmed in this study associated with e-learning system are long-on problem, access, long download time for large file, loss or forgotten passwords, and server/network failure. Some earlier studies have reported similar challenges like the ones in this study. For instance, participants in West et al (2003) indicated having technical problems. This relates to server/network failure found in this study. Vonderwell (2003) findings also indicated that interactions or social involvements found in the face-to-face classroom seemed not to develop in the online context during the 10-week course duration. The students considered online communication less personal. The low level of social interactions with the instructor and delayed feedback were perceived as a challenge to their learning. Participants in Hara and Kling's (1999) qualitative case study of a Web-based distance education course at a major U.S. university also reported lack of immediacy in getting responses back from the instructor, and as a result they felt frustrated. Some studies indicate similar results. For example, in Vonderwell's (2003) study, one reports disadvantage of an online course was the delay of immediate feedback from the instructor. One participant in the study stated that when he emailed a question to the instructor, "it might take hours, maybe a day or so before he gets an answer back for the question" (Vonderwell, 2003:84). Perceived level of expertise is another weakness identified in online learning studies. Participants in Petrides' (2002) study report skepticism of their peers' supposed expertise. Lack of a sense of community and/or feelings of isolation were other challenges learners reported in their online learning experiences. Vonderwell (2003) reports **online learning** participants indicate a lack of connection with the instructor especially "one-on-one" relationship with the instructor. One participant in the study states, "I still feel like I know a little bit about my instructor, but not the same way that I would if I was in a class. I don't know much about her personality at all" (p. 83). Other studies found similar results. For example, Woods' (2002) in a study conducted at the Regent University, Virginia Beach, Virginia on how much communication is enough in online courses. By exploring the relationship between frequency of instructor-initiated personal email and learners' perceptions of and participation in online learning; reported that online learners report feeling isolated from faculty as well as other learners in the online courses they had taken. All this reports support the findings on the challenges of e-learning identified in this study.

The fact that students are generally satisfied with the overall quality of content of the system has implication for e-learning. This is because more students will become interested and thereby increase the number of student offering courses via the platform. Aside of this, the finding will also draw the attention of the school authority to the issue of e-learning quality assurance. Furthermore, the need to upload more relevant content on the system by the teachers and the e-learning administrators at the university will be considered very imperative.

Limitation of the Study

There is no perfect study without shortcomings. This is true of this study. Some of the limitations in this study are worth mentioning: the use of force choice has limitations on the operationalisation and the results of the survey. It is observed conceptually that neutrality may be valuable and legitimate response by the respondents in this study towards questions.

Furthermore, **satisfaction with e-learning** could be measure at various perspectives such as administrative, IT personnel, staff, students, etc. This study only focuses on measuring satisfaction only from the students' perspective left alone others. Although this study utilized validated and usable measures, the data collected were from self-reported measures that have their obvious limitations. Future studies on this topic could add interview schedules to augment data collection from questionnaires. The data collected were from students (one level of determining e-learning satisfaction), future studies could include teachers/lecturers, administrative staff and IT personnel. These limitations notwithstanding, this study has contributed significantly in demonstrating the influence of system quality, content quality, service quality, teaching and learning effectiveness in determining satisfaction with e-learning.

CONCLUSION

This study has examined the users satisfaction with e-learning and the results have demonstrate that some identifiable dimensions such as perceived usefulness, ease of use, service quality, content quality and teaching effectiveness were associated with **e-learning satisfaction**. This study and results are very important in view of the fact that many tertiary educational institutions the world over have put in place or have plan to put e-learning system in place in their various schools.

RECOMMENDATIONS

Aside of the fact that perceived usefulness, ease of use and other dimension identified in this study are good predictors and determinant of e-learning satisfaction, some challenges were as well identified as associated wit the use of e-learning at the University of Botswana. In the light of these challenges, it is recommended that the University of Botswana should increase the access to e-learning possibly to allow both on campus students and those leaving outside campus an opportunity of using e-learning anytime and anywhere. The issue of network/server failure should also be considered. The University should consider increasing the bandwidth of the system to allow it to work faster than before. The technical staff needs to increase their support services to the e-learning users. This will go a long way to settle the challenges of loss and forgotten password. Students as well need to play their own part regarding this challenge. They need to be master of their password as soon as they are given. This can be done by writing it somewhere to keep reminding them of it or in the alternative store the password in the memory of their cell phone. There is no doubt about the fact that University of Botswana is a giant stride among the great Universities in Africa as far as implementation of e-learning is concerned. Upon this premix other African Universities that have not put e-learning system in place should considered it necessary and a matter that needs urgent attention. Governments of African countries as well are called upon to assist universities in the implementation of e-learning. By so doing, we are bridging the digital chasm which has been existing between advance and developing countries.

Furthermore, for University of Botswana students to fully utilize the e-learning system they need to be provided a toll free number. This is to assist most of them who usually call for assistance. There is also the need to provide Internet connection to the e-learning platform **(WebCT)** facilitators so that users are helped on daily basis. It will help WebCT administrators to monitor WebCT from their homes since they are never sure if the system is working of-campus especially during weekends. The educational technology unit in charge of e-learning at the university should develop a rubric for online instruction. In other words, the unit should come up with the criteria for evaluating an online course for the purpose of content quality. This will assist in guiding online designers to develop quality course that will enhance their teaching and learning. Moreover, for the smooth running of WebCT, Edu Tech should consider having its own IT person responsible for **WebCT** in order to attend to its issues (problems, challenges and complains) in a timely manner.

REFERENCES

Abrami, P. C., Cohen, P. A., & d'Appolonia, S. (1990). Validity of student ratings of instruction: what we know and what we do not. *Journal of Educational Psychology*, *82*(2), 219–231. doi:10.1037/0022-0663.82.2.219

Bitner, M. J. (1990). Evaluating service encounters: the effects of physical surroundings and employee responses. *Journal of Marketing*, *54*(2), 69–82. doi:10.2307/1251871

Bolton, R. N., & Drew, J. H. (1991). A longitudinal analysis of the impacts of service changes on customer attitudes. *Journal of Marketing*, *55*(1), 1–9. doi:10.2307/1252199

Carliner, S. (2005). *CMSs versus LMSs*. Retrieved February 25, 2008 from ASTD Worldplace Learning & Performance http:www.astd.org

Clark, C. (2002). E-learning Strategy document. Version 2.1 (20.5.2002). *Information Technology Policy Committee.* Chair, e-learning steering Group of ITPC. Retrieved May 12, 2007 from http://www2.warwick.ac.uk/insite/forum/archive/elearning/stategydocument/

Clerkin, M. J. (2004). *Practice: A comprehensive plan for preparing online students*. Needham, MA: The Sloan Consortium. Retrieved May 5th, 2008 from http://www.sloan.org/effective/details.asp?SS_ID=133

Delone, W. H., & Mclean, E. R. (1992). Information system success: The quest for the dependent variable. *Information Systems Research*, *3*(1), 61–95. doi:10.1287/isre.3.1.60

Delone, W. H., & Mclean, E. R. (2003). The Delone and Mclean information system success: A ten years update. *Journal of Management Information Systems*, *19*(4), 30–36.

Doll, W. J., & Torkzadeh, G. (1989). The measurement of end-user computing satisfaction. *MIS Quarterly*, *12*(2), 259–274. doi:10.2307/248851

Drury, D. H. (1998). A hierarchical structural model of information systems success. *Infor,* (3). Retrieved January 15, 2008 http://findarticles.com/p/articles/mi_qa3661/is_199802

Feldman, K. A. (1976). The superior college teacher from the student's view. *Research in Higher Education*, *5*, 243–288. doi:10.1007/BF00991967

Garity, E. J., Glassberg, B., Kim, Y. J., Sanders, L., & Shin, S. K. (2005). An experimental investigation of web-based information system success in the context of electronic commerce. *Decision Support Systems*, *39*(3), 485–503. doi:10.1016/j.dss.2004.06.015

Gatian, A. W. (1994). Is user satisfaction a valid measure of system effectiveness. *Information & Management*, *26*(3), 119–131. doi:10.1016/0378-7206(94)90036-1

Gibbons, A., & Fearweatther, P. (2000). Computer-based instruction. In T. S. Fletcher, J. (eds.), *Training & retraining: A Handbook for Business, Industry, Government, and the Military* (pp. 410-442). New York: Macmillan Reference.

Hussein, R., Abdu-Karim, N. S., Mohamed, N., & Ahlan, A. R. (2007). The influence of organizational factors on information system success in e-government agencies in Malaysia. *Electronic Journal of Information Systems in Developing Countries EJISDC*, *29*(1), 1–17.

Israel, G. D. (2003).*Determining sample size*. Retrieved from February 10, 2007, http://edis.ifas.edu

Ives, B., Olson, M. J., & Baroudi, J. J. (1983). The measurement of user information satisfaction. *Communications of the ACM*, *26*(10), 785–793. doi:10.1145/358413.358430

Kim, K. J., Liu, S., & Bonk, C. J. (2005). Online MBA students' perceptions of online learning: Benefits, challenges, and suggestions. *Internet and Higher Education, 8*, 335, 344.

Leaderer, A. L., & Mendelow, A. L. (1988). Convincing top management strategic potential of information systems. *MIS Quarterly*, (Dec): 525–534. doi:10.2307/249127

Nunnally, J. C. (1978). *Psychometric Theory*, (2nd ed.), New York: McGraw- Hill.

Oates, J. (2007). *Researching information system and computing*. Los Angeles: Sage Publications.

Oliver, R. L. (1993). Cognitive, affective, and attribute bases of the satisfaction response. *The Journal of Consumer Research*, *20*(3), 418–430. doi:10.1086/209358

Palm, J. Colombet, Sicotte, C. & Degoulet, P. (2006). Determinants of user satisfaction with clinical information system. *Amia Annu Symposium Proceedings,* (pp. 614- 618).

Petrides, L. A. (2002). Web-based technologies for distributed (or distance) learning: Creating learning-centered educational experiences in the higher education classroom. *International Journal of Instructional Media*, *29*(1), 69–77.

Quinn, C. N., Alem, L., & Eklund, J. (2005).*A pragmatic evaluation methodology for an assessment of learning effectiveness in instructional system*. Retrieved March 15, 2008 from http://www.testingcentre.com/jeklund/Interact.htm

Seddon, P. B. (1997). A Respecification and Extension of the DeLone and McLean Model of IS Success. *Information Systems Research*, *8*(3), 240–253. doi:10.1287/isre.8.3.240

Seddon, P. B., & Kiew, M.-Y. (1994). A Partial Test and Development of DeLone and McLean's Model of IS Success. In *Proceedings of the International Conference on Information Systems*, Vancouver, Canada, (pp. 99-110).

Squires, D., & Preece, J. (1999). Predicting quality in educational software: Evaluating for learning, usability and the synergy between them. *Interacting with Computers*, *5*(11), 467–483. doi:10.1016/S0953-5438(98)00063-0

UAEU Laptop Project (2004). *Laptop Project: Students' Survey Analysis report*. Al-Ain, UAE: UAEU.

UBel. (2002). *E-learning.* An Unpublished paper, University of Botswana, Gaborone.

University of Botswana. (2007). *WebCT Report*. Botswana: Center for Academic Development,University of Botswana.

University of Botswana. (2007). Facts and Figures. *University Fact Book*.

Uys, P. M. (2003a). E-learning at the University of Botswana: A Bi-Annual Reflection on Progress. *Lonaka Bulletin,* (pp.5-12). Centre for Academic Development, University of Botswana, Special Edition on educational . *Technology (Elmsford, N.Y.)*, (March): 2003.

Uys, P. M. (2003b). E-learning at the University of Botswana: A Bi-Annual Reflection on Progress. *Lonaka Bulletin,*(pp. 5-12). Centre for Academic Development, University of Botswana, Special Edition on educational Technology.

Vonderwell, S. (2003). An examination of asynchronous communication experiences and perspectives of students in an online course: A case study. *The Internet and Higher Education*, *6*, 77–90. doi:10.1016/S1096-7516(02)00164-1

Wang, Y. S. (2003). Assessment of learner satisfaction with asynchronous electronic learning system. *Information & Management*, *41*, 75–86. doi:10.1016/S0378-7206(03)00028-4

Wang, Y.S; Wang, H.Y & Shee, D.Y.(2007). *Measuring e-learning systems success in an organizational context: Scale development and validation.*

Westbrook, R. A., & Oliver, P. R. (1991). The dimensionality of consumption emotion patterns and consumer satisfaction. *The Journal of Consumer Research*, *18*(1), 84–91. doi:10.1086/209243

Xiao, Li., & Dasgupta, S. (2002). Measurement of user satisfaction with web-based information systems: An empirical study. *Eight Americas Conference on Information Systems* (pp. 1149-1155).

Zacharias, P., Vasslopoulou, K., & Poulymenakou, A. (2002). *Designing online learning courses: Implications for usability*. Retrieved March 15, 2008: http://wwwjapit.org/zacharias_eta102.pdf

APPENDIX

E-Learning Satisfaction Questionnaire

Perceived Usefulness

1. I use e-learning system to communicate information with colleagues and tutors.
2. I use e-learning system to share my general and specific knowledge.
3. E-learning system is useful.
4. E-learning system does everything I would expect it to do.

Perceived Ease of Use

5. E-learning system is user friendly.
6. E-learning system is easy to use
7. E-learning system is efficient
8. E-learning system use is effortless.

Perceived Service Quality

9. E-learning system provides a proper level of online assistance and explanation
10. E-learning support staff are always available for consultant.
11. E-learning support staff provides satisfactory support to users using the e-learning system.
12. E-learning support staff responds in a cooperative manner to suggestion for future enhancement.

Perceived Content Quality

13. E-learning system provides up-to-date content
14. E-learning provides content that exactly fits my needs
15. E-learning system provides useful content.
16. E-learning system provides sufficient content.

Perceived Teaching Effectiveness

17. E-learning system enhances good organization of subject matter.
18. E-learning system brings about effective communication between learners and tutors
19. E-learning system raises enthusiasm for the subject matter and teaching.
20. E-learning system makes teaching and learning flexible.

Overall, how satisfied are you with your use of e-learning system at the University?

1. Adequately satisfied
2. Satisfied
3. Moderately satisfied
4. Less satisfied
5. Dissatisfied

Challenges of E-Learning

Use the response format: Most Often, Often, Rarely.

How often do you faced with the under-listed challenges when using e-learning system?

1. Log on problem
2. Loss/forgotten password
3. Network/server failure
4. Access problem
5. Long download time for large adobe and PPT files.

Chapter 12
A Case Study Analysis of the Use of Online vs. Proctored Final Exams in Online Classes

Stuart S. Gold
Walden University, USA

EXECUTIVE SUMMARY

This case study examines the results of an effort by a large regionally accredited institution to assure the integrity of its online final examination process. The question of whether the student outcomes achieved when administering an entirely online final exam are comparable to the outcomes achieved when administering proctored final exams for online (elearning) university classes is the primary focus of this study. The results of an analysis of over 100 online courses and 1800 students indicate that it is possible to establish processes and procedures that allow the results achieved by students on their final exam to be comparable irrespective of whether the final exam is proctored or is a fully online examination.

BACKGROUND

The online learning program evaluated in this study is a regionally accredited, university offering a range of undergraduate and graduate degree programs to students in both online and face-to-face formats. The university offers programs in business, management, and technology specifically directed toward working adult professionals. Online courses from each of these programs were included in the study.

This study examined the question of whether the student outcomes achieved when administering an entirely online final exam are comparable to the outcomes achieved when administering proctored final exams for online (elearning) university classes.

A secondary purpose is as stated by James, McInnis and Devlin (2002) the question is whether on-line assessment is having an influence on the quality of learning. This study directly addressed the issue of the need for online universities to employ processes that will scale to allow for effective management of large numbers of online course takers.

This question directly addressed the issues of 1) the ongoing and dynamic growth of online university offerings and 2) the need for online universities to employ processes that will scale to allow for ef-

DOI: 10.4018/978-1-60566-942-7.ch012

fective management of large numbers of online course takers.

The study was a large-scale study that incorporated data from 100 online elearning courses and over 1800 students across the full range of undergraduate course offering at the institution studied. Through an analysis of archival course records and student final exam grades, the researchers were able to conduct statistical analyses of the data for a sample of 50 courses in each group (online and proctored final exams), a total of 100 courses and in excess of 1800 students.

SETTING THE STAGE

Enrollment in online courses has been growing at an extremely fast rate for the past several years and is projected to continue this growth for the foreseeable future. One of the issues this dynamic growth has created is the scalability of the internal management processes and systems within the university. Processes that were time-tested and worked well for a few hundred students tend to encounter problems when the student population increases to thousands and tens of thousands of course takers per term. The university studied is a classic example of such rapid growth.

As working business professionals participate in distance learning at increasingly higher rates, it is important to identify specific instructional technology that can scale readily to support this increasing population of course takers and provide positive outcomes for these students. Institutions of higher education are actively expanding or implementing online education programs to meet this burgeoning trend. The need to identify instructional technology that supports the increasing number of online course takers becomes increasingly important to the success of such programs.

LITERATURE REVIEW

This study focused on one of the key administrative and educational issues affecting eLearning, the scalability of the final exam process. Wellman and Marcinkiewicz (2004) state that "as educators adopt online instructional techniques, one of the challenges they face is assessing learner mastery of course content." James, McInnis and Devlin (2002) stress that if lower-order learning becomes the result of online assessment, then the gains made in efficiency, staffing and cost savings may be offset by a drop in the quality of the outcomes achieved.

The final exam process design assured that final exams were administered consistently and included comprehensive coverage of the entire course. All final exams included a wide range of questions, covering both lower level and higher-level cognitive skills as defined in Bloom's Taxonomy (Bloom et al, 1964).

Traditional assessment techniques are costly and time consuming efforts, which an online course management system should be able to alleviate, if the results of the online process can be trusted (Rowe, 2004). The issue of trust has been a significant factor in slowing the implementation of online testing. While plagiarism has been the a focus of many online programs there has been much less attention paid to other problems related to the issue of dishonesty in online assessment (Rowe, 2004).

It is important to remember that cheating on final exams is far from a new phenomenon and certainly not a situation, which is unique to the case of an online examination. Bushweller (1999) cites statistics stating that 70% of American high school seniors admit to having cheated on an in-class exam. Further 95% of the students who did admit to having cheated said, they were never caught. Numerous other authors support the perspective that cheating on exams is not a phenomenon unique to the online environment, including Cizek (1999) who makes the point that cheating increases with student age. This is a significant issue for online programs, which focus on the adult learner population.

This continues to be an intensely debated issue today. Boltuc (2008) made the point that he believed it was clear that additional anti-cheating measures would be required in online education. He went on to state that while we could also have people taking exams for others in some colleges (especially those with large sections and lots of anonymity) it would be difficult in most colleges to have a middle-aged Ph.D. in a given discipline take an exam *in lieu* of a typical undergraduate student. This is not the case online. In contrast, McCluskey (2008) took the position that you can cheat online but that you can also cheat in class. "At the beginning of each semester I never asked my students for a photo ID or fingerprints as they walked into class and answered "here!" when I called names. Should congress require an ID scanner for class? It is the same thing.... In my entire college career many years ago no one asked me to prove who I was. Why now and why online learning? The answer is that we are the future and we have always been scary. (McCluskeey 2008)"

Wellman and Marcinkiewicz (2004) state, "that there is paucity of research examining the impact of proctored versus un-proctored testing..." Quilter and Chester (2001) emphasize that few formal research studies examine the relationships between online communication technologies and teaching and learning and reaffirm that research with empirical documentation is lacking. Boltuc (2008) concluded that we should allow market forces, meaning here the choices of colleges, to regulate the use of means and that some soft regulation in terms of ends is actually needed, or at least inevitable.

CASE DESCRIPTION

This study examines the relationship between the modality used to administer a final exam and the student outcomes achieved on the final exam in online university courses. The core question is, whether the student outcomes achieved when administering an entirely online final exam are comparable to the outcomes achieved when administering proctored final exams for online university classes.

Until 2004, all final exams for online students, at the institution of interest, were delivered in a proctored setting. This required the student to obtain an approved proctor. Faculty submitted final exams to the university administration, copied manually and then a copy was sent to the proctor for each student. After the student completed the exam, the proctor returned the completed and properly validated final exam packet to the university to be copied and filed. The completed exam copies were subsequently mailed to the faculty for grading. Once the grading was complete, faculty members made copies of all exams for their own records and returned the graded final exams to the university along with an end of term grading package. In 2004, the university administration decided, that the copying, express delivery, temporary workforce requirements and inherently time consuming and error prone nature of this process required it to be changed.

In spite of the risks, the institute decided in 2004 to move to a purely online form of final exam for all classes, eliminating the proctor and the perceived safeguards that a proctor may provide in terms of academic integrity. In place of the proctor, the university took several important steps to address the integrity and quality of the final exam process.

The institution rewrote all final exams and subsequently reviewed by the academic management group and departmental Deans as appropriate. The use of question pools was encouraged to randomize the questions and the order in which questions appeared to the students. To reduce the possibility of students engaging in various forms of online activity, which could degrade the integrity of the process, special software, was implemented within the course management system. This software prevented the student from using the internet or other sources while taking the exam, though the course text was available as a hard copy reference.

Additional steps taken by the university to address the broader range of concerns were as follows:

- Length of Exam = 3 hours maximum allowed, exam was automatically closed.
- Each exam had to include a number of both objective and essay questions.
- For undergraduate exams a minimum of 30% essay, and for Graduate no more than 20% objective
- Each exam was peer reviewed to ensure that final exams assessed the mastery of each course objective.

Bloom's Taxonomy was used as the faculty guide to writing questions that effectively measured students' ability to employ critical thinking skills versus memorization of information. The goal was to include questions that addressed each level of Bloom's Taxonomy (Bloom et al, 1964).

This project analyzed the results at the end of the first year of the utilization of online exams by the subject institution. It advanced the research by providing an objective comparison of two of the more commonly used modalities for administrating final exams in online university courses and by utilization of data obtained directly from the database. This combination added a new dimension to the body of knowledge in this area.

METHODOLOGY

The study employed a non-experimental quantitative methodology utilizing archival data stored in the university course databases. We employed random sample to select the data for analysis and that data was then analyzed utilizing standard statistical techniques and concepts.

T-test for Independent Samples

The *t*-test is the most commonly used method to evaluate the differences in means between two groups. In this study, the *t*-test can be used to test for a difference in actual student raw test scores between Online Exams vs. Proctored Exams. Theoretically, the t-test can be used even if the sample sizes are very small, as long as the variables are normally distributed within each group and the variation of scores in the two groups is not reliably different. As mentioned before, the normality assumption can be evaluated by looking at the distribution of the data by performing a normality test.

The *p*-level reported with a *t*-test represents the probability of error involved in accepting our research hypothesis about the existence of a difference. Technically speaking, this is the probability of error associated with rejecting the hypothesis of no difference between the two categories of observations (corresponding to the groups) in the population when, in fact, the hypothesis is true.

Sample

The research project included a total of 100 courses and approximately 1800 students that together provide a detailed analysis of the topic. Courses were selected using random sampling techniques.

Archival data collected from the course management software databases included: (a) student final exam grades and (b) final exam delivery modality. The database recorded actual raw exam grades only and did not include any information related to end of course student surveys or student satisfaction. The use of the actual raw exam grade eliminated possible validity issues related to the issuance of a final course letter grade.

It was not possible from the available data to determine the level of student satisfaction with a course or to relate the level of student satisfaction to the method of final exam administration employed. Data analysis was performed using appropriate statistical techniques. The consistent nature of the final exam structure and process throughout the university and the wide range of disciplines and number of classes included in the study served to increase internal validity.

SUMMARY OF FINDINGS

There was not a statistically significant difference in the average final exam grade achieved for students in courses utilizing proctored final exams vs. the average final exam grade achieved for students in courses utilizing online final exams.

Table 1. Average final exam score

Modality	Average Score
Online	73.8070
Proctored	73.8838

Table 2. Statistical analysis, student's t-test

t	Sig. (2-tailed)
-.029	.977

The analysis was performed using student's t-test for comparison of the group means of the final exam raw scores. The analysis was not significant at the .01 or .05 levels. This result was verified by an Analysis of Variance test which yielded similar results.

The conclusion was that the method of final exam administration was not a significant factor in determining the average grade achieved on the final exam.

The average values for the final exam scores are shown in Table 1 The t scores, ANOVA results and related significance levels for the comparison of the group mean values are shown in Table 2. Table 3 shows the Analysis of Variance results.

The averages were surprisingly close, almost identical; clearly, there is no statistically significant difference between the two groups in this respect. However, the average does not tell the entire story. Upon further analysis, the nature of the grade distributions proved to be somewhat different.

To provide another picture of the data, box plots (Figure 1) were calculated for each sample. As can be readily observed from the box plots, the proctored final exam grades exhibited a more tightly grouped distribution in comparison to those recorded from the online final exams. The proctored exam group had fewer courses with final exam scores toward the bottom of the range. The result of this tighter grouping is that the very low scores do not fit in statistically with the rest of the proctored exam sample and are considered outliers in that group even though they would fit readily into the online exam sample. (Trochim, 2001)

As a next step, the outliers were removed from the data and the analysis was redone. When the

Table 3. Analysis of variance

	Sum of Squares	df	Mean Square	F	Sig.
Between Groups	.147	1	.147	.001	.977
Within Groups	17008.013	98	173.551		
Total	17008.160	99			

Figure 1.

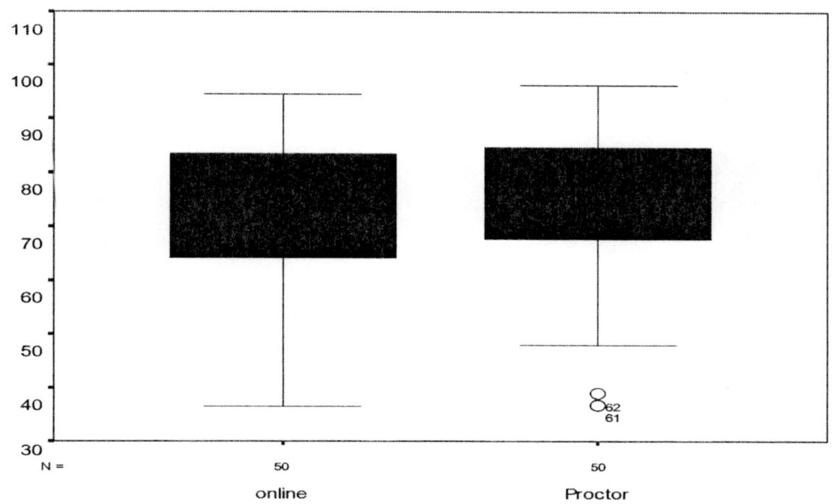

outliers were removed from the proctored final exam group that changed the sample average from 73.8838 to 75.3842. This revised average value was still not statistically different from the mean value for the online sample. The conclusion is that the online final exams appear to be allowing for a somewhat greater degree of variation in the average class scores than was present under the prior system. This is an area which may merit further research, however that research was outside the scope of this study.

In both groups the range of final exam class average scores was extremely broad with class averages ranging from approximately 36% to 94% for online exams and from approximately 48% to 96% for proctored exams (excluding the outlier values). If we include the outliers in this analysis the data for the proctored exams ranges from approximately 37% to 96% which is much closer to the values seen for the online courses.

The results relate well to earlier research by Smith and Dillon (1999) who refer to the media/method confound, a concept stating that the technology alone does not cause the effect, rather it is the combination of the technology and the way the technology is employed that impacts student outcomes.

Implications

The research study results suggest that the university was able to construct a set of online final exams, which were generally equivalent to the prior proctored final exams in terms of student outcome achieved on the test. The intent was not to validate the existing proctored exams, or to determine their efficacy in a pure online mode. The university instead decided to focus on creating a revised set of online exams that would yield similar results.

The analysis indicates an opportunity for further research by gathering a larger sample of classes in specific disciplines or courses. This would allow for a more granular analysis incorporating course specific and instructor specific variation. Another opportunity for research is in the analysis of the distribution of grades in an online vs. proctored exam environment.

The use of packaged course management software is a relatively recent development in the history of instructional technology. Faculty and student use of software facilitated communication tools will continue to evolve over time and new software features for managing online testing will be developed,. This factor represents a potential limit on the external validity of the study to generalize these findings to different course management systems or to future upgraded releases of the course management system used in the study.

The course management system database contained data focused solely on the final exam grades and did not incorporate any information related to end of course surveys or student satisfaction. The use of archival data from the course management system limited the opportunity to study factors such as the level of student satisfaction with the course.

Similarly, the experience level of the student with the technology may have impacted the internal validity of the results. Incorporating a wide mixture of courses ranging from beginning to advanced levels served to mitigate this effect.

As a single university setting was used in the research, there was no control group against which to measure the results of the research. This may limit the external validity of the study and the generalization of findings to other institutions and other forms of course design and use. This issue was mitigated by the fact that the collected data is similar to data provided by the course management systems in use at many universities.

There may be indirect relationships supported by multiple factors impacting the final exam outcomes including student perceptions and attitudes toward the final exam modality employed. The analysis of these indirect relationships was outside the scope of the proposed study.

CURRENT CHALLENGES FACING THE ORGANIZATION

Since the time of this study, the institution has continued to evolve its online environment and offerings. They decided to move to an e-book delivery method for all course texts. That decision in turn had unforeseen consequences on the final examination process and procedures.

The software employed by the institution to prevent students from leaving an online exam until it was submitted for grading also prohibited the student from having access to any e-book online course materials during the exam. Since many of the course offerings specified that the final examination would be open book, and had both course content and final examinations that were created based upon the presumption that the student would have access to a text during the final exam, this left the university in a difficult situation. That situation required either a change in the structure of both course content and the final examinations in dozens of courses or a change in the manner in which the final examinations were administered.

The university administration decided that the faculty will amend the final examination process to allow students the ability to view an e-book as needed during the exam. It was felt, that this approach would have minimal impact on the integrity of the final exam process while avoiding a lengthy and costly course redevelopment process, however, the decision, and its ramifications, if any, on the integrity of the final examinations remains to be determined. That determination requires an analysis, which scholars will need to do at a future date once there, is sufficient data available for a valid statistical analysis.

REFERENCES

Boltuc, P. (2008, July22). Chronicle article focuses on distance-ed academic integrity. *Message posted to Sloan-C Electronic mailing list.*

Bushweller, K. (1999, April). Generation of cheaters. *The American School Board Journal, 186*(4), 24–32.

Cizek, G. J. (1999). *Cheating on tests: how to do it, detect it, andprevent it.* Mahwah, NJ: Lawrence Erlbaum.

James, R., McInnis, C., & Devlin, M. (2002). *On-line assessment.* Retrieved April 25, 2005, from http://www.adelaide.edu.au/myuni/instruct/planning/assessment.html

McCluskey, F. (2008, July 22). *Chronicle article focuses on distance-ed academic integrity.* Message posted to Sloan-C Electronic mailing list.

Quilter, S. M., & Chester, C. (2001). *The relationship between web-based conferencing and instructional outcomes.* Retrieved July 31, 2003, from Wilson SelectPlus database.

Rowe, N. C. (2004, Summer). Cheating in online student assessment: beyond plagiarism. *Online Journal of Distance Learning Administration, 7.* Retrieved April, 15, 2005, from http://www.westga.edu/~distance/ojdla/summer04/olt72.html

Smith, P., & Dillon, C. (1999). Comparing distance learning and classroom learning: conceptual considerations. *American Journal of Distance Education, 13*(2), 6–23. http://www.ajde.com/contents/vol13_2.htm Retrieved February 23, 2003. doi:10.1080/08923649909527020

Trochim, W. M. (2001). *The research methods knowledge base* (2nd ed.). Cincinnati, OH: Atomic Dog Publishing.

Wellman, G. S., & Marcinkiewicz, H. (2004, December). Online learning and time-on-task: impact of proctored vs. un-proctored testing. *Journal of Asynchronous Learning networks, 8.* Retrieved April 26, 2005, from http://www.sloan-c.org/publications/jalnv8n

Chapter 13
Sharing Insights:
Teachers' Problems and Accomplishments in their Online Day-to-Day Teaching

Carmen Pérez-Fragoso
Universidad Autónoma de Baja California, México

EXECUTIVE SUMMARY

The case presents an analysis of the postings of a group of online teachers from a Mexican public university as they confront the challenges and rewards of their day-to-day teaching activities. They commented on their problems and accomplishments in a discussion forum during one semester. The problems included academic-administrative issues, difficulties of students in the appropriation of the platforms and the self-regulation of their learning, time management, negotiation and penalization of tasks delayed and other pedagogical concerns to the lack of institutional support. The findings suggest that the problems that online teachers face share specific characteristics and, according to the teachers, are mostly due to the pedagogical relationship being technologically mediated. Through the analysis, the author hopes to illustrate the complex technological, organizational and cultural issues that accompany online teaching and learning, and how the institution and the individual teachers dealt with them.

BACKGROUND

Online teaching and learning, especially in developing countries, creates the opportunity to reach more students who otherwise might not be able to access higher education. However, the requirements for successful online teaching and learning involve complex technological, organizational and cultural issues that are sometimes difficult to address. The case presented in this chapter deals with the day-to-day online activities of a group of teachers from a Mexican public university. The teachers discussed their daily experiences related to their online teaching during one semester, using a discussion forum. The analyses were carried out in three phases, following the institution's academic calendar. The teachers were teaching in six different undergraduate programs, using three different platforms. The results suggested that the problems encountered by the teachers were

DOI: 10.4018/978-1-60566-942-7.ch013

surprisingly similar regardless of the discipline of the course being taught, and despite the wide range of disciplines: from engineering to sociology and they had common problems from phase to phase. The teachers discussed their concerns and accomplishments, evidencing a high level of commitment towards online learning throughout the semester. Unfortunately, even five years after the study period, institutional and administrative problems remain.

The Universidad Autónoma de Baja California (UABC) is a Mexican state-wide public university founded in 1957 (Piñera, 2006). It is a multi-campus university that performs teaching and research activities at all levels (technical, bachelor, master and doctorate degrees), and it hosted 24,408 students working towards 65 degrees in 2004. At that time, the university had 976 full-time teachers, 97 half-time teachers and 2,917 lecturers according to the UABC Commission of Planning (UABC, 2003). The university is managed with a president appointed for a four-year term by a governing body.

The UABC has been recognized at the national level for its efforts in continually improving its services. The need for Mexican institutions of higher education to respond to the globalization pressures began in earnest in 1993 after the North American Free Trade Agreement was signed (Aboites, 1999). The federal government established norms to certify the quality of the higher education programs following recommendations posed by the United Nations Educational, Scientific and Cultural Organization (UNESCO) in relation to all higher education teaching personnel (UNESCO, 1997). Recently, the International Organization for Standardization in Mexico certified that the university was offering high quality managerial and academic laboratory services (ISO 9001: 2000, Gaceta UABC, 2007, p. 3). In addition, two national accreditation bodies have positively evaluated the study programs, including the ones that participated in this case (UABC, 2008, p.89): The Committee for the Accreditation of Higher Education (COPAES for its initials in Spanish, COPAES, 2009) and the Inter-institutional Committees for the Evaluation of Higher Education (CIEES, for its initials in Spanish, CIEES, 2005).

SETTING THE STAGE

The World Conference on Higher Education (UNESCO, 1998) was instrumental in setting the pace of development of higher education at the global, regional and national levels. The document emphasizes the necessity to offer continuous access to education to all members of society "by creating new learning environments ranging from distance education facilities to complete virtual higher education institutions and systems, capable of bridging distances and developing high-quality systems of education, thus serving social and economic advancement and democratization as well as other priorities of society" (UNESCO, 2002, p. 88). In Mexico, the National Association of Universities and Institutions of Higher Education, ANUIES for its initials in Spanish, endorses online teaching and learning in higher education (ANUIES, 1999). Due to the increasing demand for higher education, ANUIES is developing a virtual university system to complement the present system (Ruiz, 1997; ANUIES, 2002). These international and national policies have begun to affect state universities and are currently influencing budgets and universities' priorities (Rodríguez, 2002).

The UABC institutional policy regarding online courses is not consistent. On the one hand, through the official documents and policies, the UABC promotes online teaching and motivates teachers to deliver their courses entirely or partially online (UABC, 2000). The strategic planning document for 1998-2002 included a new Program of Distance Education which indicated that the institution considered among its priorities the use of ICT in regular courses. Management actions geared to promote the use of Information and Communication Technologies (ICT) in the classroom were: the licensing of a commercial platform (Virtual-U™, Harasim, Clavert, & Groenboer, 1996); the incorporation of the BSCW (Basic Support for Cooperative Work, Bentley *et al*, 1997); the investment in the development of its own platform (UABC Virtual, Luna *et al*., 2002); the teaching of diverse courses on software geared to the preparation of Web pages; the creation of contests for online educational resources and online courses with software and hardware prizes. In addition, the institution awards teachers with merit pay by including "innovation in teaching" as a category for consideration. Altogether, the institutional support seems strong.

On the other hand, the decision to prepare and deliver an online course is entirely personal on the part of the teacher. This fact has implications. Although the merit pay system rewards "innovation in teaching", the commitment of the institution to support teachers' innovation is not specified, and varies according to the policies of the teachers' own school or faculty. Consequently, the institutional policies, as far as the use of the technologies in the classroom are concerned, are not homogeneous. ICT use is dependent upon the directors of the schools and faculties, and they may or may not permit online coursework. Nevertheless, the university has some enthusiastic and innovative teachers, and they are prepared to experiment and overcome the obstacles that result from this situation. It is thanks to them that the UABC initiated its first online courses. The results were encouraging and motivated other teachers and administrators, generating an institutional interest to support this modality.

These first online courses were offered on an experimental basis in 1997, basically using email and web pages. At that time, the university offered Mathematics I for undergraduates, an elective course for graduate students in Education, and two teacher-training workshops on how to use email and online activities for teachers from all campuses. From 1998, with the simultaneous licensing of the commercially developed Virtual U and the development of UABCVirtual platform, online courses incremented.

Typically each course follows a prescribed program that is developed to be taught face-to-face. Often the same teacher who developed that program then adjusts it to work online, so the online teachers have taught the face-to-face course prior to teaching it online. In the case that follows, only one teacher had not developed the program and did not choose the platform herself. The teachers present the converted program to the faculty director and their academic committee, and once it is accepted, they upload their materials in their preferred platform. The teachers follow their online program according to the university dates to initiate, evaluate, and finish the courses.

CASE DESCRIPTION

At this university, different disciplines provide online courses, both compulsory and elective, and they are delivered through different platforms to undergraduate students. As researchers, we wanted to see what, if any, problems the teachers faced when teaching online. Therefore, the teachers were asked to contribute to a discussion forum named "Online Teaching" during the semester from February to June 2003 using the platform UABCVirtual. This period is pertinent for study because it is situated at a time when only few teachers had experience with online teaching, and their example and mentorship proved important for the less experienced teachers. All the online teachers were voluntarily teaching in this modality, and subsequently online teaching has become compulsory in a number of instances within the university. Finally, many of the problems discussed by the teachers during this study period continue to be unresolved.

This chapter presents the analysis of the postings in that discussion forum of a group of nine online teachers. The teachers were teaching with three different platforms: five used Virtual U™; two used UABCVirtual (our university platform); and two used their own webpage and BSCW for handling homework and discussion forums. The number of students per course ranged from 5 to 39. The courses were from six different bachelor programs: engineering, sociology, computer science, educational sciences, business administration, and informatics.

The purpose of the teachers' online forum was to discuss their experience while teaching their online courses, specifically: 1. Accomplishments of the week; 2. Problems encountered; and 3. Noteworthy matters at the individual level. The analyses were carried out in three phases, based on the academic calendar: the initial phase covered from the beginning of the courses until the students' deadline to drop the course without academic penalty (4 weeks); the second phase covered the continuation of the course until the Easter holidays (6 weeks); and the last phase covered the consolidation and closure of the course (6 weeks). As a verification procedure (Lincoln & Guba, 1985), at the end of each phase, another forum was opened; it was called "Discussion of the Discussion", where the teachers read, gave their opinions and showed their agreement with the interpretations of the researcher. As the researcher[1], I participated only in this second forum; if I had any doubts regarding their postings, I asked them in this forum at the end of the phase under study. The participation in both forums was voluntary and therefore was not homogenous: the teachers made 75 comments in the first phase, 63 in the second phase, and only 17 in the last phase. All comments were analyzed using content analysis (Weber, 1990; Krippendorff, 2004).

The teachers' postings were grouped thematically. Two main categories emerged: Factual information (reference to specific facts or events), and Personal perceptions (opinions or judgments evidencing conclusions from actual or past experiences). The factual information category was subdivided into: Problems (difficulties encountered during the course processes), and Actions (the ways they chose to treat the problems and activities related to teaching and communication with their students). From the category of Personal Perceptions, two subcategories emerged: On their students' performance (personal inferences regarding students' participation on the course), and On their own teaching performance (personal reflections on their own participation and experience as teachers). See Figure 1.

Figure 1 shows the number of commentaries for each category. It is worth noting that the fewest commentaries addressed problems related to the students' use of the platforms, indicating that the technologies used soon were transparent to the students. Most of the discussion centered on problems of how to prepare pedagogically sound online activities, and how to keep the students engaged throughout the course. Thus the teachers' primary concerns focused on how to transfer their knowledge of good pedagogical practice to this new modality. It is interesting to note that the teachers reported differences in the students' reactions to online group tasks depending on which semester they were enrolled in, with the more advanced students showing fewer problems for working in teams. Surprisingly, the teachers faced similar problems regardless of the discipline of the course being taught, and despite the wide range of disciplines: from engineering to sociology. The problems differed from phase to phase throughout the semester, but again, the teachers' experiences were largely uniform as the phases progressed. In general, the teachers discussed both their failed and successful practices when dealing with problems in each phase of the course, emphasizing their successful practices for the benefit of the novice teachers. Such consistency of findings allows for well-grounded recommendations in addressing the current challenges facing an institution such as this public university in its efforts to effectively deliver online teaching.

Figure 1. Categories and subcategories of analyses of the discussion forum and showing respective frequency of teacher commentaries

Categories												
	Factual information						**Personal perceptions**					
	Problems			**Actions**			On their students performance			On their own teaching performance		
Subcategories	**Administrative**	14		**Administrative**	9		**Previous performance**	8		**Differential participation**	6	
	-Lack of information		4	-Use of diverse media to update themselves		9	-Participation		4	- Related to face-to-face teaching		6
	-Registry system		6				-Study habits		2			
	-Group unknown		4				-Marks		2			
				Pedagogical	49							
	Pedagogical	29										
	-Self-regulated learning		4	-Alternative ways of making tasks		3	**Time related**	3		**Online experience**	9	
										-Platforms		3
	-Reading and writing		6	-Supports for advancing tasks		5	**Autonomy**	3		-Teaching		4
										-As a student		2
	-Following instructions		4	-Improving instructions		8				**Satisfactions**	9	
							Expectations	4				
	-Generation of peer interactions		3	-Group work		5				**Frustrations**	5	
	-Individualization		1	- Design of integrative activities		2	**Satisfactions**	9				
	-Evaluation		3	-Optional activities		2				**Group feelings**	4	
	-Homework's delays		4	-Gradual rigor		6						
			4	-Socialization of group achievements		3						
	-Platform appropriation	6		-Continuous support		4						
	Personal											
	-Organization		2	- Stimulus & sanctions		7						
	-Coordination		1									
	-Errors committed		3	- Rules enforcement		4						
	TOTALS	49			58			27			33	

Phase 1: Teachers' Discussion During the First Four Weeks

The first phase of the online courses taught by these teachers covered the first four weeks, and their commentaries were centered on two themes: the academic-administrative problems they faced and the students' difficulties in using the platforms and self-regulating their learning. Overall, teachers resented the lack of administrative support regarding information on online courses, which caused them many difficulties in starting their courses on time. According to the teachers, in order to be able to begin their online courses on time, they require special institutional support in two related areas: dissemination of the course's basic information (title, teacher, teachers' email address, course's electronic address), and recognition of the teachers' efforts and "generation of the culture of online courses [T2][2]". Teachers, in general, thought that their efforts regarding the activities to teach online are undervalued. For example, one teacher said: "[T4] I think that there has been a great effort on the part of the teachers that have accepted the challenge to change the modality of the course. Although this could be a theme for a thesis, I dare to say that it is easier to create from scratch an online course than adapt an existing one designed for face-to-face teaching". Another teacher felt that "[T2] apparently we are not taken very seriously yet".

The teachers discussed the limitations of the registry system and the problems it caused them. At the UABC, students have the opportunity to enroll for the courses during the first two weeks of the semester, in other words, after courses have begun. For online teachers, this creates problems with the students' course list. The students that enrolled once classes had begun were not in the original lists, and they "don't know their classmates or the electronic address of their teachers or their courses [T4]". The teacher does not write to these students because the teacher does not know about them. As one teacher put it: "we do not know that there is a student in search of his or her group [T3]". Another related problem concerns the lack of knowledge regarding the actual number of students enrolled and the maximum number of students accepted. As one teacher explained: "For my fortune or my disgrace, the group enrolled was very numerous and was divided into two after three weeks: a face-to-face group and another one online…, and some students that were no longer online continued to be registered in the platform and vice versa [T5]." The teacher could not update her group participants because the central administration had to do it.

Regardless of the mode of teaching, whether online or face-to-face, in order to open the courses the university requires that they have a time schedule and a classroom assigned as well as a minimum number of students enrolled. The registry system was designed to control the allocation of time and space in order to avoid overlaps in the face-to-face courses, thus schedules have to refer to working hours during the week or Saturdays and courses cannot be delivered on Sundays. This causes problems to online participants since rigid scheduling is irrelevant. To resolve this, one experienced teacher told the others she had learned "to register my online courses on Saturdays, only for administrative purposes [T3]"

To compensate for the administrative problems, most of the teachers, especially the more experienced ones, programmed the academic activities of their courses after the second week, and dedicated the first week to establishing contact with the students and solving logistical problems. They also programmed training sessions on the use of the platforms, some including face-to-face sessions with non-compulsory attendance. Some teachers established practice sessions with the most used platform functions (how to send assignments, post comments, etcetera). One teacher established Instant Messenger schedules to give support to the students. Another asked the students to open an alternative email account (besides the university one) in order to assure communication.

The comments about the students' problems focused on two topics. The teachers remarked that students had insufficient skills to use the platforms effectively. In addition, the teachers commented on students' sudden confrontation with the new exigencies of online learning, not only in relation to the self-regulation of their learning, but also in relation to reading and writing.

The teachers found considerable differences between the students who had not taken online courses and those that already had experience. They learned to use the platform gradually, according to the teachers, and they required different supports for different lengths of time based on their previous experience, number of semesters at university, study area, and tools used. As one teacher explained to the rest: "I think it is important to keep their experience in perspective, and not to forget what it means to be there for the first time [T3]". In general, the teachers continued to offer technical support well beyond the period of the initial training sessions: "I have opted for individual follow-ups and continuous technical support [T2]", said one, and "I give personal tutorials [T7]" said another. Another teacher explained the students' lack of attendance to his non-compulsory training sessions saying, "The majority of my students have visited my web page from previous courses or they know other classmates that have taken the course. Besides, most of them come from the computing area and they are like fish in water [T4]". These commentaries demonstrate the differences in the students' attitudes and technical skills, depending on the disciplinary area and semester of study.

The problems of technology use were relatively easy to solve, but the problems related to the students self-regulation of learning, as well as those related to online reading and writing required more diversified educational strategies. Students' reactions to the new demands of online learning were varied. One teacher explained: "In this second week, the students resent the change from a face-to-face to an online course; they miss the teacher's traditional role and, above all, they suddenly feel the great responsibility of taking charge of their own learning [T4]". Another teacher echoed that perspective, saying that "the [students'] great enthusiasm with which they entered now turns and shows them a reality of much more self-regulated [personal] work [T5]".

Those kinds of comments are evidence that the students' preparation for online courses must not be reduced to the technical aspects, as these courses demand communicational and self-managerial skills that are not necessarily developed in face-to-face courses. In order to promote self-regulation of learning, teachers used different teaching strategies. For example, one teacher gave different options to do the task, and presented students with different materials, handbooks and tutorials. Other teachers encouraged experienced students to submit tasks before the due date as a strategy to motivate the less experienced ones to try to accelerate their pace as well. Another teacher detailed the strategy that worked best for her course after three years: "I present them with the course purpose, skills to develop, methodology, evaluation, and bibliography when I frame the course in our face-to-face initial session. …I include an explanatory guide with the components and requirements of each task, and I add completed examples. [Among their tasks] the students have to present an integrating project that they have to develop in stages. This project is done in teams, and that also helps as they support each other [T7]".

The demands of the students' day-to-day learning activities evidenced reading and writing problems, particularly among the novice online students. One of the teachers' complaints was that students "do not read their peers' contributions; they tend to only send their own opinion [T1]". In online courses, to read --and to make sense of-- the participants' contributions is an essential part of their learning work, but they are not accustomed to this. Since this activity is not very common in their conventional classes, the students require structured support specifically to discover and develop their own communicative potential.

In order to support the development of the necessary reading and writing skills, teachers modeled the responses and began their feedback with a positive comment in order to stimulate the students' confidence to continue posting. One teacher discussed the importance of "being there, participating all the time as one of the group [T8]", and giving emotional support throughout the course. In order to make sure the instructions of the activities were clear, teachers wrote them differently when using the discussion forums or the chat. One teacher had a student as her monitor and she verified with her the way she understood instructions.

The first four weeks seem to be more difficult for the participants of online course than face-to-face ones. For online teachers, they involve the resolution of a series of administrative and pedagogical problems, and for the students, they entail the sudden confrontation with new requirements that are different from those experienced in face-to-face courses. Nevertheless, the teachers' commentaries showed great commitment with their online teaching activities, and the students' dropout rate in the nine courses was only 7.5% (20 out of the 267 students' chose not to continue the course during the grace period).

Phase 2: Mid-Section of the Courses

The teachers' postings during the semester's 5th to 10th weeks showed different kinds of concerns. There were hardly any comments regarding problems with the administrative processes or with the platforms, except for two teachers who received their final students' enrollment list after a great delay. During this second phase, the commentaries revolved around changes in the pace of the development and delivery of activities and showed an array of teaching strategies that the teachers used to negotiate (and penalize) delays. The teachers perceived the appropriation of the tools by the students as a gradual process that required continuous reinforcement, and the teachers began to be more demanding (and strict) in their requirements. For example, in order to grade the projects and send feedback to the teams, one teacher rejected all other means of submitting completed projects and he sent feedback to the teams only through the platform. This way, he said, the projects were available for all other teams to review. Another teacher offered two final opportunities to try the platforms' tools: "I sent them to the technician so they can see if they are doing things incorrectly, and I inform them that the next assignment will only be reviewed and marked on the platform. Generally, the problems are solved, but if there is recidivism, I lessen the opportunities… It must be very clear how the work is done on this modality and the use of the supports [T6]".

However, the majority of the commentaries in this stage referred to the difficulties of generating a more significant interaction among the students themselves and with the contents of the courses, on the one hand, and on the other, the need for posting information related to individual and collective accomplishments. Some teachers also perceived that the students' difficulties were related to the type of activity. One teacher pointed out: "What I want is that they give their opinion, but because of the diagrams involved [in the course material], they are finding it difficult to express themselves [T6]". Other teacher noticed a change of pace when the students approached certain tasks: "I have noticed that some students, the same way as in previous semesters, relatively abandon the activities when they have to begin the tasks related to the development of the major project [T4]".

The teachers perceived that the students had difficulties regarding their work in groups and yet they also found it to be rewarding. Apparently this practice of group work is less common in traditional face-to-face courses. "Only a third of the students has managed to interact to the extent of working in groups [T8]" said one teacher. In the online courses, the students "begin with a new dynamic to work on these projects. The students like the experience of being able to share their work in progress with other teams, an experience, they said, which is not practiced frequently in face-to-face courses [T1]".

Regarding teamwork, the number of team activities included in the courses varied considerably. In one course 90% of activities had to be done in teams. This particular teacher assigned a person in charge of each team to encourage participation. Those leaders then assigned the responsibility to initiate the activity to a different person each time and the rest were to support and work from his/her initial ideas to encourage participation. After the group reached a consensus, the leader posted their commentary in the general forum. The teacher wrote: "This way, I promote everybody's contributions and critiques to the central ideas [T1]".

Generally, the teachers perceived that the students from the advanced semesters were better at working in teams. For example, a teacher of freshmen students wrote: "I have tried to put them in virtual teams so they can review their writings in order to develop their critical reading skills… The confusions began… [They asked] if they have to make specific time appointments … [and] that they prefer to work individually, etcetera [T8]". In contrast, the teacher with an advanced semester course said: "They enjoy working in teams... They are so well integrated by the last semesters that they get together any way...; It may help that I allow them to form the groups the way they want. They decide who and how many will work in their teams, so I have teams ranging from 2 to 8 persons [T7]". In addition, the teachers perceived more self-regulated learning from advanced semester students: "These guys are in their seventh semester, and I believe that they have realized that one of the skills that they will use as professionals is self-teaching. They have participated and contributed with materials that they investigated, and two of them initiated criticizing constructively the others' contributions, which has given a sort of climate of competition, and everybody has begun to evaluate the aspects contributed by the rest of the group [T1]".

Another theme that emerged concerning interacting with students was the need to make information visible regarding individual and group achievements. One teacher commented: "In order that students realize their performance in relation to the performance of the group, I published the averages of everybody. I believe that this is a necessary measure in order to motivate the students that are working hard, and to hurry along the students that are behind [T9]".

Publishing information on the group's performance should include, according to the teachers, information related to the "class attendance". For an online course, attendance is translated as the continuous participation required to "be present" in the courses. One teacher reasoned, "In face-to-face courses, we generally take into account the attendance in order to give or withdraw the right to take the exams; if a student missed classes constantly, he or she loses the right to be evaluated. In my case, I take into account the students' accesses to the platform's tools, and obviously their participation, in order to give them the right to be evaluated [T1]".

In summary, the analyses of the commentaries in this phase showed that there was an intense pedagogical dynamic, showing a diversity of strategies used by the teachers to assure that the learning activities were accomplished within the planned dates. Given that one of the attractive features of online courses is indeed the possibility of completing the activities at one's own pace, the negotiations with (and sanctions against) the students because of delays in the delivery of assignments are almost continuous in this period. Also, the less experienced teachers raised the topics of how to determine the adequate length of time to do the activities and how to integrate the contents learnt. Interestingly, there was one teacher that had a motivated, hard working group: "They continuously ask me to review their exercises. I cannot keep up with their pace [T3]"

Phase 3: Consolidation and Closure of Courses

The third phase of the study covered the final six weeks of the courses. This was an intense period of activity for both students and teachers. The teachers discussed the motivation teaching strategies used to engage the students in the final projects including designing optional activities. There were also many comments regarding their own satisfaction and they shared comments from their students both positive and negative.

The teachers discussed their different pedagogical techniques to motivate their students. For example, they personalized the learning activities: "I reviewed and marked the optional activities done by the students, awarding them extra points. I believe that with this tactic the students felt more freedom within the activities of the course because they could now leave out doing one compulsory activity that is not compatible with the characteristics of their projects, or simply that they cannot or don't want to do [T2]".

Nevertheless, the most numerous commentaries referred to the students' delays in the delivery of assignments. One teacher remarked, "Six students struggled a lot this week in order to update their assignments… On the one hand the course represents an advantage to them since the schedule is flexible, but on the other hand, the review work gets accumulated for me [T2]".

Some experienced teachers are less flexible in accepting late assignments. For example, one teacher explained his deadlines saying: "I do not give more time than what was initially allocated. All the activities of the course are already programmed, and the students must manage their time. Because I have delivered this course several times, I have calculated well the time to prepare and hand in the assignments... In addition, the students know that there is a mailbox entitled 'pending assignments' where they can deposit one of their missing ones, but receive a lower mark. Altogether there are 7 activities. In order to pass the course they must have at least a mark higher than 60% and 4 activities done, otherwise they have to repeat the course [T9]".

Thus, the teachers demonstrated differences regarding the rigor and specification of the rules implemented in the online courses. "Of course I penalize them with a lower mark for sending their assignments late [T2]". Another teacher said that some students that "have not given signs of life except to complain because they cannot enter the site, cannot find the readings, cannot unzip the files (ha!) now feel that they are up to their necks in work and are sending desperate messages seeking 'special treatment'… For them, there will be few opportunities, because I think that, even online, we must help them, but not devalue the students that have met the requirements in deadlines and forms [T8]".

In this last phase, the teachers' commentaries on their personal satisfaction regarding the work of their students are more numerous than in the previous ones. For example: "23 of 35 students have jumped the barrier of passing from theory into practice; finally, the problem that had been present during a month was solved [T2]", or "The concept maps done by some students have pleased me a lot [T8]".

Teachers also commented on their own errors. For example, one teacher explained: "I uploaded lots of good exercises and assignments, but the problem was that I did not set them up suitably, and I did not prioritize which were basic (mandatory) and which were complementary (optional). Problem: some students were scared, others were paralyzed, and others were lost [T5]". Another said, "The first time that I prepared an online course I overdid the number of links for them to review (and show them how up-to-date my course was). I wanted lots of work from the students, and they had to put lots of effort into it, and so did I in reviewing all their work. When I delivered the course the second time, I reduced the workload [T7]".

Lastly, but equally important, the teachers commented on their solidarity and collective identity of being online teachers throughout the phases. Typical commentaries of this phase were: "The participation in this kind of forum has been very important to me; belonging to a group of teachers who share their opinions is exciting, and this is more so because the forum refers to online teaching [T2]". Such expressions of solidarity and satisfaction with online teaching indicate the value of this type of discussion for supporting online teachers, even if such a finding was not anticipated in this study.

Summarizing the results, the comments posted by the teachers suggest that the problems they faced at each stage of the semester, the same as when teaching face-to-face, are different. However, the differences are accentuated for online teaching and acquire specific characteristics, perceived by the teachers as mostly due to the pedagogical relationship being technologically mediated. One teacher even commented on the difficulties to locate the origin of the problems: "[T6] When [these problems] show up, one wonders if it is because of the teacher, or the learning materials, or the platform […]. If it were a face-to-face class, it would be much more clear". When discussing the problems faced by the students, the teachers emphasized the need to develop their students' academic skills without assuming that the students were already autonomous learners. The teachers sought ways of offering ongoing and varied support in the three phases of the courses. They shared their doubts and suggestions, as well as their positive and negative experiences. Regardless of their previous experience teaching online, all the teachers made comments that show that they enjoy teaching online and derive great satisfaction from using this modality.

CURRENT CHALLENGES

The teachers' participation in this forum, where they discussed their successes and challenges while simultaneously teaching online courses, was rich and helpful for them. Some of the teachers commented that being a part of an online community made them proud. Unfortunately, the university institution does not recognize their efforts as it should. In this respect, the results agree with Salmon (2006) when she states that it is important to recognize and reward individual academics for their successful teaching and celebrate teachers who have made beneficial changes. For the teachers that participated in the forum, the personal satisfaction derived from online teaching was their reward, but if the university plans to offer online degrees, one challenge definitely is to find ways to motivate and prize online teachers.

From the teachers' perspectives, there is a consensus on the challenges presented by online teaching. From the pedagogical point of view, they are conscious that the students need to be active and engaged in their learning processes (Pallof & Pratt, 1999, 2001). Consequently, the teachers developed strategies to keep students united and interested as well as an array of exercises to personalize their paths as much as possible. From the social point of view, the teachers discussed how they could emphasize the emotional part of learning, being conscious of the need to make an "extra effort to humanize the [virtual] environment" (Pallof and Pratt, 2001, p. 32). From the technological point of view, they agreed on the need to promote positive examples and models to support the appropriation of the technologies. As Proulx (2001, p. 143) stated: "Each individual develops a certain level of technical proficiency, so that this 'feeling of competence' allows him to move with greater or lesser flexibility within the virtual environments". Thus they viewed the development of the students' feelings of competence as a challenge requiring their continuous attention.

It should be reiterated that in the first phase of the courses the most numerous problems were administrative and they must not be overlooked. The teachers were conscious of the importance of a good beginning, and they yearned for better logistical support so that the courses could begin on time. As easy as this sounds, this has proven to be a very tough challenge. Such administrative problems apparently are similar to the ones facing higher education institutions that offer distance education programs in other countries, at least when they first offered them. As Moore (1994) describes, part of the explanation resides in the inertia that accompanies prevailing procedures in face-to-face administrative systems. According to Moore, this prevents changes due to the way traditional university education is conceived. Berge (1998) reported that half the obstacles reported by his sample of online teachers were related to the organizational culture. Procedures get institutionalized so that they are "the way things are done at this institution" (Berge & Muilenburg, 2001), and change is difficult. At present, some measures are being taken at the university to make the course registration system suitable for both online and face-to-face courses.

Another big challenge regarding online learning today is the creation of clear policies that assure the quality of the courses without interfering with teachers' individual freedom to choose tools and pedagogical models. Recently, some teachers have begun using Moodle (Dougiamas & Taylor, 2003) with success, and the university has a Blackboard licence (Blackboard, 2009). Institutional training and technological support for the online teachers is also crucial. Specifically, the university needs to continue offering teacher-training courses on the pedagogical uses of Blackboard, especially when teachers first initiate their online teaching.

In synthesis, online teaching of undergraduate courses in this public university in Mexico presents challenges for teachers and the institution. The university is working to solve the administrative and delivery problems, but many institutional challenges persist. Further, the teachers are conscious of the challenges that online teaching and learning presents to them, as individuals and as a group and they, too, continue to seek solutions. Such optimism is warranted because the institutional conditions for online teaching are very flexible, and that is a big asset. Teachers may choose to use any pedagogical technique, and even create their own. They can also look for support in their colleagues and their social relations, although those supports are entirely informal and, as one teacher explained, an organizational culture does not exist for online courses. Nevertheless, the freedom to design and deliver online courses leaves an ample margin for maneuverability, and the teachers profit from this situation to experiment with innovative practices.

REFERENCES

Aboites, H. (1999). *Viento del norte: TLC y privatización de la educación superior en México*, (2nd ed.). México, DF: UAM / Plaza y Valdés.

ANUIES. (1999). *La educación superior hacia el siglo XXI: Plan Maestro de Educación Superior Abierta y a distancia.* XIV Asamblea General Ordinaria de ANUIES. Universidad Autónoma del Estado de Hidalgo.

ANUIES. (2002). *Hacia la construcción de un sistema virtual para la educación superior*. Informe de la tercera reunión nacional de educación superior a distancia. Cozumel, Quintana Roo.

Bentley, R., Appelt, W., Busbach, U., Hinrichs, E., Kerr, D., & Sikkel, S. (1997). Basic Support for Cooperative Work. *International Journal of Human-Computer Studies*, *46*, 827–846. doi:10.1006/ijhc.1996.0108

Berge, Z. (1998). Barriers to online teaching in post-secondary institutions: Can policy changes fix it? *Online Journal of Distance Learning Administration*, 1. Retrieved from http://www.westga.edu/~distance/ojdla/summer12/berge12.html

Berge, Z., & Muilenburg, L. (2001). Obstacles faced at various stages of capability regarding distance education in institutions of higher learning. *TechTrends*, *46*, 40–45. doi:10.1007/BF02784824

Blackboard (2009). *Blackboard web site.* Retrieved from http://www.blackboard.com

CIEES. (2005). La evaluación interinstitucional de la educación superior en México. Comités Interinstitucionales para la Evaluación de la Educación Superior. Retrieved from http://www.ciees.edu.mx/ciees/publicaciones.php

COPAES. (2009). Programas educativos de licenciatura y técnico superior universitario reconocidos por su calidad. Enero 2009. Consejo para la acreditación de la educación superior, México, D.F.

Dougiamas, M., & Taylor, P. (2003). Moodle: Using Learning Communities to Create an Open Source Course Management System. In D. Lassner & C. McNaught (Eds.), *Proceedings of World Conference on Educational Multimedia, Hypermedia and Telecommunications 2003* (pp. 171-178). Chesapeake, VA: AACE.

Gaceta, U. A. B. C. (2007). Reconocen a UABC por su calidad certificada. Gaceta UABC, num. 195, 6 de octubre de 2007. Available from http://www.uabc.mx/gaceta/gacetas/g195.pdf

Harasim, L., Calvert, T., & Groeneboer, C. (1996). Virtual-U™: a web-based environment customized to support collaborative learning and knowledge building in post secondary courses, (pp. 120-127). In *Proceedings of the 1996 International Conference on Learning Sciences*, Evanston, Illinois.

Krippendorff, K. (2004). *Content Analysis: An Introduction to Its Methodology*, (2nd. Ed.). Thousand Oaks, CA: Sage.

Lincoln, Y., & Guba, E. (1985). *Naturalistic Inquiry*. Beverly Hills, CA: Sage.

Luna Serrano, J., Pérez Fragoso, C., Morita Sandoval, O., & Ruiz Rodríguez, A. (2002). Evaluación de una herramienta computacional para la educación en línea. VIII Congreso Internacional de Informática en la Educación, La Habana, Cuba.

Moore, M. (1994). Administrative barriers to adoption of distance education. *American Journal of Distance Education*, *8*, 1–4.

Pallof, R., & Pratt, K. (1999). *Building Learning Communities in Cyberspace: Effective Strategies for the Online Classroom*. San Francisco, CA: Jossey-Bass Publishers.

Pallof, R., & Pratt, K. (2001). *Lessons from the Cyberspace Classroom: The Realities of Online Teaching*. San Francisco, CA: Jossey-Bass Publishers.

Piñera Ramírez, D. (2006). *Los primeros cincuenta años de educación superior en Baja California*. México, DF: Miguel Ángel Porrua-UABC.

Proulx, S. (2001). Usages de l'Internet: la « pensée-réseaux » et l'appropriation d'une culture numérique In E. Guichard (Ed) *Comprendre les usages de l'Internet*, (pp. 139-145). Paris: Éditions Rue d'Ulm.

Rodríguez Gómez, R. (2002). Continuidad y cambio de las políticas de educación superior. *Revista Mexicana de Investigación Educativa*, *8*(14), 133–154.

Ruiz Durán, C. (1997). *El reto de la educación superior en la sociedad del conocimiento*. México, DF: ANUIES.

Salmon, G. (2006). Above the parapet for e-learning: Recognition, reward and celebration. *International Journal of Learning Technology*, *2*(4), 342–356. doi:10.1504/IJLT.2006.011339

UABC. (2000). Plan de Desarrollo Institucional 1999-2002. Mexicali, BC: Universidad Autónoma de Baja California.

UABC. (2003). Informe de Rectoría 2003. Comisión de Planeación, Universidad Autónoma de Baja California.

UABC. (2008). Informe de labores 2008. Mexicali, B.C.: Universidad Autónoma de Baja California.

UNESCO. (1997). *Recommendation concerning the Status of Higher-Education Teaching Personnel*. Date of adoption November 11, 1997. Retrieved from http://unesdoc.unesco.org/images/0011/001102/110220e.pdf#page=32

UNESCO. (1998). *World Declaration on Higher Education: Higher Education in the Twenty-First Century: Vision and Action*, Paris, Fr., 5 - 9 October 1998. Retrieved from http://www.unesco.org/education/educprog/wche/declaration_eng.htm

UNESCO. (2002). *Open and Distance Learning. Trends, Policy and Strategy Considerations*. Paris: Division of Higher Education. Retrieved from http://unesdoc.unesco.org/images/0012/001284/128463e.pdf

Weber, R. (1990). *Basic Content Analysis*, (2nd. Ed.). Newbury Park, CA: Sage.

ADDITIONAL READING

Goodyear, P., Salmon, G., Spector, M., Steeples, Ch., & Tickner, S. (2001). Competencies for online teaching. *Educational Technology Research and Development*, *49*(1), 65–72. doi:10.1007/BF02504508

Harasim, L., Hiltz, R., Teles, L., & Turoff, M. (1995). *Learning networks: A field guide to teaching and learning*. Cambridge, Ma.: The MIT Press.

Hiltz, R., & Turoff, M. (2002). What makes learning effective? *Communications of the ACM*, *45*, 56–59. doi:10.1145/505248.505273

O'Leary, R., McFarlane, A., Timmis, S., Greenwood, S., et al. (2003). E-Learning Strategy (Rep. No. 1.9). Bristol, UK: University of Bristol.

Pallof, R., & Pratt, K. (1999). *Building Learning Communities in Cyberspace: Effective Strategies for the Online Classroom*. San Francisco, Ca.: Jossey-Bass Publishers.

Pallof, R., & Pratt, K. (2001). *Lessons from the Cyberspace Classroom: The Realities of Online Teaching*. San Francisco, Ca.: Jossey-Bass Publishers.

Rosenberg, M. (2001). *E-learning. Strategies for delivering knowledge in the digital age*. New York, N.Y.: McGraw-Hill.

ENDNOTES

[1] I want to thank my colleague Karen Englander for her invaluable support and insightful comments about this case.

[2] All quotes refer to teachers' comments and they refer to the teachers by number.

Chapter 14
The Effects of E-Learning on African American Males: Three Case Studies

Tammy J. Graham
The Citadel School of Education, USA

Stephenie M. Hewett
The Citadel School of Education, USA

EXECUTIVE SUMMARY

The chapter examines the experiences of three African American males who were placed in an electronic learning (e-learning) classroom in a rural secondary school. The three case studies provide detailed descriptions of the young men's backgrounds, educational experiences, and academic achievement results before the implementation of e-learning. Furthermore, the case studies detail their academic achievement results and dispositions during the e-learning process, pitfalls of their e-learning program, and lessons learned from the implementation of the program. It is the authors' hope that educators and business professionals will utilize the information and lessons learned in this chapter when planning and implementing e-learning classes and trainings in order to enhance e-learning experiences for African American males.

INTRODUCTION

In this age of blogging, texting, electronic mailing (e-mailing), instant messaging, and computer gaming, one school in the Southeastern United States decided to pilot a version of electronic learning (e-learning) with students who are in danger of being dismissed from school. These students are repeating ninth grade for at least the second time. They have a history of numerous discipline referrals, and the e-learning program is their last hope for earning enough credits to graduate from high school.

This chapter follows the accomplishments and difficulties of three African American males participating in the pilot e-learning program.

BACKGROUND

The high school selected for these case studies has approximately 430 students and is located in the southeastern section of the United States, approxi-

DOI: 10.4018/978-1-60566-942-7.ch014

mately 20 miles from the closest city. It is a rural farming community with four elementary schools, one middle school, and one high school. The high school population is 99% African-American, with approximately 82% of the students living in poverty.

The high school is considered a low performing school with high dropout rates, high retention rates, high suspension and expulsion rates, and low test scores. However, the school has made major changes to improve its state report card rating in the past few years. In 2005, the school was rated as an at-risk school with an at-risk growth rate, indicating that the school failed to make adequate progress towards the state's 2010 performance goal. However, in 2008, the school was rated as an average school with an excellent growth rate, indicating that the school exceeded the expected level of progress toward the state's 2010 performance goal (SC Department of Education, 2009).

In order to decrease the 2.7% school district dropout rate, each high school in the county was issued access to the Apex Learning Academic Curriculum, a standards based online offering of courses to high school students. The district granted individual schools the freedom to decide how to implement the e-learning program. Because 18% of the school of study's students are overage for their grade level, the school decided that the optimum use of the Apex e-learning program would be for credit recovery. Students were enrolled in a variety of different classes, including math (Algebra I and Geometry), science (Physical Science and Biology), English (English I and II), and social studies (Global Studies I and Geography). Each course consists of units, lessons, and activities. A typical lesson has activities that include "practice, readings, journals, labs, discussions, projects, web explorations, reviews, and both computer- and teacher-scored assessments" (Apex Learning, 2009). The Apex Learning Academic Curriculum provides active learning experiences, in which the students read, watch, listen, write, and discuss to gain better understanding of the concepts presented. Different learning styles are addressed through the images, movie clips, sound clips, animations, charts, and graphs. The goal is to allow students that are overage and unsuccessful in the traditional classroom the opportunity to work through an alternative, self-paced instructional program in order to earn high school graduation credits.

Because the high school has a limited number of computers in classrooms as well as a limited number of computer labs that must be reserved in advance, a special classroom was converted and named "the virtual classroom" for the e-learning program. Student computers with internet access and head phones were placed in a classroom that is located in an annex of the school building. Additionally, a teacher's computer was added in order for the teacher to upload and reset student lessons and assessments.

The administration has been unable to hire a qualified teacher who is interested in taking responsibility for the class on a full-time basis. As a result, during the first semester, the principal appointed a different faculty member to supervise each ninety-minute class. A behavior intervention specialist directed the first ninety-minute class. She appeared to be the only one of the three teachers to have a positive relationship with all three case study students. Despite the relationships, the teacher indicated that the students did not want her help, and she was not observed to offer help.

Since most of the students have extreme difficulty with math, a retired math teacher supervised the second ninety-minute class. She has a reputation for being a strict teacher and does not put up with any misbehavior in the classroom. The three students responded to her when she spoke but typically did not ask her for help when needed. However, by mid-semester, one student started asking for her assistance, and she responded positively to his requests. As a result, this student passed his Algebra I class by the end of the first semester.

A math teacher who is relatively new to the school supervised the last ninety-minute class. He reported that some of the students were frightening to him. He wrote the majority of the behavioral referrals that the case study subjects received during the first semester. Seven of the fourteen referrals were issued due to students being late to class.

At the beginning of the second semester, the principal hired a substitute teacher to supervise each virtual class; however, after only one week, the substitute teacher moved to another classroom in order to cover for a teacher who left on maternity leave. The principal promptly hired another substitute teacher to supervise the virtual classes. The three case study students appear to work quietly for him and follow his directions but do not typically ask him for academic assistance.

SETTING THE STAGE

Eight students (7 males and 1 female) were selected to participate in the e-learning program. All eight of the students are African-American, have been in the ninth grade for two or more years, and have multiple discipline referrals. The students and their parents were notified that this alternative to the traditional classroom setting is the last option for the students to remain in high school. Without this credit recovery program, the selected students will be unable to earn enough high school credits to graduate from high school by the time they reach their twenty-first birthday. Additionally, summer school will not be offered due to the lack of funding.

This lack of funding has created many of the academic difficulties for these students. For example, 39% of the teachers have provisional certificates, as noted on the 2008 school report card. Furthermore, the school has limited technological resources. Additionally, 14% of the students in the school were retained in 2008, and 8% were suspended or expelled for violent or criminal offenses (SC State Department of Education, 2009). The e-learning program offers an alternative to the traditional educational programs that were unable to meet the students' needs.

In addition to the school of study's difficulties, current research suggests startling data related to the education of African American males (Edelman, Holzer, & Offner, 2006; Graves, 2006; Greene & Winters, 2006; Wynn, 2005). African-American males lead the country in dropout rates, special education diagnoses (Attention Deficit Disorders, Learning Disabilities, and Mental Retardation), lower GPA's, lower college attendance rates, and lower college graduation rates (Greene & Winters, 2006). Additionally, a report by Petit and Western (2004) acknowledged that in 1999, 21% of African American male high school dropouts were in state or federal prisons, compared to 2.9% of their white male high school dropout counterparts. Moreover, a study conducted by the Justice Policy Institute (2002) between 1980 and 2000 revealed that approximately 3 times as many African American males were incarcerated as were attending college.

In too many American classrooms, African-American males are not reaching their full potential. Many high school classrooms are using the "old school" high school design of students working in their seats and staying quiet for fifty to ninety minutes at a time, listening to a teacher or doing seatwork. Limited visual aids are utilized, and physical activity is restricted. The results of the "old school" classroom settings are dismal for many African-American males.

The instructional strategies for some classrooms are designed to teach to one learning style, offering little support for many males. "Instructional practices that offer little benefit or intervention can lead to increased learning anxiety for Black students, which in turn can lead to wider gaps in achievement, culminating in Black students turning off to learning, disciplinary problems, and/or Black students dropping out" (NEA, 2005, p. 2). Additionally, Tatum (2006) suggested that

African American male students sometimes deal with self image problems by displaying behaviors such as "acting tough" and "dissociating from school" that may be misinterpreted by teachers and administrators and subsequently lead to grade retentions and suspensions (p. 44).

With the notable deficits in achievement for African-American males, educators have begun to focus more attention on the learning styles and preferences of African-American males. Alternative forms of education are being explored. Educational technology in the form of e-learning offers one such alternative to the traditional classroom. E-learning offers a self-paced style of individualized instruction that reaches the three learning styles of auditory learners, visual learners and kinesthetic learners by providing auditory readings of texts and visual aids in the form of video clips, charts, and diagrams (Kruse, 2004). Dr. Simone Kruger, a researcher at the Edge Hill University's SOLSTICE Center for Excellence and Teaching and Learning, found that "e-learning has a dramatic heightening effect on students' ability to sit back and reflect both on their own role and capabilities as a 'learner' compared with students who solely study through traditional classroom and textbook means" (Anonymous, 2007, p. 1).

In his article, *The Benefits and Drawbacks of E-Learning*, Kevin Kruse (2004) indicated that there are numerous advantages of e-learning for students, including the following: availability of the instruction at home at any time of the day, reduction of stress and increase in personal satisfaction because of the self-pacing, engagement of the learner by "pushing them through the assignments rather than pulling them," and improvement in confidence levels of students (p. 1). In the same document, Kruse concluded that schools and school districts also benefit from e-learning by saving money through the purchase of e-learning licenses versus teacher salaries, reducing learning times for students by 40% to 60%, decreasing dropout rates, providing consistent delivery of content, providing expert knowledge of the content without having to search for highly qualified teachers, and offering a fair, objective means of assessment and completion of the courses. (Kruse, 2004)

Due to the possible advantages of e-learning for students, schools, and school districts, an average-size school district located in the southeastern section of the country decided to purchase an Apex Learning license to give to all high schools in the district. The instructional design of e-learning correlates with best practices of instruction. Best practices in instruction include involving students in active learning, providing feedback on performance, allowing sufficient time on task, offering activities with high expectations, and utilizing a curriculum that respects diversity of learning and worldviews.

E-learning provides opportunities for active learning by including activities that require critical thinking, application of course content, and construction of personal knowledge of the concepts. Students receive immediate feedback on their work, revise, and review, allowing an expansion of the content and process of learning. The e-learning objectives and standards are comparable to in-class objectives and standards without the distractions of student disturbances. Students also realize that the e-learning curriculum is equally as challenging as a traditional classroom, but many times, due to the lack of distractions, e-learning produces better outcomes. Finally, the e-learning curriculum is designed to offer a variety of ways to learn and apply course concepts (Billings & Connors, 2006).

The APEX e-learning program is an asynchronous digital learning program that provides opportunities for all types of learners to learn the concepts of math, English, social studies, sciences, world languages, and electives. As a credit recovery program, school districts and/or schools can buy a license for students to access all of the state required graduation courses through the Apex digital curriculum. Students earn high school credits by completing each e-learning class. For each course, the Apex e-learning program consists of content reading passages similar to a textbook, a speaker link that will orally read the passage to the student through a set of headphones, and a video or PowerPoint offering visual images connecting with the content information in the passage. Each class has an average of 750 pages, 1,000 images, 250 multimedia tutorials, 250 interactive exercises, 50 computer-graded assessments, and 85 vetted web links (Apex Learning, 2009, par 2). Students take a diagnostic assessment before starting each section of the course. If the student scores 70 or above, the student can proceed to the next section without having to complete the assignments for that section. However, if the student scores 69 or below, he/she must complete the assignments within the section that serve as study guides for the section assessments. Once all assignments and assessments within a course are completed at a level of 70 or above, the student receives a high school credit for the completed course.

CASE STUDY DESCRIPTION

Three African-American males participating in the e-learning program were selected for this case study. They were selected because they have been participants of a federally funded grant program since they were in the seventh grade. Their academic progress, including grades and state test scores, has been monitored for over three academic years. Additionally, background information, attendance rates, and discipline issues have been noted over the past three years. All three of the students have had difficulties in traditional classrooms throughout middle school and high school. For example, one student was sent to an alternative school, a high school for students who have been suspended from their home school due to behavior issues, and is in his first year back in a traditional high school. The students' names have been changed to protect their anonymity.

Case Study One: James

James is a seventeen-year-old male who is at least the second generation who has grown up in the community. He currently lives with his mother. He has an older sister who is known as a community prostitute and still lives in the home. He has a younger brother who has experienced serious mental health issues and has been hospitalized for at least six months in the mental facility of a local hospital. James' father is known as a "crack head" and has been imprisoned multiple times. When James' grandmother died and left her house to James' father, he stripped it for copper wiring and other valuable materials. He sold the materials for money to buy crack. He lived in the shell of the house before being arrested for crack.

James' mother reports having no contact with James' father due to his history with crack. She states that she has "never been a user." She currently works for a cleaning company contracted with the local district to clean the schools after school hours. She cleans in the high school that James attends. She is very willing to attend parent conferences concerning James and his brother and their academic progress. She has difficulty with transportation because she does not own a car. The family has been seen walking fifteen miles from home to obtain food for dinner. They live down the road from the high school in a trailer that was donated by a local community organization. Their original trailer was condemned.

When James was in sixth grade, his sixth grade class was reportedly so challenging that all but one of the sixth grade teachers quit. It was difficult to assess the sixth graders' mastery of required objectives due to the numerous changes in teachers. As a result, the students were promoted to the next grade level.

In the seventh grade, James had a positive school experience due to a teacher who cared about him. The teacher had high expectations and worked hard to keep him focused on the academic assignments. She offered the structure

James needed to stay focused and complete assignments. She also had a great relationship with James' mother. Test scores showed improvement, although he still scored below basic, and his grades also improved. Although he did not pass all subject areas, he attended summer school and was promoted to the eighth grade.

Eighth grade was academically disastrous for James. A separate class for the "bad boys" was established in which James was included. His disruptive behaviors of playing and talking resulted in numerous in-school and out-of-school suspensions. The teachers were not as caring as he had experienced in the seventh grade. His behavior became so disruptive that the principal often allowed him to leave class to install pencil sharpeners in classrooms. James took pride in that work and accomplishment. However, his grades suffered, and his test scores plummeted. James was promoted to the ninth grade despite his lack of academic success.

James attended the traditional ninth grade academy in his school. Only students in the ninth grade for the first time are allowed to be in the academy. It is based on the teaming concept. Unfortunately, James displayed the same maturity level and behaviors that he demonstrated in the eighth grade. He played too much in class, laughing and talking and exhibiting no impulse control. Teachers repeatedly wrote referrals for nonviolent offenses. James had numerous out-of-school suspensions. He was once arrested for attempting to steal a truck, run over the owner, and fight the deputy sheriff. He entered the juvenile justice system and spent the last part of the ninth grade in the juvenile detention center. He is currently under probation.

Case Study Two: Andrew

Andrew is also a seventeen-year-old product of the community. His parents grew up in the area, as well. Andrew lives with his mother, and little is known about his father.

He has an older sister who lives at home.

Andrew's mother has expressed concerns about his academic progress. Additionally, things have not always gone smoothly between Andrew and his mother. At the end of the previous school year, Andrew ran away from home and was placed in a foster home for a limited time.

Academically, Andrew has been slightly more successful than James. His behavior is too playful and talkative, but he has more impulse control. Andrew has excellent manners and a pleasant attitude. He is very respectful to teachers; however, he tries hard to be accepted by peers who are also having academic difficulty.

Andrew repeated two grades before attending the seventh grade; however, he was promoted numerous times despite of his lack of academic success. He was part of the sixth grade that "ran off" all but one of the teachers. He and James had the same seventh grade teacher. The structure of that classroom benefitted Andrew, and his academic progress improved that year.

In eighth grade, Andrew was a member of the all male "bad boy" class that was designed to get the "trouble makers" out of the regular classes so that students who wanted to learn could do so without being disrupted by the "bad boys." This classroom environment was as disastrous for Andrew as it was for James. The math teacher had an accent that Andrew could not understand. Additionally, his other teachers had difficulty with classroom management, wasting much of their instructional time taking care of student disruptions. There was very limited academic instruction implemented in any of the classes, and test scores plummeted.

Like James, Andrew attended the traditional ninth grade academy. However, at the end of the school year, he ran away from home and missed his end-of-year exams. As a result, he did not pass 9th grade. Each year, Andrew has improved on the state assessment, but he has never reached academic proficiency.

Case Study Three: William

William is a seventeen-year-old who lives with his mother. He has some contact with his father. There is no information available on siblings, but it is believed that he is the only child of his mother.

William's mother has expressed concerns for her son that are similar to those expressed by James' and Andrew's mothers. For example, she reported that William does not listen to her and does what he wants to do despite her objections. Additionally, she indicated that William has difficulty staying focused enough to complete his school assignments.

William is quiet and displays an angry attitude. He is confrontational when approached and immediately portrays a tough guy. This persona frightens some of his teachers. At least one teacher has indicated that William is not welcome in his classroom.

William attended the same middle school as James and Andrew in the sixth (the "bad class") and seventh grades. However, during his eighth grade school year, because of repeated offenses, including fighting and bringing a weapon to school, the school administrators and district school board assigned William to an alternative middle school for students with behavior issues. While attending the alternative school, William was expelled for slamming a door on a teacher's hand and breaking the teacher's fingers.

Unlike James and Andrew, William did not attend the traditional ninth grade academy at his school. Instead, he attended the alternative school, which did not have an academy in place. He scored below proficiency on all parts of the state assessment for all of the middle school years.

Class Day

This year, James, Andrew, and William all started in the repeater ninth grade class (not the ninth grade academy). They were in a homeroom with students who do not have the required number of credits to be promoted into the tenth grade. However, early in the 2008-2009 school year, the school principal moved James, Andrew, and William to the virtual classroom, after teachers and administrators discussed their failure to succeed in the traditional ninth grade classroom. The school day starts at 7:20 a.m., with the students coming into the classroom. The classroom is located, along with the in-school suspension room, in an annex of the school building. Announcements from the principal come over the intercom, along with the Pledge of Allegiance to the United States flag, and a moment of silence. Once the intercom is silent, the students sit at their assigned computers, boot their academic programs, and slip on headphones. The students are allowed to bring MP3 players, CDs and other music devices to listen to music while they complete their assignments. The rule is that the music cannot be heard by the teacher or other students.

The students are participating in three virtual class periods that are ninety minutes each. They work through two classes (3 hours) before taking a break. They then go to an elective class and lunch before returning to the virtual classroom for the last ninety-minute class.

Before beginning their virtual classes, the students were briefly shown how to start their programs and maneuver from study guides to assessments on the computer screen; however, they were not taught basic computer skills nor given direction on the number of lessons they needed to complete within a certain timeframe. As a result, the students initially struggled with the e-learning program. However, midway through the first semester, a certified reading specialist and a certified special education specialist began monitoring the students' progress. They currently meet with the students once per week and help them develop weekly and monthly goals. They also help them develop plans to meet those goals.

RESULTS TO DATE

Although the students continue to have difficulty with the traditional school rules, each student has made academic progress in the e-learning classroom. The quality of the students' work has been far above what was completed in the previous school years. Additionally, the students have been observed to spend more time on task than they previously did in the traditional classroom. The students do not devote their time and energy to criticizing each other, as was observed in the traditional classrooms. A review of school records indicate that the three students have had fewer discipline referrals than they had in the traditional classrooms, and the offenses have been minor as compared to the previous years.

At the end of the semester, after the case study students had been working in the virtual classroom for thirteen weeks, James had completed two classes, basic math and reading. He was unable to complete the Algebra I, Geography, or Biology classes, and he completed only a limited amount of work in each class. James reported enjoying the virtual classroom because of the lack of disruptions. He also liked being able to listen to music as he worked on the assignments. However, James needed more guidance to complete the virtual classes. He had a tendency to skip around from subject to subject without following a plan to get any of the subjects completed. Although the software program provided opportunities for James to listen to the reading passages, he did not take advantage of that aspect of the program. He skimmed over important reading passages that may have helped him complete the virtual classroom program.

Although school records indicate that James had fewer discipline referrals than in the past, during the first semester, he was arrested for an incident in the community, and he spent some time in the county jail. During the middle of the second semester, due to what was considered limited progress in the e-learning class, school administrators decided to move James to the GED program. James is reportedly struggling with academics in his current setting. He wants to attend classes at a local technical college in order to become a barber; however, he needs at least a 9th grade education in order to be accepted into the college.

Andrew has had the most positive results out of the three case study participants. He completed five classes by the end of the first semester and is currently working on Geometry, English II, and World History. He reportedly likes the virtual classroom because of the lack of distractions. Andrew realizes that he tends to get into trouble by following other students, and the virtual classroom provides distance from the temptations.

Andrew sometimes voluntarily attends Saturday school in order to complete his assignments. Additionally, he responds very positively to volunteers who visit the e-learning classroom once weekly. With the help of the volunteers, he developed a plan to work on one virtual class per class period.

Andrew appears interested in learning and motivated to complete his work. In fact, Andrew's grades were high enough for him to play on the school's varsity basketball team during basketball season, and he currently plays on the school's baseball team. His plan is to complete enough high school credits to enter the fall 2009 school year as a junior. Additionally, he hopes to graduate from high school and attend a local college.

William completed Geography, Physical Science, and Algebra I by the end of the first semester. He works on one course at a time and is currently working on English II. Reportedly, he likes virtual classes better than traditional classes because the computer lets him know the assignments that he has to do without bothering him while he is working. William also likes to listen to music while he works. However, the teacher often has to remind him to refrain from singing and to turn his music down so others cannot hear it.

William could benefit from individual assistance as he completes the assignments, but he typically responds that he does not need help and can do it on his own. Although he is friendly to volunteers, he prefers to work alone. He hopes to complete another course by the end of the school year, but absences and suspensions sometimes hinder his ability to complete work. In the future, William wants to attend truck-driving school. However, his parents want him to complete high school before becoming a truck driver.

PITFALLS AND LESSONS LEARNED

Report cards and completed credit hours indicate academic progress for these young men. Additionally, the number of discipline referrals for these students have decreased. However, the e-learning classroom has encountered barriers to making it completely successful. When the e-learning program first began, the students were given limited guidance. As a result, the case study students faltered. For example, when taking the diagnostic assessments, if the students scored below seventy, they often attempted to memorize the correct answers on the assessments before retaking the final test instead of completing the study guide assignments. This sometimes worked, as some of the questions were the same as the diagnostic tests. In math, the students quickly figured out that they could ask the e-learning program for the correct answer and then print out the answer and the problem. Consequently, the students went to the final assessment and just matched the problems with the answers from the practice test without ever knowing how to solve the problems. Due to limited teacher monitoring, the students developed a system that outsmarted the program.

The teachers graded assessments and facilitated the computer technology but did not typically offer one-on-one assistance or instruction on the coursework. Due to this classroom environment, there appeared to be a lack of camaraderie between the students and the teachers. As a result, the students appeared to be reluctant to ask for assistance from the teachers, and some of the teachers appeared reluctant to offer assistance overtly. The young men preferred to work on their own without asking for help. Dr. Simone Kruger, researching at the Edge Hill University's SOLSTICE Center for excellence in teaching and exploring, studied part-time student experiences of e-learning at the University and noted that males typically found that e-learning made it easy for them to "do my own thing" (Anonymous, 2007, p.1).

Another problem observed was that students were not given a timeline to complete lessons. As a result, some students completed lessons sporadically.

Additionally, some students concentrated on one subject area and ignored other subject areas. As a result, the students had numerous lessons from certain subject areas to complete in a limited amount of time. Moreover, the students appeared to lack basic computer skills. As a result, it appeared to take them longer to go through computerized lessons. They seemed to click the computer mouse repeatedly without direction. They were unsure of the links to take them to the correct assignments, and they were unable to locate the information that they were trying to access in a timely manner.

CURRENT CHALLENGES FACING THE ORGANIZATION

There are several major challenges facing the school in the initial stages of the e-learning program implementation. One challenge with the program is insuring that the students master the skills necessary to help them pass the minimum competency tests required by the state for high school graduation. Although the students are passing their self-paced computer assessments, some of their strategies may prevent them from doing well on standardized assessments. Another challenge is helping students obtain computer literacy skills before introducing them to e-learning. However, the most pressing issue is the lack of a qualified teacher to facilitate the e-learning class. The teacher needs to be experienced in working with students who have academic difficulties. The teacher also needs to be computer literate and able to problem-solve when computer difficulties arise. One major requirement for a teacher, in order to ensure success of these students, is the ability to build positive relationships with the young men. This essential role may help the students master the skills to pass the state exit exam (minimum competency test).

The challenges of the school can be addressed by conducting a needs-assessment of the school and its students. Additionally, strategic planning may benefit the school. Determination and hard work will also be essential in addressing the challenges.

CONCLUSION

Research suggests alarming statistics related to the education of African American males. As a result, school systems have experimented with various strategies to assist the students. However, Edleman, Holzer, and Offner suggested that numerous proposed solutions to address the problem of African American male school disengagement have been unsuccessful (2006, pp. 1-2).

School administrators in one rural high school implemented an e-learning program in order to help African American students who were unsuccessful in the traditional classrooms. Overall, results from three case studies appear promising. The students demonstrated greater academic achievements in the e-learning class than in traditional classrooms. They also experienced fewer discipline referrals. While one of the three students recently transferred to a GED program, two of the three students are steadily completing high school credits. In fact, one student's academic scores have improved enough for him to play two varsity sports.

The computer offers active, engaging instruction and activities that many males enjoy, and the e-learning program is self-paced and meets more than one learning style of students. As suggested by Newkirk (2006), "Cutting literacy learning off from students' media-immersed lives makes school an alien and unappealing place. For schools to effectively teach literacy, they should work with, not against, the cultural tools that students bring to school" (p. 64). However, as results from the three case studies indicate, students who participate in e-learning programs continue to need guidance from teachers. When paired with support from a qualified teacher, e-learning programs have the potential to offer valuable learning experiences for students who are unsuccessful in traditional classrooms.

REFERENCES

Anonymous. (2007). *Women get more than men from e-learning.* Retrieved November 12, 2008, from http://findarticles.com/p/articles/mi_qa5402/is_200703/ai_n21284857

Apex Learning. (2009). Retrieved March 17, 2009, from http://www.apexlearning.com/courses.html

Billings, D., & Connors, R. (2006). *NLN living book: best practices in online learning.* Retrieved November 20, 2008, from www.electronicvision.com/nln/chapter02/index.htm

Edelman, P., Holzer, H. J., & Offner, P. (2006). *Reconnecting disadvantaged young men.* Washington, DC: Urban Institute Press.

Graves, E. G. (2006). Saving our young black men. *Black Enterprise, 36*(6), 10.

Greene, J. P., & Winters, M. A. (2006, April). *Leaving boys behind: Public high school graduation rates* (Civic Report No. 48). New York: Manhattan Institute for Policy Research. Retrieved September 10, 2006, from http://www.manhattan-institute.org/pdf/cr_48.pdf

Justice Policy Institute. (2002). *Cellblocks or classrooms: The funding of higher education and corrections and its impact on African American men.* Washington, DC: United States Justice Fund. Retrieved September 9, 2006, from http://www.soros.org/initiatives/justice/articles_publications/publications/cellblocks_20020918/coc.pdf

Kruse, K. (2004). *The benefits and drawbacks of e-learning.* Retrieved November 12, 2008, from http://www.e-learningguru.com/articles/art1_3.htm

National Education Association. (2005). *Focus on Blacks*. Retrieved September 18, 2006, from http://www.nea.org/assets/docs/mf_blacksfocus05.pdf

Newkirk, T. (2006, September). Media and literacy: What's good? *Educational Leadership, 64*(1), 62–66.

Pettit, B., & Western, B. (2004). Mass imprisonment and the life course: Race and class inequality in U.S. incarceration. *American Sociological Review, 69,* 151-169. Retrieved September 16, 2006, from http://www.princeton.edu/~western/ASRv69n2p.pdf

SC Department of Education. (2009). *The State of South Carolina 2008 Annual School Report Card.* Retrieved February 26, 2009, from http://ed.sc.gov/topics/researchandstats/schoolreportcard/2008/

Tatum, A. W. (2006, February). Engaging African American Males in Reading. *Educational Leadership, 57*(8), 44–49.

Wynn, M. (2005). *Empowering African American males: Teaching, parenting, and mentoring successful black males.* Marietta, GA: Rising Sun Publishing.

Chapter 15
"Cross Talk": The Connected Stance of One Successful Student's Online Interactions

Susan J. Wegmann
University of Central Florida, USA

EXECUTIVE SUMMARY

Asynchronous online discussions can be complex and fruitful mimicking their face-to-face counterparts in undergraduate college classes. However, some researchers note a discrepancy in substance and interest levels between online and face-to-face discussions. This chapter describes the interactions of one thriving student in an asynchronous online course. It analyzes the student's interactions with his peers, and uses these interactions to provide ways that online instructors can structure courses to optimize genuine and engaging online discourse. Additionally, it suggests that students and instructors who assume a Connected Stance show a depth of learning within the computer-mediated framework. Finally, it provides a unique format for analyzing online discussion boards.

"We had pretty significant 'cross talk' going on." A comment from one online student at the end of an online course

BACKGROUND

Can rich engaging interactions occur online? What about "cross talk," or students communicating with each other about various topics, in a short amount of time and space? As universities follow the global trend to increase online delivery of classes, researchers have investigated good practices in androgogy (Greene, 1998) and whether online interactions are as robust as face-to-face interactions can be (Wegmann, & McCauley, 2007; King & Doerfert, 1996; Mondada, 2006; Ruan & Beach, 2005).

This case study is a picture of one student, in one section of a class in a large urban university in the South eastern United States of America. The university is within the top ten universities in the United States, in terms of undergraduate student enrollment. There has been a steady increase of online course offerings since the university's first offering in 1997. The university was founded in 1968 and

DOI: 10.4018/978-1-60566-942-7.ch015

offers over 200 degree programs for undergraduate and graduate study, along with over 20 doctoral programs. The university faculty and staff have attracted over $122 million in research funding. There are over 1300 international students from 141 countries at the university. The university has an extensive program to help instructors develop online courses, including extensive training to develop effective interaction. Each online instructor spends one semester (45+ hours) attending a face-to-face course that supports the creation of an online course. In it, 10+ hours are devoted to course structure and ways to increase students' interaction.

The course where the data originates was housed in the College of Education. This course is taken by all students who are studying to be middle and high school content area teachers. The undergraduate course was an exploration of Content Area Reading Strategies, targeting middle and high school future teachers. Students in this particular section ranged from English, Social Science, Math, Science, Health, Physical Education, to Music majors. I wrote, designed, and delivered the course content fully online, based on previous courses I had taught. For each of the 10 lessons, students were expected to read a chapter from the assigned textbook. In total they were to complete six activities, four quizzes, and seven discussion boards, where they were expected to respond to an open-ended prompt as well as reply to their peers. A Discussion Board Rubric (See Appendix A) was used to evaluate each discussion board entry. It highlighted five aspects of each initial posting: a. content of initial response, b. depth of initial response to lesson question(s), c. content of reactions to peers, d. depth of reactions to peers, and e. mechanics of initial responses and peers' reactions. Students were given this rubric at the beginning of the course. They were evaluated and assigned points to each discussion posting, based on the rubric.

There were 55 students in this two-section class. This manuscript is the result of a focused look at one participant in a much larger research study. The other students' stances, or the ways participants used their language, were tallied and used as a comparison for the focused case study student. The research is true participant observation (Spradley, 1979), as the instructor of the course was also the researcher. In-depth member-checks were conducted by asking the participant numerous e-mail questions throughout the analysis. All parts of the analysis were done after the academic semester was completed.

Many online classes make use of discussion boards, on which students can interact with their peers, the content, and the instructor. As a co-constructed place in an online class, discussion boards can offer interactive possibilities, whether synchronous (i.e. real time) or asynchronous (i.e. not real-time). But, a discussion board may or may not elicit engaging discussions and interactions. Therefore, what are the elemental aspects of discussion boards that encourage students to wonder, challenge their peers, initiate their own topics and participate in ways that show they are deeply interacting with their peers, the instructor, and the content?

This case study sheds light on one aspect of effective discussion boards by examining one successful Master's of Education student's interactions in an online Content Area Reading course. In particular, the researcher analyzes the discussion boards, student tracking, course emails, and student grades using discourse analysis techniques. The analysis of all students in the course was beyond the scope of this study. Instead the researcher chose to deeply analyze one participant.

SETTING THE STAGE

This case study relies on the theoretical underpinnings of two lenses: the transaction theory of reading (Rosenblatt, 1996) and discourse analysis research methodologies (i.e. Mehan, 1998; Cazden, 1988, and Britton, 1993). These two threads work together to reveal understandings about reading and learning online. They also help to describe optimum amounts of structure that will encourage cross talk.

Transactional Theory of Reading

The first theoretical strand builds on the work of Louise Rosenblatt and her notions of the Transactional Theory of Reading (Rosenblatt, 1982, 1996, 2005). Since most content delivered online is read, the transactional theory of reading explains some of the complicated internal processes that occur while students progress through an online course. Rosenblatt purports that all people assume a mental position on a spectrum that ranges from *efferent*, or reading for the purpose of a later event such as a test, to *aesthetic*, or reading for an "in the moment" feel, for pure pleasure.

Rosenblatt also reveals that our *linguistic experiential reservoir* (LER), or the experiences and language we have learned, influences and engages the position we will assume on the efferent/aesthetic stance. Linguistic Experiential Reservoirs are as varied and personal as fingerprints and can be enlarged and focused by teachers who are willing to investigate students' interests and prior knowledge. Rosenblatt purports that each reading event produces a unique *evocation*, or understanding of the text. This evocation is different at each reading event, even if the text remains the same, due, in part, to the LERs influence.

While reading course content online, students assume a position or stance somewhere between the efferent and aesthetic, influenced by the way the online course is structured. This stance is made apparent by what learners write on discussion boards and how they approach course content. Even though it is different in many ways, online reading has some similarities to "hard-copy" reading. For example, in both platforms as words are decoded, or interpreted, students: a. assume a stance on the efferent/aesthetic spectrum, b. activate their LERs (most often subconsciously), and c. create a unique evocation. However, particular influences on each of these three aspects of reading are different online. In the first component, stance is influenced by the online instructor's choice of text and structure, as well as the students' familiarity and comfort with technology. Some students report an automatic efferent stance when reading online due to their lack of expertise with computers. This makes the online reading experience labored and focused only on how the content might be used for a future activity (i.e. test or assignment). However, if students are comfortable with the learner-computer interaction (Moore, 1991), they will be more likely to assume an aesthetic stance, or one in which the reading event is savored and students can engage more deeply with the content. Studies have shown that an aesthetic stance while reading is more likely to result in deep student learning and memory of the content (Gill, 2008; Many, 1991; Rosenblatt, 2005). Thus, an optimal stance while reading, whether online or hard-copy, is more aesthetic.

Students' LERs are also affected when reading online. Online texts can include extra-textual information like sidebars, Internet links, possibilities for further reading on other websites, and other links to scaffold LERs at the exact point of need. So, when reading online, students' LERs are more likely to be enlarged if they take advantage of the extra information readily available. However, the extra-textual information likely to surround online reading may serve as a distraction to some students, who will not likely take advantage of enlarging their LER. Finally, some students report feeling a *transactional distance*, or disconnect that occurs when the learner and teacher are separated geographically (Stein, Wanstreet, Calvin, Overtoom, & Wheaton, 2005). This transactional distance affects the evocation because the reader may be adversely psychologically affected if their feelings of transactional distance are severe. This adverse reaction to reading when online affects the evocation because it shadows the reading event, making an aesthetic stance nearly impossible.

Analyzing Discourse Online

In face-to-face courses, discourse is analyzed by transcribing the spoken word, assigning moves, and analyzing them for content and affect. Online discourse measures language-in-use as well by looking closely at asynchronous (or not in "real time") discussion boards and course e-mails. This case study borrows from the respective oral discourse analysis works of Courtney Cazden (1988) and Hugh Mehan (1979, 1998), and moves the field of online discourse analysis forward by suggesting an optimal way to analyze online discourse.

The works of Cazden and Mehan inform the study by suggesting that we can understand what people are saying by analyzing their possible moves, or reasons why participants choose particular words. According to Mehan (1979) and Cazden (1988), oral discourse patterns in most face-to-face classrooms consist of an Initiate, Respond, and Evaluate, or IRE, pattern. In the IRE pattern teachers *Initiate* a question or topic, students merely *Respond* to teacher-generated topics, and teachers *Evaluate* students' ideas. The IRE pattern is prevalent, yet not optimal for classroom discussion (Mehan, 1998), because it limits the role of the student to that of responding to teacher-directed questions.

The discussion board is used extensively as the vehicle for interactions in an online course (Burnette & Buerkle, 2004; Waltonen-Moore, Stuart, Newton, Oswald, & Varonis, 2006). Interactions and opportunities to interact online (including discussion boards) do not occur without planning —they need to be an intentional part of the instructional design (Althaus, 1997; King & Doerfert, 1996; Smith, 2005; Zhang, Perris, & Yeung, 2005). Instructors designing online courses need to ensure that students have ample opportunity to respond to each other on discussion boards and in group activities (Ruan & Beach, 2005).

Additionally, all discourse sequences online are mediated by the computer. Computer-mediated communication (CMC) (Mondada, 2006), or communicating with others using technology as a go-between, is a complex relationship between teachers, students, content, and the interface (or computer, in this instance). Distance educators have defined four types of online interaction: (1) learner-teacher, (2) learner-content, (3) learner-learner, and (4) learner-interface (Hillman, Willis, & Gunawardena, 1994). This last aspect of interaction is unique to distance education and greatly influences online students and instructors.

One benefit of asynchronous online interactions is the reflection time that students have before formulating their written responses. This increased time to reflect may result in profound ideas and may be missing from face-to-face interactions. Certain personality types may benefit from online asynchronous interactions. Lin et al. (2005) conducted a study of asynchronous online environments as they relate to psychological types. Participants reported that the time they were given to reread and clarify their thoughts was valuable. Their findings were driven by personality types and the ability of learners to feel a "co-presence" in the online class.

More recently, researchers have begun investigating online discourse using discourse analysis techniques (Williams & Humphrey, 2007; Wegmann & McCauley, 2007, 2008). These investigations have used traditionally face-to-face discourse analysis as a lens for understanding online discourse. Rather than an IRE pattern online, a more optimal way of using language online has been described as using a *Connected Stance* (Wegmann, & McCauley, 2008). The Connected Stance is characterized by higher-order thinking skills (Bloom, 1956), or those in which teachers and students wonder, question, and do much more with their language than the IRE pattern. A Connected Stance occurs when students show high participation plus high engagement with the content. On the other hand, the *Disconnected Stance* is one in which teachers and students enact what Bloom would call low-level thinking

Figure 1. The Connected Stance or the nexus of high engagement plus high participation

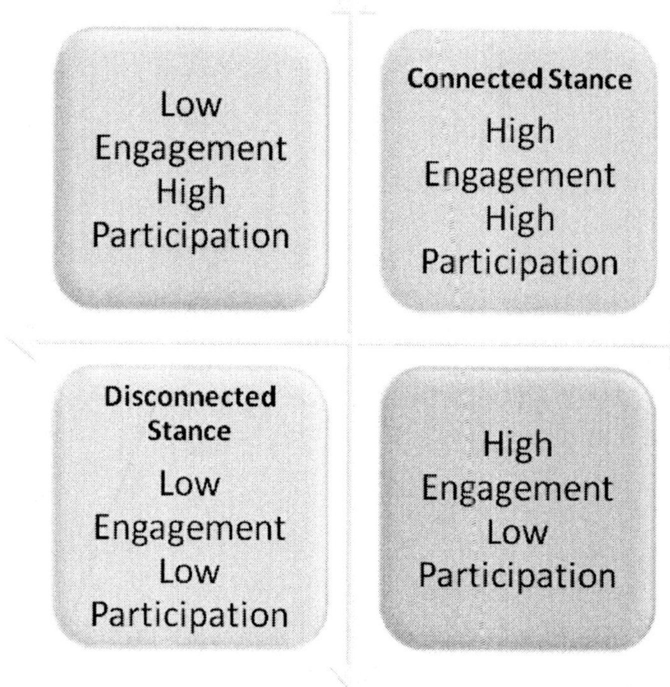

(1956) and merely answer questions or respond to teacher-generated questions. Figure 1 shows the Connected and Disconnected Stance, in relation to students' engagement and participation levels. Appendix C displays further information about the variables in the Connected Stance.

In an earlier study of online interactions (Wegmann & McCauley, 2008) the following moves were found in online discussion board posts:

1 Introducing a new topic
2 Sharing opinion
3 Sharing beliefs
4 Connecting to other readings
5 Connecting to their own experiences
6 Connecting to their own classroom
7 Connecting to their own thinking
8 Building rapport
9 Suggesting a new organizational theme
10 Revealing their own struggles
11 Responding to other peer's question
12 Giving information
13 Giving advice
14 Connecting to previous thought
15 Questioning
16 Giving an example
17 Sharing "grand idea"
18 Challenging peer
19 Connecting to course content
20 Using humor

Moves such as these serve to define what students are doing with their language during a discussion board thread. In a two-university, two-class study of discussion board entries, Wegmann and McCauley found that students who used a wide range of these moves were more deeply engaged in the content of the course (2008). There was also a correlation between the amount of various moves made and a meaningful amount of engagement (Wegmann & McCauley, 2008).

The Connected Stance is similar to Rosenblatt's notion of an aesthetic stance. Both denote high levels of thinking (Bloom, 1956) as well as engagement with the content. Engagement and participation are two important aspects of both stances. However, the aesthetic stance and the Connected Stance differ in several ways. A student enacting an aesthetic stance need not use a wide range of moves in order to attain an "in the moment" feel. On the other hand, this range is a critical part of identifying the Connected Stance. Also, an aesthetic stance may be assumed while reading silently or participating in a solitary activity while the Connected Stance needs social interaction to thrive. Even so, the similarities between the Connected Stance and the aesthetic stance are striking: higher-order cognitive skill use, increased motivation to complete assignments, and higher levels of satisfaction with the reading events.

CASE DESCRIPTION

Focus Case Study Participant

Seth Brockton (a pseudonym) was a History major with a Political Science/Social Science Education minor in a fully online class. He was an older than average college student, who had two children (one, 22 and the other, 10). He reported that he was "reasonably computer savvy" due to his job and taking online classes. Early in the semester, he wrote about an unfortunate message he lived with during his middle and high school years:

I was told in seventh grade that I was academically "challenged" and that lie stuck with me all the way through high school. I barely made it through high school, and without sports, (I do believe), I would have quit high school (Introductory Discussion Board, Brockton, p.3).

He referred to this struggle several times in his postings as one motivator to get a teaching degree: help other boys who struggle with academics. He earned a high A in the course and turned in exemplary work. His thoughts on discussion board postings were cogent and showed he was engaged in the course. At the beginning of the semester he revealed that he ". . . avoided Internet-based classes because I wanted the student interaction." He ended the semester by saying that, "In this class, we had pretty strong 'cross talk' going on, and it was helpful." Because he worked in the cable TV industry, I was a bit worried about what he meant by "cross talk," which, in his field, is actually audio-visual interference from other sources. However, the context of his usage shows that the "cross talk" he experienced was a desirable thing, and that he was helped by the interactions with his peers. I elicited Seth's participation for this case study after the course was completed and he had begun the next semester of coursework.

Results

This study describes one way to gauge engagement and participation. To measure participation, I tabulated the number of times Seth accessed class folders, viewed content documents, and posted messages on the discussion board. These were then compared to the average student in this section. The amount of time he spent completing assessments and the total number of hours he spent online for the course were also calculated and compared to his peers. Table 1 shows Seth's participation on these facets compared to the course averages.

Seth's total number of sessions and the number of times he viewed content files were well above the course average. He also spent more time on assessments and posted more messages on the discussion boards than his peers. In fact, he frequently replied and reacted more than the basic requirements. Seth spent 38 hours during the 10-week course online, compared to 26 hours

Table 1. Seth's participation in the course compared to the course average

	Number of times the course was accessed	Number of times content files were viewed	Number of times students posted messages on discussion board	Total hours spent on assessments	Total number of hours spent online in the course	Grade at the end of the term	% of time each stance assumed
Total (n=55)	5611	10360	2272	437 hours	1442 hours		
Class Average	102	188	41	8 hours	26 hours	89%	54% disconnected 46% connected
Seth's count	139	308	66	9 hours	38 hours	98%	21% disconnected 79% connected

average for his peers. This equated to logging in to the course roughly two times a day. Seth showed high participation in the course, based on these numbers alone.

However, the numbers of times and hours he spent on accessing course materials do not paint a complete picture of Seth's engagement in the course. For that, we turn to discourse analysis of the moves that made up what he was actually saying in his postings.

As students reply to open-ended questions or their peers, they use language to support their thoughts and participate in the discussion. How they use their language is the target of this case study, as there are numerous functions for speaking. If students merely confirm what their peers write by saying, "good job" or "I totally agree," they do not seem to be engaged in the content at hand. On the other hand, if students agree and challenge their peers to consider other options or to ask for more information, then they are more engaged. Thus, the number of different moves that students enact can illuminate their depth of engagement in the content.

Additionally, Seth's academic success in the course out distanced his peers. He earned a 98% in the course, compared to his peers, who averaged an 89%. Even though this high academic grade was a combination of many factors, he reported that the

> *. . . discussions I had on the discussion boards helped confirm what I was thinking. I had sort of a study group, because I could ask them about anything, and someone would reply. We even had "discussions" break out and then we would write about it, until we understood it. Someone said it was good to be able to write about what they were thinking. (End of term survey, Brockton, p. 5)*

Seth assumed a Connected Stance 79% of the time, compared to the class average of 46%. From his end of the term "Highlights from this Class" discussion board and interview questions, he revealed a connection between the ways he was able to relate to his peers and his good grades. He liked the "interactions with the students" and they helped him clear up "any questions I have had."

To investigate Seth's engagement with course content, I coded the moves that all students made on all discussion board posts (transcribed to 305 pages of 3 inch columns), using the earlier list of 20 moves. The discussion board topics varied according to different purposes and concepts in the class. Six topics were assigned in Seth's semester:

1. What is the role of the teacher in your content area class?,
2. Quiz Review: What is literacy? What are some seminal theorists/theories in literacy? How can teachers assess content areas appropriately? How has the history of assessment

influenced the assessments we use today? What would the following theorists say is appropriate assessment? Vygotsky, Rosenblatt? Piaget?

3. Please upload your article summary (see rubric for explanation) to the Assignment page AND this discussion board (that's your response). Then react to all of your group members about their articles. Finally, reply to all who reacted to you. Remember: Respond, React, and Reply.
4. Please copy and paste your Script Information Page into the reaction box. THEN attach your vocabulary digital story to the post by the due date. (Thursday, midnight of Lesson 6) You must use PhotoStory and post to TeacherTube. Remember to React and Reply to all peers in your group.
5. After reading Lesson 8, please post a response to the discussion board that agrees/disagrees with one of the following: a. An aesthetic stance has no place in the content area classroom, b. The main reason teachers don't teach comprehension strategies in the content area is that they are supposed to be taught in reading class, or c. Reading Guides should only be constructed by teachers. (Please don't just rely on your personal experiences, although you may choose to share some. Instead, support your answers by using theories, your textbook, and/or Internet sites) Remember: Respond, React to all peers in your group, Reply to everyone who reacted to your post.
6. Highlights from the class - What is the highlight from this class that you will take into your teaching? Here you may share personal experiences, concepts you learned, or any other bit of information you wish.

On average, Seth used 7 different moves per posting, compared to the class average of 4. His most common moves were to question, reveal his own struggles, share his opinion, and connect to his own thinking. The use of these particular moves reveals that Seth was using higher-order thinking abilities to engage with the course content. Due to his high participation and his high engagement with the course content, most often Seth enacted a Connected Stance.

For example, on one discussion board, the assignment was to summarize and reflect on a journal article. Students were to post their article reviews and react to all three of their group members' reviews. Then, students were asked to reply to all who reacted to them. (See Appendix B for further description of the complex interactions and rich involvement with the course material he showed in this discussion thread.) Seth posted his journal article review, which gave wholehearted support for using digital storytelling. Karolyne responded by challenging his positive review of digital storytelling. He replied by giving some examples to support his ideas. Table 2 shows Seth's and Karolyn's responses and details my coding of their moves.

During this exchange, Karolyne raised an interesting question about hiding behind technology. Her Connected Stance was evident by the length (138 compared to a class average of 94 words per post) and the number of different moves she made. She "challenged a peer," and "questioned" his initial response. She also gave an example as support. Instead of simply dismissing her challenge, Seth also assumed a Connected Stance, made visible by his 202 words and six moves he enacted. He agreed with her thoughts to a point, and then offered a logical suggestion for teachers to deal with the problem she posed. He answered her challenge by "sharing his own opinion" and then "giving an example" for support of his thoughts. He did not dismiss her questions; instead he tentatively (by using the term "perhaps" twice and "I guess" to sum up his thoughts) gave his opinion and supported it with his own approach to the issue. This exchange represents the high level of thinking that Seth consistently used when responding to the discussion board replies to my questions and those of his peers.

Table 2. An example of Case Study Participants' moves within an asynchronous discussion board post

Student Name	Students' posts – Discussion board #2	Moves coded
Seth	(copied his journal article review of Ohler's (2006) article in *Education Leadership* on using Digital Storytelling in the classroom.)	
Karolyne	. . . While I agree with this statement (that digital storytelling gives students "voice" that they would not have otherwise) to a certain extent, which is that students can creatively express themselves through digital mediums, I wonder if projects like this really allow students to hide even further from social interactions. By assigning a digital story as a project, students can become as creative as they want, and even show great amounts of comprehension in their work. However, as teachers, we are also to guide them through social interactions during academics and press them to develop critical thinking skills as well as social skills. How would you suggest incorporating social interaction in a digital story assignment with your students to ensure that students are discussing and debating the content with each other, and not just "telling" each other their material? (Karolyne, Discussion board 3, p. 4)	14 connecting to a previous thought 8 building rapport 2 sharing an opinion 18 challenging a peer 16 giving an example 17 sharing a "grand idea" 15 questioning
Seth	Thanks for reading my review. I do see your point regarding the possibility of some students using digital technology to "stay hidden." In fact, I do believe that with the advent of e-mails and IMs and the like, many young people prefer to use the computer vs. actual physical interpersonal exchanges. Personally, I think this is a bad thing. My opinion does not, however, seem to be stopping this reality. You ask how I might combat such a thing, and perhaps draw students into debating content? I would try to build certain of the digital storytelling assignments with a debating component. Perhaps assigning pairs of students with point, counter-point assignments. An example might work for a social studies assignment. Imagine asking two students to develop presentations in which each student has to design a campaign commercial for a presidential candidate. Perhaps the oral component could be that each student take part in a five-minute debate with an opposing student. The debate could be on a point made in the digital presentation. I guess the key would be that the teacher must stay on top of the process and not allow students to hide without at least an attempt to draw them out.	8 building rapport, 14 connecting to a previous thought 16 giving an example 2 sharing opinion 16 giving an example 17 sharing a "grand idea"

This exchange also represents a Connected Stance in full form. On the end of the term "Highlights of this class" discussion board Seth reported that, "I love the crosstalk aspect of regular classes, and in many of the Internet based classes I have had to far, that element was lacking. In this class, however, I felt like I actually was getting to know a handful of the students!" He went on to say that the crosstalk helped him resolve academic questions. "Another highlight was the interactions with the students. . . . it was helpful to have my peers there to clear up any questions I had."

Discussion and Findings

Understanding the relationship between students' participation and their engagement is critical to this case study and why this participant was successful in assuming a Connected Stance.

Finding #1: *Seth's high level of interaction* (revealed by number of times he posted and the number of words he used), *plus his engaged attitude* (revealed by the number of various moves he employed) *helped him achieve a Connected Stance and have good academic success.*

In other words, because Seth assumed a Connected Stance he was a more successful student. This Connected Stance enabled him to relate to his peers in ways that were educationally helpful: asking for help with the course content, being encouraged by his peers agreeing with him, and helping his peers understand the content. Seth's posts were complex and his interactions with peers showed rich involvement with the course material and his peers (See Appendix B). Seth enacted a Connected Stance because he participated highly and he employed a wide range of moves. Even though final grades are a weak indicator of ultimate success in an online course (since students may learn numerous things that are not evaluated), Seth's high grades in the course (98%) show that he understood the course content. In this way he was a more successful student than his peers, who averaged 89% final course grade.

Finding #2: *Students' word-counts by themselves do not reveal the complicated interactions that might be occurring.*

No doubt the type of questions that Seth was required to answer influenced the stance he assumed. He mentioned in the end of term survey that he was excited about being able to "really communicate" with his peers in the class. He reported that he liked the fact that open-ended questions (I.e. "Agree or disagree with the following statements. . . ") were mixed with more structured questions (I.e. "How can teachers assess content areas appropriately?"). He also reported that this engaged him in the course content. Seth's notions of *cross talk*, or the interactions that occur among students, revealed that Seth and his peers communicated and interacted widely and often. Simply tabulating the number of words that students use misses some important data: that of the stances they assumed while participating.

Finding #3 *Course structure plays an enormous role when creating online spaces that are open to various stances and moves.*

The types of responses that teachers encourage influence the stance that learners assume. Seth reported that he found the content and questions of this course engaging. He noted that "the questions made me think. I had to support my answers and push-back when my group members challenged me. I liked the fact that I could share my knowledge on some parts and share my opinions on other parts of the course." In effect, the structure of the course allowed Seth to tell his knowledge *and* share his opinions. However, the structure of online courses can inhibit students' assuming a Connected Stance. In mastery focused courses (i.e. Mastery for Learning, Bloom, 1956), students are expected to simply answer questions and move forward (in Seth's words, "share my knowledge"), if they have attained a pre-determined percentage rate. This structure, while efficient for short-term learning and low-level thinking abilities, is not optimal for encouraging a Connected Stance, in which students deeply engage with the content.

CURRENT CHALLENGES AND PROBLEMS FACING ONLINE LEARNING

This research contributes to our understanding of the role of the learner in online courses. Seth's high level of satisfaction (Interaction Survey results) seemed to result from his high level of participation and engagement in the course. In other words, by enacting a Connected Stance, Seth was satisfied with his participation and engagement with the course. The challenge for instructors of online courses at the university level is how to structure online courses with engaged discourse in focus. This case study describes the use of a Discussion Board Rubric and the structure of discussion boards which promote a Connected Stance.

This research project can also inform the role of the teacher in online courses, in that the structure of the discussion boards and the nature of activities assigned allowed for students to enact more than just a simple IRE pattern (Mehan, 1979). This case study participant showed that numerous possible moves can be used, when given the freedom to respond. This research indicates that both structure and student engagement opportunities can enhance online teaching and learning.

Comparing the moves to higher-order cognitive skills may be fruitful for future research. Which possible moves motivate higher levels of thinking and a Connected Stance? Which moves reduce students' chances of enacting a Connected Stance? In addition, a full investigation of students' LERs before they enter the course compared to their LERs at the completion of the course (as the LERs relate to the course content) would prove fruitful. This would require more than a pretest/post test, as LERs include more than cognitive knowledge. However, a study designed to measure students' language use plus background knowledge might reveal ways in which instructors can structure more effective online classes.

The Connected Stance that Seth assumed was a result of the structure of the course, high participation level, and a high engagement with the course. He revealed that he enjoyed the "cross talk" that he experienced. In his words on the last discussion board, "in this class I felt very encouraged all the way through. I knew that DrW wanted us to think long and deep about our own questions. That made it easy for me to participate and engage."

REFERENCES

Althaus, S. L. (1997). Computer-mediated communication in the university classroom: An experiment with on-line discussion. *Communication Education*, *7*, 158–174. doi:10.1080/03634529709379088

August 4, 2006, from http://www.ssu.missouri.edu/ssu/AgEd/NAERM/s-e-4.htm

Bloom, B. S. (1956). *Taxonomy of Educational Objectives, Handbook I: The Cognitive Domain.* New York: David McKay Co Inc.

Britton, J. (1993). *Language and Learning.* Portsmouth, NH: Boynton/Cook.

Burnette, G., & Buerkle, H. (2004). Information exchange in virtual communities: A comparative study. *Journal of Computer Mediated Communication, 9*(2). Retrieved March 5, 2009, from http://www.ascusc.org/jcmc/vol9/issue2/burnett.htm

Cazden, C. B. (1988). *Classroom discourse.* New York: Heinneman.

Gill, S. (2008, October). The Comprehension Matrix: A Tool for Designing Comprehension Instruction. [from Professional Development Collection database.]. *The Reading Teacher*, *62*(2), 106–113. Retrieved December 1, 2008. doi:10.1598/RT.62.2.2

Green, J. (1998). Androgogy: Teaching adults. In B. Hoffman (Ed.), *Encyclopedia of Educational Technology.* Retrieved March 5, 2009, from http://coe.sdsu.edu/eet/Articles/andragogy/start.htm

Hillman, D. C., Willis, D. J., & Gunawardena, C. N. (1994). Learner-interface interaction in distance education: An extension of contemporary models and strategies for practitioners. *American Journal of Distance Education*, *8*(2), 30–42. doi:10.1080/08923649409526853

King, J. C., & Doerfert, D. L. (1996). *Interaction in the distance education setting.* Retrieved

Many, J. (1991). The name assigned to the document by the author. This field may also contain sub-titles, series names, and report numbers.The Effects of Stance and Age Level on Children's Literary Responses. *Journal of Reading Behavior*, *23*(1), 61–85.

Mehan, H. (1979). *Learning lessons: The social organization of classroom instruction.* Cambridge, MA: Harvard University Press.

Mehan, H. (1998). The Study of Social Interaction in Educational Settings: Accomplishments and Unresolved Issues. *Human Development*, *41*, 245–269. doi:10.1159/000022586

Mondada, L. (2006). Participants' online analysis and multimodal practices: Projecting the end of the turn and the closing of the sequence. *Discourse Studies*, *8*(1), 117–129. doi:10.1177/1461445606059561

Moore, M. G. (1991). Distance education theory. *American Journal of Distance Education*, *5*(3), 1–6. doi:10.1080/08923649109526758

Ohler, J. (2006). The World of Digital Storytelling. *Educational Leadership, 63*(4), 44–47.

Ohler, J. (2008). *Digital storytelling in the classroom: New media pathways to literacy, learning, and creativity*. Thousand Oaks, CA: Corwin Press.

Retrieved December 1, 2008, from ERIC database.

Rosenblatt, L. (1982). The Literary Transaction: Evocation and Response. [from Professional Development Collection database.]. *Theory into Practice, 21*(4), 268. Retrieved December 1, 2008. doi:10.1080/00405848209543018

Rosenblatt, L. (1996). *Literature as Exploration*. New York: Modern Language Association of America.

Rosenblatt, L. (2005, March 1). From What Facts Does This Poem Teach You? *Voices from the Middle, 12*(3), 43–46.

Ruan, J., & Beach, S. (2005). Using online peer dialogue journaling to promote reflection in elementary preservice teachers. *Action in Teacher Education, 27*(3), 64–75.

Smith, R. (2005). Working with difference in online collaborative groups. *Adult Education Quarterly, 55*(3), 182–199. doi:10.1177/0741713605274627

Spradley, J. P. (1979). *The Ethnographic Interview*. New York: Holt, Rinehart, and Winston.

Stein, D. S., Wanstreet, C. E., Calvin, J., Overtoom, C., & Wheaton, J. E. (2005). Bridging the transactional distance gap in online learning environments. *American Journal of Distance Education, 19*(2), 105. doi:10.1207/s15389286ajde1902_4

Waltonen-Moore, S., Stuart, D., Newton, E., Oswald, R., & Varonis, E. (2006). From virtual strangers to a cohesive learning community: The evolution of online group development in a professional development course. *Journal of Technology and Teacher Education, 14*(14), 287–311.

Wegmann, S., & McCauley, J. (2007). Can you hear us now? Stances toward interaction and rapport. In Y. Inoue (Ed.), *Online Education for Lifelong Learning*. Hershey, PA: Information Science Publishing.

Wegmann, S. & McCauley, J. (2008). *Shouting through the fingertips: Computer-mediated discourse in an asynchronous environment.* Presentation at Society for Information Technology, March 3, 2008.

Williams, R. S., & Humphrey, R. (2007). Understanding and fostering interaction in threaded discussion. *Journal of Asynchronous Learning Networks, 11*(2). Retrieved December 2, 2008 from http://www.aln.org/publications/jaln/v11n2/v11n2_williams_member.asp

Zhang, W., Perris, K., & Yeung, L. (2005). Online tutorial support in open and distance learning: students' perceptions. *British Journal of Educational Technology, 36*(5), 789–804. doi:10.1111/j.1467-8535.2004.00492.x

ADDITIONAL READING

Curtis, D. D., & Lawson, M. J. (2001). Exploring collaborative online learning. *Journal of Asynchronous Learning Networks, 5*(1), 21–34.

Delucchi, M. (2000). Don't worry, be happy: Instructor likeability, student perceptions of learning, and teacher ratings in upper-level sociology courses. *Teaching Sociology, 28*, 220–231. doi:10.2307/1318991

DeRouin, R. E., Fritzsche, B. A., & Salas, E. (2005). E-learning in organizations. *Journal of Management*, *31*(6), 920–940. doi:10.1177/0149206305279815

Dobrovolny, J. (2006). How adults learn from self-paced, technology-based corporate training: New focus for learners, new focus for designers. *Distance Education*, *27*(2), 155–170. doi:10.1080/01587910600789506

Jonassen, D. H. (1999). Designing constructivist learning environments. In C. M. Reigeluth (Ed.), *Instructional-design theories and models: A new paradigm of instructional theory* (Vol. II, pp. 115-140). Mahwah, NJ: Lawrence Erlbaum Associates.

Lin, L., Cranton, P., & Bridglall, B. (2005). Psychological type and asynchronous written dialogue in adult learning. *Teachers College Record*, *10*(8), 1788–1813. doi:10.1111/j.1467-9620.2005.00542.x

Mezirow, J. (1997). Cognitive processes: Contemporary paradigms of learning. In P. Sutherland (Ed.), *Adult learning: A reader*. Sterling, VA: Kogan Page.

Moore, M. G. (1993). Three types of interaction. In K. Harry, M. John & D. Keegan (Eds.), *Distance education: New perspectives* (pp. 19-24). New York: Routledge.

Restauri, S. L. (2006). Faculty-student interaction components in online education: What are the effects on student satisfaction and academic outcomes? *Dissertation Abstracts International*, *67*(01), 602. (UMI No. 3206695)

Wegmann, S. (2009) Interactions Online. In Rogers, P., Berg, G., Boettcher, J., Howard, C., Justice, L., and Schenk, K. (Eds.) *Encyclopedia of Distance and Online Learning*. Hershey, PA: Information Science Reference.

APPENDIX A

Discussion Board Rubric

1. There are three parts to each discussion board entry: **Initial response** (where you answer the question, issue, etc. and start a new thread on the discussion board), Peer **reaction** (where you react to your group members' responses.), **reply** (where you reply to each reaction to your response). Each time you make an initial response you will be assessed on the depth, length, and content of your response, as well as your reactions and replies to others. Mechanics will also be assessed. **Worth 10 points each response/reaction/reply.**
2. Note: In general, your "initial response" is **Due Wed., midnight of the lesson week;** "reactions" are **Due Sat., midnight of the week,** and replies are due the following **Monday**.
3. Mechanics of initial responses and peers' reactions DO COUNT! Points will be taken off for any posting that has more than 3 mechanical errors.
4. Another important thing to remember is to use the Reply button when you are responding to another person's posting. **Do not create a new posting.**

Table 3.

	1	2
Content of initial response	The initial response did not focus on the question at hand and/or was posted late.	The initial response focused on the question at hand and was posted by Wed., midnight of the week it was assigned.
Depth of initial response to lesson question(s)	The response was "hurried" or weak but at least it brought up one or more new ideas not found in the text or lesson. It was less than 150 words.	The response was thoughtful and brought up at least two new ideas not found in the text or lesson. It was at least 150 words.
Content of reactions to peers	The reactions were not polite and/or did not keep to the person's topic(s) or was late.	The reactions were polite and focused on the topic(s) that the person raised, and was posted by Sat., midnight of the week it was assigned.
Depth of reactions to peers	The reaction was "hurried" or weak but it brought up one or more new ideas not found in the person's response.	The reaction was thoughtful and insightful. It raised at least two new ideas, related to the person's response.
Reply to peers' reactions	The replies were not thoughtful, nor did they answer the questions posed by peers.	The replies were thoughtful and insightful. They addressed any questions the peers raised.

APPENDIX B

Figure 2. Seth and peer group's interactions

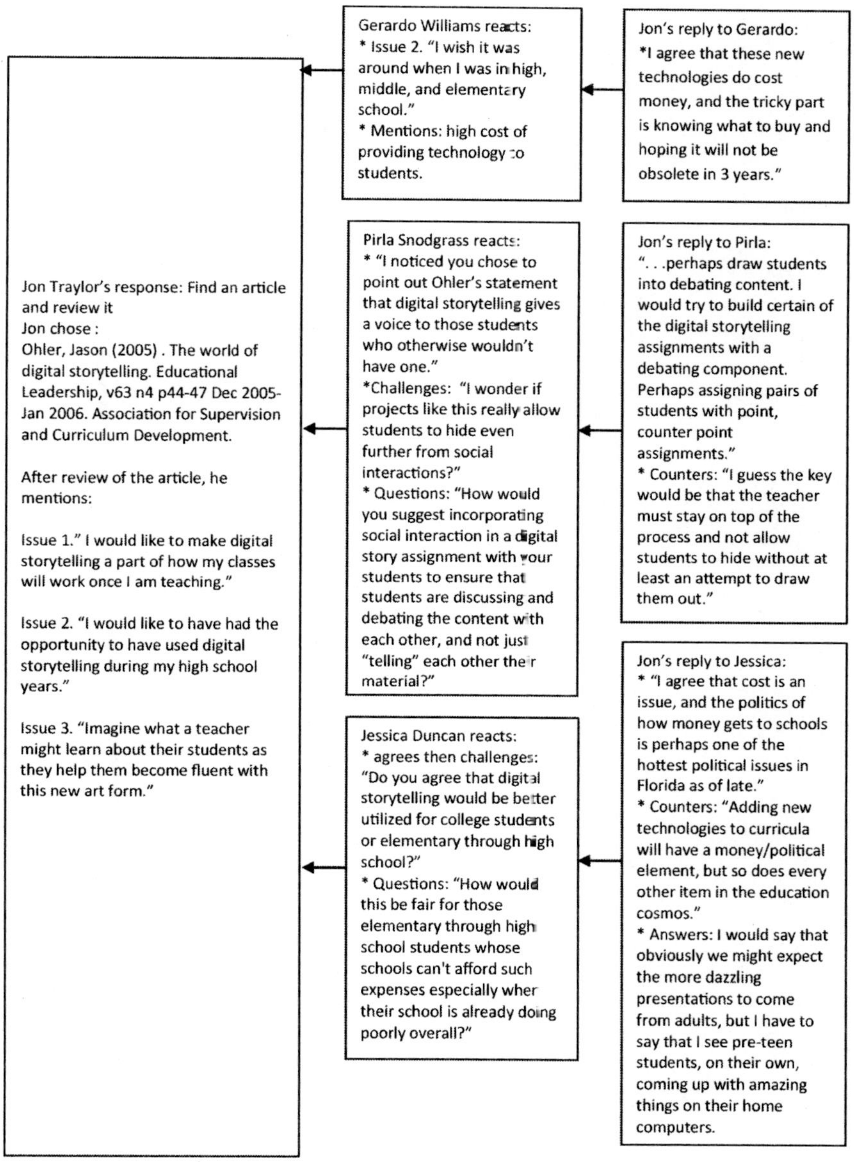

APPENDIX C

Figure 3. Relationship of efferent/aesthetic response to connected/disconnected stance

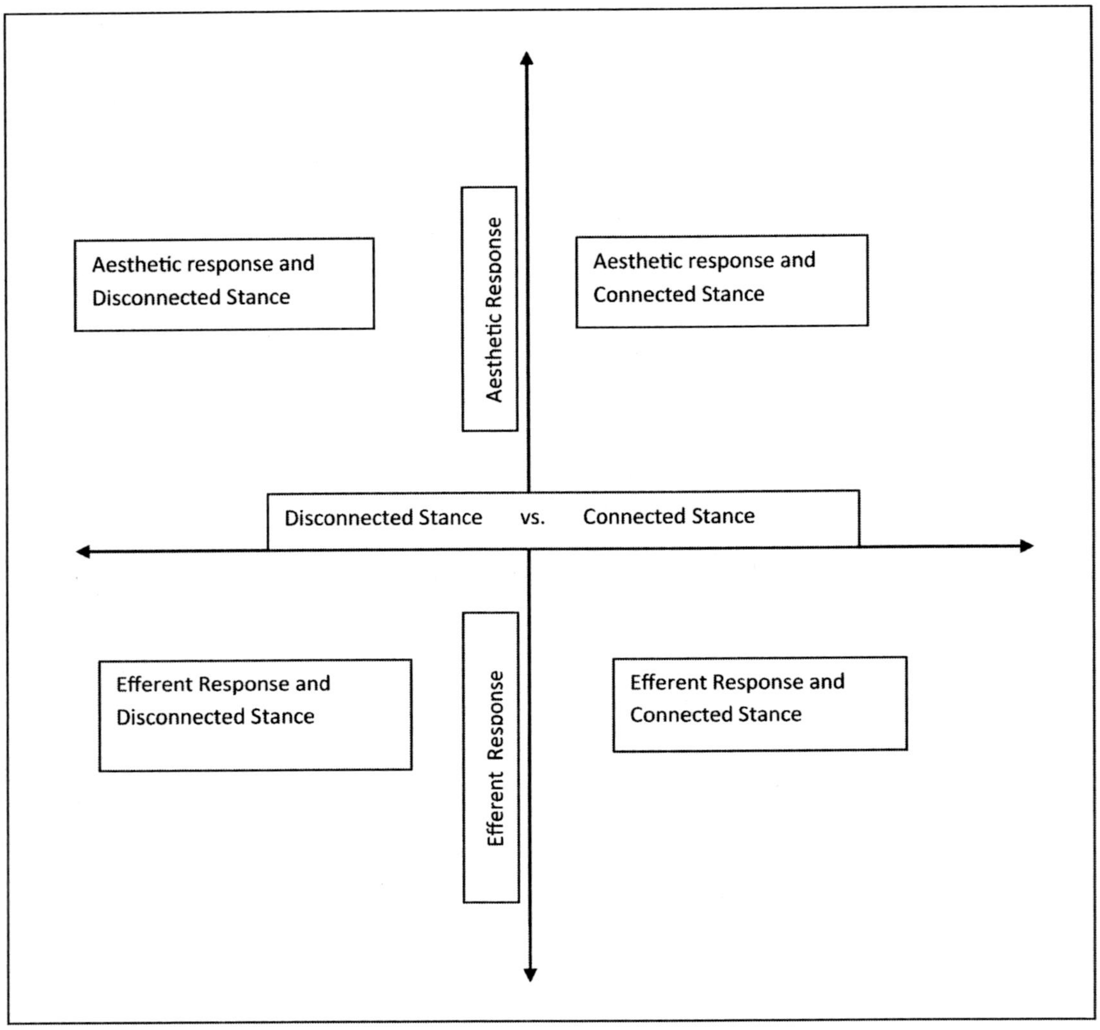

Section 4
Special Considerations in E-Learning and Development

Chapter 16
Building Quality Assessment into Online Courses Across the Institution

Michael L. Rodgers
Southeast Missouri State University, USA

EXECUTIVE SUMMARY

This case shows how a long-term, campus-wide effort balanced technological, pedagogical, financial, and political considerations to develop and implement a system for online course quality assessment at a medium-sized public university in the Midwest. The case shows how the need for an assessment system came to be recognized, and how the committee charged with creating the system arrived at a solution which took into account both course design and instructor performance. Thus, the institution now has, for the first time, a tool for improving the quality of its online courses. Moreover, it is hoped that administrators, faculty, and faculty developers will see that the quality assessment system joins a course management software suite development effort and a series of faculty training workshops in a wide-ranging list of tools for enhancing faculty competence as users of technology for teaching and learning.

THE CASE

This case study describes the events that led to the development and implementation of a system to assess the quality of **online courses** offered by a comprehensive, Master's-granting, public university in the Midwest. The University, which was founded in the 19th Century as a Normal School, enrolled 8000 students by the mid-1990's; enrollment has since grown to almost 11000, with almost 11% of credit hours now generated from **online courses**. From 1997 to 2008, the faculty grew from 380 to 400 in number. As the case study reveals, the institution's regional mission - shaped by sensitivity to the disadvantaged economic condition of much of the region - was a major factor in the development of a strong schedule of **online courses**.

While implementation of a system for **quality assessment** of **online courses** is the ultimate focus of this case study, it is useful to trace the history of

DOI: 10.4018/978-1-60566-942-7.ch016

online courses at the institution, for it is through the history that we see how the users of the assessment system – faculty and administrators – became educated in the art and science of online teaching by the development process itself. Had that education not occurred, the assessment system would have been impossible to develop, and meaningless to implement. Much like a qualifying exam in a Master's program, the institution's **quality assessment** system serves simultaneously as an indicator of how well the institution responded to the challenge of online teaching, and a predictor of its future success. The process described in the history was not perfect. Rather, it was the product of many faculty, professional staff, and administrators, serving on several committees over many years. The process was shaped by political imperatives, funding levels, and, importantly, the time that the various committee members were able to commit to the effort. Despite the many variables, the institution came to consensus on a system for assessing online course quality. Perhaps because of the many variables, this case study offers its users insight into the complex problem of online course assessment.

Creating an Online Course Capability

The following is a highly paraphrased conversation between members of the Technology Associates, a **technology** steering committee consisting of a member from each academic division, and formed to assist the University's teaching and learning Center, *circa* 1995:

The Committee Chair: "You know that we recruit students from every part of the United States and beyond, but our primary service region consists of 26 nearby counties. And some of them are among the most economically disadvantaged counties in the nation."

Committee Member #1: "Yes, I read Census data showing that some of those counties are actually losing population ...."

Committee Member #2: "Mmmm ... other than some farming, there's not much going on there." (Mumbles of agreement from Committee Members #4, 5, 6, and 7).

The Chair: "That's right, but they're in our service region, and we need to be more effective in providing them with educational opportunities." (More indications of agreement from the Committee.)

Committee Member #3: "Well, students could drive to campus, but they'd have to be awfully motivated. I once did a workshop for a school district at the border of our region; I had to drive over two hours on narrow, winding roads to get there. I felt like I'd fallen off the edge of the world!"

Committee Member #2: "I know we send faculty to teach night courses, using rooms in some of the high schools. Surely we could just increase the number of those classes."

The Chair: "A major problem is that faculty don't like going off campus to teach. Even if we pay mileage to drive there, faculty complain that driving time doesn't count for promotion and tenure. In fact, some departments won't even count off-campus courses toward faculty teaching loads. We can't afford to open a branch campus in every town, and students – especially those with jobs or families – don't have time to drive to the main campus to take courses during regular class sessions. Besides, the cost of transportation alone may prevent many students from taking courses and pursuing a degree. And that brings me to the purpose of our meeting today. The Center recently received mission enhancement funds from the State for online course development; we need to discuss how we'll use that funding to deliver **online courses** *to students."*

Committee Member #1: "I've heard of those. Students get course materials over the Internet."

Committee Member #2: "That's part of it. Students would download content, but they would also be able to check grades, communicate with the instructor, and essentially do everything that students now do in our on-campus classes. If the **online course** *only distributed content, we may as well offer correspondence courses. In an online course, students actively communicate with the instructor, using e-mail, listservs, or discussion boards on the World Wide Web." (Heads nod in recognition of the differences between* **online courses** *and correspondence courses.)*

Committee Member #3: "Even if you could find an instructor willing to teach online, how can the university support the course? Have you tried to dial into the campus network from home lately? We only have 16 incoming dialup lines, and those go down all too often. Even when they are not down, I have to redial and redial just to log in, because the lines are usually all busy. Does anyone here know an instructor who has taught an **online course**? *I don't."*

The Chair: "OK, let's follow up on that. Let's make a list of everything we need to successfully offer an **online course**.... *"*

Within a few minutes, the Committee came up with a preliminary list of resources needed to teach a course online, including:

1. Ready access to the campus network (in 1995, that meant more dialup lines)

2. A highly reliable server to deliver content to students any time of the day

3. **Course management software** *(CMS) to manage student access to the course, secure grade display, and other administrative and pedagogical functions*

4. Systems to provide services to online students comparable to those students receive on campus (for example, textbook services, library, student counseling services, and Writing Center assistance)

5. An instructor who is comfortable teaching with **technology**

Soon after the meeting, the campus began to change. Network drops were installed in faculty offices. More computer labs were opened on campus. It became much easier to dial into the campus network from home, and important resources, such as the Library's card catalog, were accessible from the home connections. A few students and faculty began to use e-mail to communicate with each other. It was obvious that Items 1 and 2 in the Committee's list were being addressed, as the institution made a major investment in computer infrastructure. How would the institution address Items 3-5?

By 1996, the **technology** committee decided that faculty "access" to **technology** was key to developing an institution-wide online course capability. But "access" meant much more than building fast networks, wiring offices with network drops, and making powerful software available. For faculty to actually *use* the resources and infrastructure required by online courses, faculty required training, support, and successful experiences with **technology**. Indeed, we came to understand that a "culture of **technology** use" among the faculty had to be developed, before online courses could become common at the university.

At the same time, the University's teaching and learning Center, in collaboration with the Technology Associates, worked to find a suitable CMS solution for the campus. Over a two-year period, we reviewed several browser-accessible systems (Albright, 1997) that were available at the time, notably Blackboard, Mallard, Web Course in a Box (WCB), and World Wide Web Course Tools (WebCT). Cost was important to our decision, but there were also concerns about customer service: would the software be customizable to our specific needs, and would faculty be able to get tech support appropriate to their "rookie status" as CMS users?

Eventually, our analysis of costs and faculty needs led us to build our own CMS, a product that has become known on campus as the Online Instructor Suite, or "OIS." While a "homegrown" CMS might not have as many features or resources as the leading commercial products, we saw great value in directly involving faculty in decisions about the pedagogy that our CMS might support: How is learning supported by linking lists of grades on a grade display page to the actual assignments and course content? What is the most effective way to carry on an asynchronous online discussion? What question formats should the online testing module support? Users were also encouraged to contribute their own ideas. Indeed, by deeply involving our faculty in the CMS development process through the equivalent of a community of practice, we hoped that access at this level would contribute to the creation of a cadre of faculty capable of teaching online.

Moreover, the modular design of the OIS – originally intended to permit piecemeal deployment of the CMS, rather than waiting for a complete product from our very small programming team – also promoted faculty understanding of the CMS software, and online teaching in general. Modules for gradekeeping, discussion, and test-taking obviously extended to the online environment activities that are common in traditional face-to-face courses. With those modules providing basic orientation to online teaching, instructors could then advance to less familiar activities and concepts, such as collection of data through HTML forms, production of streaming media, and submission of grades to the Registrar from an online gradebook.

Despite the CMS' high value as a teaching tool, we recognized that direct training was essential for producing online instructors. We therefore inaugurated a series of Technology Serving Learning Institutes[1] in 1997 (Rodgers & Starrett, 2000). The TSL Institutes were offered exclusively to faculty, and they coincided with a major effort by the University's Computing Services unit to upgrade networking and connectivity assets on campus. Although participation in the Institutes was entirely voluntary, the rapid and widespread upgrades of desktop computers in faculty offices, vastly improved Internet access, and adoption of standardized word processing and data management software, combined with small faculty development awards, free software, and other incentives to make the Institutes quite popular. By 2002, over 75% of all faculty on campus had attended at least one Institute session.

Early TSL Institutes were scheduled for late May or early June, soon after the conclusion of Spring Semester. At first, the Institutes were organized around a 5-8 day, short course model, designed to produce **technology** "experts" – faculty recognized within the academic departments for their expertise. The Center's **technology** committee planned the Institutes, including selecting topics, setting the schedule, identifying faculty to serve as facilitators, setting participant incentives, and publicizing the program. Despite their popularity with faculty, the short course model failed to leverage previous faculty experience, and some faculty found the lengthy time commitment difficult to honor. Moreover, the model did not account for demands imposed on faculty by course content, accreditation agencies, and differences in resources from department to department.

As we gained experience with the Institutes, we explored alternate models, eventually abandoning the short course model for an *a la carte* menu of sessions inspired by the "Low-Threshold Applications and Activities" approach championed by Steven W. Gilbert and the TLT Group (Gilbert, 2002). The Institutes were also expanded to include sessions in mid-August and early January. Experiments with online interest groups, communities of practice and brief lunchtime sessions ("Wired Wednesdays") complemented the Institutes. It soon became obvious that participation incentives were not needed to sustain interest, permitting a phaseout of participant incentives, from $37 per participant-hour in 1997 to $0 in 2001. Despite the incentive phaseout, participation by faculty grew from 46 in 1997 to 117 in 2000; since 2000, participation has stabilized at ~150 faculty per year, of which much less than half are first-time participants.

The consistently high turnout for the TSL Institutes – with special emphasis on the large percentage of repeat participation - was only one indication that a "culture of **technology** use" had been created on campus. In the late 1990's, a time when less than 20% of the campus community had e-mail access, faculty joined students and staff in vigorous campaigns to get more of the campus wired, and to expand connectivity for off-campus users. Universal online submission of student grades was implemented in 2002, and the demand for conversions to **technology**-enabled classrooms became so intense that friction between the "haves" and the "have-nots" became apparent, especially during bad budget years[2]. Without the imposition of any campus-wide requirements to do so, content-rich Websites were built for hundreds of face-to-face courses, and a third of the faculty began using the institution's **course management software** to securely disseminate grades to students.

Why did faculty accept **technology**? To be sure, acceptance was not a foregone conclusion: the substantial commitment of time and energy to **technology** training was seen by some as a diversion from faculty's responsibility to content within the disciplines. High-value alternate uses for the money spent on **technology** could always be cited, and, importantly, the notion of interposing **technology** between student and instructor threatened to isolate and depersonalize students. Nevertheless, faculty embraced **technology** for several mutually-supporting reasons:

1. ***Technology was the focus of one of six Priorities in the University's Strategic Plan.*** The Strategic Plan, adopted in 1996, was the result of intense collaboration by all units on campus, including every Department and College. The specific language clearly stated the role of **technology** in teaching and learning:
   - INFORMATION TECHNOLOGY
   - In order to enhance student learning by making optimal use of information technology and to develop/adopt innovative applications of technology throughout the university, we will:
   - Network the campus and give it full access to the information superhighway.
   - Extend access to information technologies to faculty, staff and students through systematic acquisition of hardware, software, and by providing training opportunities and support.
   - Utilize technology to streamline all university administrative services.
   - Maintain an up-to-date "information

technology plan" to guide the coordinated acquisition of hardware and software, personnel training, and support-service development.
- Support [the campus] Library in its transition to a more complete information-access resource.
- Extend distance learning opportunities via technology.
- (University Strategic Planning Committee, 1996)

2. ***Key administrators promoted technology for teaching and learning.*** The Provost, the Vice-President for Academic Affairs, and the Dean of the institution's general education unit provided early leadership in the effort to develop the TSL Institutes, including work to secure the mission enhancement funding.
3. ***Stakeholders in the promotion and tenure process publicized ways to "count" technology use.*** A faculty action team produced guidelines that promotion and tenure committees could use to value a faculty member's use of **technology** in all three areas evaluated when a candidate stands for promotion or tenure: teaching, professional growth, and service (ITFRR, 2000).
4. ***Students wanted a technology-enabled learning environment.*** Consistent with EDUCAUSE President and CEO, Diana G. Oblinger's presentations on the Net Generation (Oblinger, 2004), our students requested **technology**-enhanced learning environments that could enhance, but not replace, student-instructor interactions.
5. ***Faculty became convinced that technology could improve teaching and learning.*** From the start, great care was taken to ensure that the TSL Institutes balanced **technology** and pedagogy. From the selection of session facilitators from among faculty who actually used the **technology** in teaching, to the use of Chickering and Gamson's landmark *Seven Principles for Good Practice in Undergraduate Education* (Chickering & Gamson, 1987) as a framework for planning and session design, the Institute sessions presented **technology** as a tool to support good teaching. Many sessions, such as "Using Questioning and Reflection to Lead to Critical Thinking on the Web" so strongly emphasized online pedagogy that **technology** was little more than a footnote. Pedagogical insights, rather than the **technology** itself, or even institutional benefits (for example, using **technology** to open new markets to the institution, manage higher enrollments, or save money), remained the Institutes' focus.

A Harvest of Online Courses

When the TSL Institutes were inaugurated in 1997, the goal was to develop in the faculty the ability to develop and teach **online courses**. That year, no **online courses** were taught at the University. Two years later, the only **online course** offered was taught by an early adopter who certainly did not need the help of the Institutes to develop and teach an **online course**. Indeed, many of the early Institutes were designed to help faculty become comfortable with **technology**, as a gateway to **online course** development. Thus, sessions were offered that featured software which could support both online instruction and traditional face-to-face instruction: PowerPoint and Photoshop were two especially popular examples. By 2003, however, the number of **online courses** had grown to 150, and the University was able to offer its first fully online baccalaureate degree. Nearly all of the **online courses** were developed by Institute participants. The number of **online courses** continued to grow, to more than 250 by 2007, with approximately 160 **online courses** offered in any semester. By 2008, even the University Honors Program included **online courses**. In addition, many Web-enhanced ("blended") courses are regularly taught by Institute participants.

The **online courses** were well-received by the institution's students and the public. Internal student satisfaction surveys showed slightly better retention rates than in face-to-face courses, and students showed a high degree of comfort with the online environment, as suggested by the low (25%) percentage of students disagreeing or strongly disagreeing with the statement, "I am better able to visualize the ideas and concepts taught in this course than I am in face-to-face courses" (Rodgers and Lockhart, 2008). Students also appreciated the greatly improved flexibility in scheduling. Not only were students able to fit courses into busy work and family schedules, but scheduling classes became easier, because there were fewer time conflicts among the courses needed for a degree. Moreover, implementation of **online courses** served to boost enrollments in upper-division and graduate courses that, as face-to-face courses, would sometimes be cancelled for low enrollment. The high visibility of the online course program gave prospective students and parents a sense that the University was on the "cutting edge" of **technology** – a new position for the University, which had a long history of underfunding **technology**.

University administrators also found reason to celebrate the **online course** movement. Offering online sections has opened up some courses to entirely new clientele. For example, in one course that meets a programmatic need in the Nursing Program, and also meets a general education science requirement, almost 80% of students in the face-to-face sections are freshman Nursing majors, but in the online sections, over 80% are third- or fourth-year students in a major other than Nursing. Obviously, the online format is steering busy upper-class students into the course. The online format also provides benefits in Summer Session scheduling: moving courses online allows students who leave town for the summer to continue to take courses from the University, without the uncertainties associated with transfer credit.

A Maturing Perspective

The University was by no means alone in its quest for **online courses** in the 1990's. However, the University was careful to distinguish its development efforts from some other institutions' efforts through careful coordination with enduring institutional goals and values, as identified in strategic planning exercises. A solid reputation for high-quality teaching was repeatedly cited by surveys, consultants' reports, and other activities associated with the University's strategic planning efforts. Consequently, **online courses** were from the very beginning intended to exemplify quality instruction, and a stated goal of the online course development effort was to produce courses "equivalent" to face-to-face courses bearing the same catalog number. Indeed, student transcripts at the University have never made a distinction between **online** and face-to-face courses. Many courses that served as prerequisites to other courses were developed for online delivery, a sign that "**equivalence**" was taken seriously. Certainly, the "**technology** with pedagogy" approach taken by the TSL Institutes served as a testament to the notion that "**equivalence**" was possible through high-quality teaching in **online courses**.

Despite the focus on quality of instruction, competitive pressures to be first to market, and uncertainty associated with rapidly evolving **technology** led the University to pursue a "numbers first" strategy in the early days of the **online course** development effort. This strategy was reasonable, in light of the need to compete for resources on campus, and to maintain a high degree of visibility for the newly-developed **online courses**. Moreover, a strategic decision to focus on the number of **online courses** developed would promote efforts to build entire online degree programs. The Provost, Academic Affairs Vice President, the Director of the office charged with administering and promoting online education, the Director of the Teaching and Learning Center, and other advocates of **online courses** within the Administration, all seemed to be willing to rely on the long-standing institutional value of teaching excellence, coupled with the Institutes' emphasis on pedagogy, to generate **online courses** that were "good enough" to credibly represent the University, and attract students over the years during which the catalog of **online courses** was being built.

In essence, online course quality was left to instructor discretion: there was no system in place to comprehensively assess **online course** quality. Other than a direct comparison of the content, there was no way to gauge whether or not the goal of "**equivalence**" had been met. While Department Chairs were generally supportive of faculty efforts to develop and teach **online courses**, many Chairs came to understand that neither the online courses themselves, nor the performance of the instructors assigned to teach those courses, could be adequately assessed using existing mechanisms for evaluating instructor performance[3]. Of course, the Chairs would have no problem judging course content; the problems arose with the need to determine how well course **design** aligned with generally recognized **best practices** for **online course design**. Also, some Chairs required guidance assessing how well the technologies around which the course was built – the **course management software**, social networking tools, multimedia, calculation and graphing tools, etc. – "fit" the content.

At the same time that Chairs were beginning to express the need to assess course quality, marked increases in students' technology sophistication provided additional impetus to develop an assessment system. Increasingly, students sought online courses that took advantage of advanced tools, and especially, meaningfully engaged learners with the content, the instructor, and with other students. We soon realized that we needed an instrument that could be applied to each **online course**.

Developing an Online Course Assessment Tool

Responding to the manifest need for assessment, the Provost charged the University's teaching and learning Center with the task of finding or developing an appropriate procedure for assessing **online course** quality. Within the Center, the task was turned over to the Technology Associates, so as to take advantage of the group's expertise with **technology** for teaching and learning. Because the Technology Associates was a steering committee comprised of faculty from the Library and each School and College in the University, selection of the group was politically prudent, as it was expected that the assessment procedure would be a single, generalized assessment process applied to all online courses at the University. A single **rubric**, applied campus-wide would minimize charges that courses from specific Departments are being targeted for extra scrutiny. Adopting a single **rubric** would simplify training of faculty and Chairs, and, importantly, provide the teaching and learning Center with a wealth of data that could be studied, for the purpose of improving all **online courses**. The process would not duplicate existing measurements of learner outcomes: assignments such as exams and papers would still measure individual student learning. Nor would the process assess the performance of degree programs; that function would be left to standardized exams, accreditation committees, exit surveys, and other traditional measures. Rather, the Technology Associates sought a process that would provide evidence for the quality of each **online course** as a learning environment: was the course **design**ed to present content in a meaningful, readily accessible way? Did the course support student engagement with content, the instructor, and, where appropriate, other students? Was the pedagogy sound? Did students have ready access to support services (tech support, library, academic support services)?

From the beginning, the Technology Associates realized – as did the Chairs - that merely adapting existing mechanisms for assessing the quality of face-to-face courses to **online courses** would not work: a one-page checklist completed by a senior faculty member during a scheduled visit to a probationary faculty member's classroom might suffice to identify instruction in dire need of remediation, but a 50-minute visit to an **online course** - especially when the course is taught asynchronously – could not be expected to give the deep insight into a course's **design** and delivery that the Technology Associates believed necessary for meaningful assessment.

Research into existing **online course** assessment models revealed a rich literature, much of which was grounded in student evaluation of instruction. A published literature review (Ali, 1998) and the IDEA Papers Website (The IDEA Center, 2008) helped us to better distinguish between our existing online student evaluation of instruction, hosted by the IDEA Center, and the proposed assessment system. Student evaluations are useful because they record perceptions of course **design** and teaching practices from a client's perspective; our new system would view the same **design** and practices from a colleague's perspective. For the new assessment system to have value, the faculty and administrators who would use it **must** be informed by awareness of sound pedagogy, as applied to online teaching.

We had concluded that the assessment **rubric** should be designed for use by colleagues who have insight into quality teaching. But, the teaching and learning Center's earlier work to help promotion and tenure committees know how to "count" faculty **technology** use (ITFRR, 2000) left the Center and the Technology Associates strongly affirming the scholarship inherent in activities to develop and teach courses in the new and largely unfamiliar online environment. Happily, this view coincided with Ernest Boyer's broad view of scholarship, as set forth in *Scholarship Reconsidered* (Boyer, 1990). Boyer's name commands a special degree of respect on campus, largely because he was an early (1980's) supporter of our University Studies Program. Thus, the notion that **online course** quality should be assessed by insightful colleagues seemed reasonable.

A survey of existing models for **online course** assessment revealed several efforts. We specifically sought models grounded in the research. Like our effort, some were centered at academic institutions, such as California State University, Chico's **Rubric** for Online Instruction (ROI) (California State University, Chico, 2006), and the University of Texas' Criteria for Evaluating **Online Courses** (DIIA, 2008). Others, such as the **Online Course** Evaluation Project (OCEP) (Monterey Institute for Technology and Education, 2006) and **Quality Matters** (MarylandOnline, Inc., 2006), were offered by educational organizations. The Technology Associates found the **Quality Matters** (QM) peer review process most attractive for its **rubric**'s thorough coverage of course **design**, its solid grounding in research, a "faculty-centered" approach to assessment that places the instructor on the review team, and the training of review teams trained to consider the course from the students' perspective. A survey of the extensive literature review (MarylandOnline, Inc., 2005) used to construct the QM **Rubric**'s Review Standards found it to be quite credible, although reconstructing the links between specific articles and review standards proved to be very time-intensive. The Technology Associates were so impressed with QM that four members of the committee requested funding from the teaching and learning Center to take the QM online training. The Center's Director enthusiastically covered the training fees. The training provided many valuable insights into **online course** assessment that were often interjected into the committee's discussions. The training also underscored a critically important fact about the QM **rubric** that was already known to several on the committee: **the QM rubric intentionally addresses course design only; instructor performance is not assessed**.

The stated needs of both the Department Chairs and students led the Technology Associates to an assessment tool much broader than the QM **rubric** – one that would probe instructor performance and offer examples of discipline-independent **best practices** for the items that eventually comprised the draft **rubric**. The 26 items were organized into four broad categories (Technology Associates, 2008):

- **Course overview, introduction, and learning objectives (competencies)**, which generally serve to make course organization and learning objectives clear to students,
- **Assessment and measurement**, which measure student progress toward stated learning objectives,
- **Resources, materials, and learner support**, which explore how well the instructional materials selected for the course support the learning objectives, and
- **Course technology and security**, which gauges how effectively the **technology** used in the course supports instruction and promotes student interactivity.

Like the QM **rubric**, the Technology Associates' document distinguished items on the basis of their importance to course quality: all 26 items were classified as either "Required" or "Recommended."

From Development to Implementation

The draft **rubric** was formally presented to Department Chairs at a special, invitation-only TSL Institute session in January 2008. The Deans' Council saw the document in Spring 2008. The **rubric** was well-received, but both groups offered suggestions for refinement. Interestingly, the **rubric** development process revealed some significant differences in opinion regarding the **rubric**'s nature and function. The Deans wanted a **best practices** document, not a peer-review **rubric**, as the Technology Associates envisioned. The Deans also asked for a list of **best practices** for teaching content within the disciplines. The Technology Associates balked at that request, on the grounds that adding such a list would make the **rubric** unwieldy, and would usurp the established role of faculty in the disciplines to decide how discipline-specific content should be taught. The Chairs requested the addition of a checklist, consisting of a synoptic version of each item, characterization of each item as "Required" or "Recommended", and a checkbox for the evaluator's use, to indicate whether each item was present in the course. The checklist was added is a way that suggests to the reader that the document can be used with or without the checklist. Chairs could thus easily employ the document as a formalized assessment instrument, and faculty could use the document as a self-study tool, or a guide during course renovations. Faculty were also provided with a preview of the document at a May 2008 TSL Institute session (Rodgers, 2008).

The final version of the document was disseminated campus-wide in mid-2008. Although the document is available at the teaching and learning Center's Website, faculty are invited to participate in TSL Institute sessions that investigate the 26 items in detail, and provide faculty with guidance on using the document in an **online course** assessment. The sessions will also explore ways to maximize the value of a course assessment for faculty who will stand for promotion or tenure.

Conclusion

At the University, there is a sense that the **online course** has found its place on the list of course delivery methods best suited to meet the needs of our students and the institution. New **online course** development continues, but at a slower pace than before, with greater attention being given to learning outcomes, student satisfaction, and wise management of resources. I suspect that other institutions are experiencing similar changes. It is our hope that this case study helps faculty, administrators, and professional staff at other institutions, who are ready to embark on the next phase in the evolution of the **online course** – a phase which focuses on matching increasingly powerful **technology** with good pedagogy and effective **design**, to create learning environments that serve students extremely well.

Perhaps as significant as the development and implementation of the **online course** assessment system is the system's value as a faculty development tool, potentially as important to the work of building online teaching skills as the decision to involve faculty in the formation of the Online Instructor Suite, and the effort to offer the Technology Serving Learning Institutes. Our assessment system's nascent importance as a validation of teaching effectiveness for promotion and tenure decisions, and its intended use by post-probationary faculty, promise to spread precepts of good pedagogy to all corners of the institution. It seems that the **online course** initiative has provided a focal point for participatory faculty development: through software specification (OIS development), **technology** training with a pedagogy emphasis (TSL Institutes), and now, application of **best practices** (the assessment system), faculty can hardly miss the points of contact between online teaching and good practice.

REFERENCES

Albright, M. J. (1997). Web Course Authoring and Management Systems. *MC Journal: The Journal of Academic Media Librarianship, 5*(1), 14–23.

Ali, D. L. (1998) *Appendix 4: Literature Search to address some commonly expressed questions regarding student ratings of instruction*. Retrieved March 5, 2009 from http://www.ucalgary.ca/UofC/departments/VPA/usri/appendix4.html

Boyer, E. L. (1990). *Scholarship Reconsidered.* San Francisco: Jossey-Bass.

California State University. Chico (2006). *Rubric for Online Instruction*. ROI Rubric for Online Instruction. Retrieved March 5, 2009, from http://www.csuchico.edu/celt/roi/

Chickering, A. W., & Gamson, Z. F. (1987). Seven Principles for Good Practice in Undergraduate Education. *AAHE Bulletin, 39*(7), 3–7.

Division of Instructional Innovation and Assessment. (2008). *Criteria for Evaluating Online Courses*. The University of Texas at Austin. Retrieved March 5, 2009, from http://www.utexas.edu/academic/cit/howto/resources/assessment/evallearning.html

Gilbert, S. W. (February 2, 2002). The Beauty of Low Threshold Applications. *Syllabus*. Retrieved November 25, 2008 from campustechnology.com/articles/38981_2/. See also www.tltgroup.org/ltas.htm.

Information Technology Faculty Roles & Rewards Action Team (ITFRR). (2000). *Information Technology Promotion, Tenure, and Merit Guidelines*. Retrieved November 25, 2008 from cstl.semo.edu/ITFRR/.

MarylandOnline, Inc. (2005). *APPENDIX 1: References. Quality Matters Inter-Institutional Quality Assurance in Online Learning*. Retrieved March 5, 2009, from http://www.qualitymatters.org/Rubric.htm

MarylandOnline, Inc. (2006). The Quality Matters Rubric. *Quality Matters Inter-Institutional Quality Assurance in Online Learning*. Retrieved November 25, 2008, from http://qualitymatters.org/Rubric.htm

Oblinger, D. G. (October 20, 2004). *Educating the Net Generation*. Featured Speaker Address at the Educause 2004 Conference, Denver, CO.

Rodgers, M. (May 28, 2008). The Online Class Rubric and You. *TSL 2008 Summer Institute.* Retrieved November 25, 2008 from http://cstl.semo.edu/institute/2008Summer/Rubric/Rubric.htm

Rodgers, M., & Lockhart, F. (2008). The Dog Ate my e-Homework: Excuses, Academic Honesty, and the 21st Century Learner? *TSL 2008 Summer Institute.* Retrieved November 25, 2008 from http://cstl.semo.edu/institute/2008Summer/Edog/edogschedule.htm

Rodgers, M., & Starrett, D. (March 2000). *Technology Serving Learning Summer Institutes.* Paper presented at the Syllabus Conference, Orlando, FL.

Technology Associates. (2008). *Online Course Best Practices Checklist.* Center for Scholarship in Teaching and Learning, Southeast Missouri State University.

The IDEA Center. (2008). *IDEA Papers.* Retrieved March 5, 2009, from http://www.theideacenter.org/IDEAPapers

University Strategic Planning Committee. (1996). *The Southeast Missouri State University Strategic Plan.* Office of the Provost, Southeast Missouri State University.

ENDNOTES

1 Colloquially named "TSL Institutes" or simply, "Institutes" on campus, and in this case study.

2 The post-911 economic downturn, for example.

3 For example, the yearly classroom evaluations of probationary faculty conducted by senior, tenured faculty from the same Department.

Chapter 17
Case Study of the CUForum @ CUHK

Peter Jakubowicz
The Chinese University of Hong Kong, Hong Kong

EXECUTIVE SUMMARY

In contrast to the formal school setting where learning is often linear, structured and controlled (be it online or face-to-face), for the 'net generation,' (Google, MySpace, MSN, YouTube and Yahoo) learning is often incidental and a sense of 'fun' is frequently of great importance. Such students' learning is often non-linear, unstructured and explained well by the tenets of Anderson's theory of online learning. This research discusses the benefits of fostering non-linearity in an online learning environment. A case study of an online business communication course at a university in Hong Kong is used to illustrate the importance of non-linear online learning by demonstrating how participants in this course adopted learning approaches that are consistent with, and a reflection of, the theory of online learning. Qualitative data from complete sets of online communication (including focus group interviews) collected over a one-semester, tertiary level course conducted at a university in Hong Kong are analyzed. The findings show that Chinese-speaking learners' online interactions, categorized into three broad areas (cognitive, affective and social), demonstrate that interactivity is a key feature of an online learning environment. Its nature is exposed and discussed, not least the finding that for the participants in this study, learning was incidental and a sense of 'fun' was important. The study suggests ways in which online theory can contribute to, as well as help in, understanding this phenomenon and makes recommendations for future research.

DOI: 10.4018/978-1-60566-942-7.ch017

BACKGROUND

Drawing from more than ten years experience in using Web-based courses at The Chinese University of Hong Kong (CUHK), the author aims to explore the application of the CUForum (a course management platform, similar to WebCT and Noodle, and developed by the Information Technology Services Centre (ITSC) at CUHK in enhancing *interactivity* in an online learning environment. Online learning, according to Anderson (2004), is defined as

the use of the Internet to access learning materials; to interact with the content, instructors, and other learners; and to obtain support during the learning process, in order to acquire knowledge, to construct personal meaning, and to grow from the learning experience (p.5).

A one-semester, thirteen week (Sept. – Dec. 2004) online business communication course for 3rd. year students studying in the Faculty of Business Administration in a tertiary-level institution in Hong Kong is used here as the database for this study. This course is offered for final year students and consists of advanced levels of business communication, namely, writing proposals and case studies, giving presentations, writing business-related documents such as resumes, cover letters, memos and emails. In addition, this course prepares the students to enter the workforce by showing them job search techniques and interview skills. All online correspondence on the CUForum is done in English, with the specific objective of improving the students' English language skills. This Web-based course is conducted both in a computer lab (with every student having access to a computer) and outside of the classroom through a high speed Internet connection called WiFi (available 24 hours/day, 7 days a week). The course plan stipulates that 10% of the final grade is accorded to interaction on the CUForum (i.e. the grade is based on the quality and quantity of the student messages posted and replied to). For the majority of the students in this study, aged between 19 – 23 years old, Cantonese is a mother tongue (90%) and the remainder has Putonghua as their mother tongue (10%). This particular course was chosen at random from a wide range of courses that use online learning as being a representative sample of a specific business communication course using the CUForum.

This investigation focuses on the concept of *interactivity* and considers how users interact with computers in a Computer-Supported Collaborative Learning environment (CSCL) to promote learning (Slavin, 1990; Cooper, 1992; Lai, 1993; Nardi, 1996; Johnson & Johnson 1998; Kekkonen-Moneta and Moneta 2001; Phelps and Ellis 2003; Roskams, 1998; McConnell, 2000; Napierkowski, 2001; Sheard and Markham, 2005; Tu and McIsaac 2002). Walther (1996), states that *interactivity* is the key to communication, and *the concept of interactivity is the key to online communication* (Author's italics). If there is no interaction in an online environment between the participants (student/student, teacher/student), then the CUForum (the online platform under study) remains an empty shell.

In order to study online interactions on the CUForum (student /teacher messages) and to understand the processes that enhance learning, an in-depth analysis using NVivo V.1.3 of the Messages section was conducted. The Messages section was chosen for its rich text-based data (transcribed in over 600 pages of text) and created by the participants (students and teachers) during a one-semester, 13-week business communication course (Sept. – Dec. 2004).

The nature of *interactivity* in a virtual context may or may not differ substantially and qualitatively from that in a non-virtual world, and this needs to be explored further, together with its impact on learning. For example, the lack of sensory stimulation and face-to-face meetings in online learning, the opportunity for private and public messages, the risk of 'flaming' (defined as sending angry or inflammatory messages, either by e-mail or newsgroup posting - Source: http://www.computeruser.com) and so on, are important aspects of *interactivity* and these will be discussed further in this paper.

In particular, in order to have online learning avoid simply being an extension of the transmission model of communication (Chandler, 1994) in another medium using technology, or as Hammer (1990) states, 'paving the cow paths,' the nature of *interactivity* not only *per se,* but in a virtual environment, needs to be discussed. The issue here is that online learning risks being a bolt-on to existing paradigms of teaching and learning, and it is important to connect online learning with newer, more effective forms of teaching and learning *per se*, i.e. regardless of the virtual environment, and then to apply these, where relevant, to virtual learning worlds. In other words, there is a serious pedagogical dimension to the study in that it explores this dimension in relation to *interactivity* both virtual and non-virtual. Hence this study looks at relationships between online *interactivity*, online collaboration and the specifics of the learning process.

In order to address the issue of interactivity, and specifically how Chinese-speaking students interact online, it is important to look at some of the literature on this subject. There is a scarcity of research that focuses on how Chinese-speaking adult students work in, and manipulate, online learning environments (Sheard and Markham 2005; Kekkonen-Moneta and Moneta 2001; Tu and McIsaac 2002). Furthermore, considering the current more powerful Web-based applications included in learning platforms, it becomes even timelier to understand the scope and nature of online interactions as they are seen from Chinese-speaking students' perspective/experience and to explore the *cognitive, affective and social* dimensions of learning.

The nature of online interactivity appears to be quite different from face-to-face interactivity (Anderson, 2004; Jakubowicz, 2006; Kekkonen-Moneta, *et al.* 2001; Tu, *et al.* 2002) and this central theme in the current research has not been explored very much in a Chinese context. The problem is one of under-research and this will be addressed in future studies.

Anderson (2004), writes in his chapter entitled 'Toward a Theory of Online Learning' that 'theory allows—even forces—us to see the "big picture" and makes it possible for us to view our practice and our research from a broader perspective than that envisioned from the murky trenches of our practice' (p. 33). Before considering any theory of online learning, it is important to provide a definition of online learning. Online learning, according to Anderson, is defined as

the use of the Internet to access learning materials; to interact with the content, instructors, and other learners; and to obtain support during the learning process, in order to acquire knowledge, to construct personal meaning, and to grow from the learning experience (p.5).

Anderson starts with a general assessment of how people learn (i.e. What are the specific attributes of learning?). This is based on the research done by Bransford *et al.* (1999). Following that, he provides an assessment of the unique characteristics of the Web, which enable it to enhance the aforementioned generalized learning contexts. Anderson then proceeds to discuss six forms of interaction and he outlines how these interactions engage and support, learners/teachers. Finally, he presents a model of e-learning, which he posits is 'a first step toward a theory in which the two predominant forms of e-learning—the collaborative and independent study modes—are presented' (p. 34). How do people learn? Is there a difference between learning face-to-face and learning online? Garrison and Shale (1990) have argued that online learning could be considered a subset of learning in general. Therefore, issues relevant to how people learn in general are relevant to how they learn in an online environment. Bonk and

Reynolds (1997) suggest that higher-order thinking in Web-based contexts could be developed by creating online learning materials that challenge the learner, link new information to old, and provide meaningful knowledge to the participants. Kozma (2001) argues that computers are required to bring simulations and real-life models to learners so that they can interact with thoes models and simulations. Further, Ring and Mathieux (2002) posit that online learning should be authentic (i.e. students should learn in a context, whether it be a classroom context or a workplace context). Bransford *et al.* (1999) [cited in Anderson, 2004] provide evidence 'that effective learning environments are framed within the convergence of four overlapping lenses. They argue that effective learning is learner centered, knowledge centered, assessment centered, and community centered' (p. 35). As Anderson states, 'Discussing each of these lenses helps us to define learning in a general sense, before we apply this analytical framework to the unique characteristics of online learning' (p. 35). How are the four overlapping lenses defined? Bransford *et al.* (1999) [cited in Anderson, 2004] offer the following definitions:

1. **Learner-centered** learning 'includes awareness of the unique cognitive structures and understandings that the learners bring to the learning context' (Anderson, p. 35). Further he states that, 'the learning environment respects and accommodates the particular cultural attributes, especially the language and particular forms of expression, that the learner uses to interpret and build knowledge' (Ibid, p. 35).
2. **Knowledge-centered** learning does not happen in a content vacuum; it is always grounded in a particular discipline or knowledge base For example, a novice mechanic does not learn to become an expert mechanic outside of a garage or other work/school-related environment. Techniques and thinking skills (and their teaching thereof) have to be grounded in a particular knowledge domain, according to McPeck (1990). In the same vein, Bransford *et al.* (1999) argue that 'effective learning is both defined and bounded by the epistemology, language, and context of disciplinary thought' (cited in Anderson, p. 35). Each discipline or field of study has a specific world view; a world view that offers, in most cases, quite a unique way of comprehending and discussing knowledge.
3. **Assessment-centered** learning, according to Bransford *et al* (1999), should focus more on formative assessment rather than summative assessment. They claim that by doing the former type of assessment learners are motivated, informed and are provided with rich feedback both from their peers as well as their teachers. Online learning offers a wide variety of opportunities for assessment such as online quizzes with instant feedback, peer review of documents, creation of online CVs using templates, etc. More importantly, by using formative assessment, learners are encouraged to assess their own learning and reflect upon what they have learned. Can online learning assessment in the workplace be more effective? Anderson (2004) argues that online learning assessment is enhanced because most workplace adult online learning provides opportunities that 'are project and workplace based, that are constructed collaboratively, that benefit from peer review, and that are infused with both the opportunity and the requirement for self-assessment' (p. 38).
4. **Community-centered** learning offers the crucial component of the social aspect of learning, especially in an online learning environment. This echoes Vygotsky's (1978) concept of social cognition and the creation of new knowledge collaboratively. Lipman (1991) expands on this concept to include the notion of communities of inquiry and

Wenger's (2001) communities of practice. Members of a community of practice challenge and support each other, and in the process, create a relevant and effective knowledge base. Many researchers on online learning highlight the positive aspects of online community creation (Harasim *et al.*, 1995), but there are also its distractors who argue that there are difficulties associated with online communities, most notably the lack of attention and participation (Mason & Hart, 1997). Hine (2000) posits that evidence from studies of the Internet show that when the community is located in a virtual space, there is a lack of a specific geographical location and the lack of a clear identity (i.e. a participant may choose to use an alias and thus be anonymous). These two factors – mutual presence in time and space – and the fact that there is an absence of physical contact (including body language and social presence) may hinder the creation, and sustainability of, an online learning community.

What is the definition of 'interaction' or 'interactivity?' How important is interaction (or interactivity) in an online context? To answer some if not all of the aforementioned questions, Anderson (2004) discusses six forms of interaction: 1) Student-student Interaction; 2) Student-teacher Interaction; 3) Student-content Interaction; 4) Teacher-teacher Interaction; 5) Teacher-content Interaction; and, 6) Content-content Interaction. He proceeds to outline how these interactions engage and support learners/teachers. Anderson (2004) states that 'although interaction has long been a defining and critical component of the educational process and context, it is surprisingly difficult to find a clear and precise definition of this concept in the education literature' (p. 43). For the Net generation everything is interactive! People interact with their PDAs (Personal Digital Assistants), send text messages on their mobile phones that have an Internet connection, programme their LCD TV monitors, work on their notebook computers, play online games, etc. The term interaction could very well describe all of these ways of interacting, and this is why this concept is so difficult to define. Anderson (2003) discussed a variety of definitions of interaction at length, but for the purposes of this discussion, the following definition is useful. Interaction is defined by Wagner (1994) as 'reciprocal events that require at least two objects and two actions. Interactions occur when these objects and events mutually influence one another' (p. 8).

In the educational context, as opposed to the popular culture context, Lipman (1991) argues that interactivity is fundamental for the creation of learning communities, Similarly, Wenger (2001), posits that interactivity plays a crucial role in the formation of communities of practice. In constructivist learning theories (Jonassen, 1991), the value of peer evaluation and assessment (i.e. the value of another person's perspective) is highlighted, and it is through interactivity that a learner acquires knowledge.

How is online interaction different from face-to-face interaction? What are some of the characteristics of online interaction? As mentioned earlier, in face-to-face interactions, participants can not only see the other person's body language, detect an individual's attitude, but can also hear the interlocutor's speech. In online interaction, on the other hand, these physical cues are absent. Online interaction does however afford certain modalities that are specific to the Web medium, as opposed to a regular face-to-face classroom situation. Currently, Web-based learning includes features such as, video streaming (virtually all media outlets today have the capacity to offer video clips; YouTube specializes in videos produced by the public and Yahoo is promoting its video capabilities in addition to its text-based content); audio streaming (this includes podcasting in its many forms); learning world languages (one can learn basic Chinese via the Internet); obtain grammar help, especially in English (websites have a variety of interactive exercises with instant correction and assessment, templates for CV writing, etc.); and finally, create Web logs or "blogs" (Notess, 2002). The user-friendly course-content management systems, built into Web delivery systems such as WebCT and Moodle (and the CUForum), have created a learning environment where learners and teachers can update and create new course content virtually overnight, and from anywhere/anytime. All of the aforementioned learning features of the Web were inconceivable a mere decade ago when the bulk of the information was text-based. The assumption made in much of the early work on the educational use of the Internet (Harasim, 1989; Feenberg, 1989) was that communication was asynchronous and interaction was text-based. Online education today is considered much more than access to a global information database, and according to Anderson (2004), the main benefit of the 'Web for educational use is the profound and multifaceted increase in communication and interaction capability that it provides' (p. 42).

Thus the main focus of this study is to examine what happens during interactions, what the contents of an interaction are, what and who is interacting with what and whom to produce, promote and support learning (e.g. directly or indirectly), and how this may be enhanced.

SETTING THE STAGE

The central theme of Chinese student *online interactivity* in this study is addressed at various stages. The phenomenon in question raised here is explored by the relevant methodologies, using specific methods and instruments, namely a close analysis of the 1224 messages posted on the CUForum by 3rd. year business communication students and the instructor during the period of Sept. – Dec. 2004. The data was analyzed using NVivo V.1.3. Following that a focus group was videotaped and the data was transcribed. In addition, the 1224 messages were investigated for evidence of linguistic features that are particularly prominent when Chinese-speaking students write in English. Then the data from the end-of-term questionnaires that were administered in Dec. 2004 was scrutinized. Finally, consistent with the methodology of a qualitative analysis, the technique of participant observer was used in this study. A detailed examination of these methodologies is beyond the scope of this chapter, nonetheless it was done in order to gain a fuller understanding of the CUForum data and to answer the research questions outlined below. The analysis was performed of the *cognitive* (including *linguistic*), *affective* and *social* factors that determine how Chinese-speaking students interact and learn online.

The main research question(s) which guide this study are:

1. What is the nature of *interactivity* in an online programme of business communication in English?
2. How does *interactivity*, in an online business-writing programme, enhance *Chinese-speaking adult students'* learning (specifically, business communication in English) from *cognitive* (including *linguistic*), *affective* and *social* perspectives?

There is extensive research on the Chinese learner [as elaborated below] (Bond, 1991; Biggs, 1999; Watkins and Biggs, 1996; Watkins and Biggs, 2001), however research on how the Hong Kong Chinese student learns online is relatively sparse (Huang and Liu, 2000; Kekkonen-Moneta, 2002; Moneta 2000; Morrison, 2006; Sheard and Markham 2005; Thompson and Ku, 2005; Tu and McIsaac 2002). In light of this, this study aims at rectifying this omission. One of the important features of this research lies in the fact that it is one of the first studies that uses an online business writing course to examine Hong Kong Chinese-speaking student interactions (in an environment where English is used as the medium of instruction).

As mentioned above, this section examines the Chinese learner in the Hong Kong (and mainland Chinese) contexts, including the learning styles of the Chinese learner. The learning styles of Chinese students appear to be qualitatively different from their Western counterparts (Bond, 1991; Cortazzi and Jin, 1996; Flowerdew, 1998; Ho, 1991, 1996; Huang and Liu, 2000; Kekkonen-Moneta, 2002; Kennedy, 2002; Merryfield, 2003; Nelson, 1995; Nunan, 2004, 2005; Sheard and Markham 2005; Thompson and Ku, 2005; Tu, 2002; Watkins and Biggs, 1996). This section outlines some of the main differences in learning styles, points out areas where Western researchers may have misunderstood the nature of the Chinese learner, offers a glimpse into the way Chinese-speaking students, coming from a Confucian-heritage culture (CHC), can best be understood, highlights the 'paradox of the Chinese learner' and offers a glimpse into why the notion of the Chinese learner, as understood by Western researchers, may be a misinterpretation and misrepresentation of this particular learner.

A number of prominent researchers in the field (Bond, 1991; Cortazzi and Jin, 1996; Flowerdew, 1998; Fung, 2004; Ho, 1991, 1996; Kennedy, 2002; 1996; Nelson, 1995; Watkins and Biggs, 1996) have noted that learners from a CHC background often subscribe to certain principles embedded in Confucian traditions. In Confucian philosophy, faithfulness (i.e. filial piety and social relationships, including the acceptance of the distance between teacher and student), and propriety (i.e. the concept of 'face' and 'self-effacement') are key Confucian values. These values have an influence on the learning styles of Chinese learners. For example, when the teacher asks a question in a class of Chinese students, there is complete silence. No student will volunteer an answer. Confucian norms influence and even prevent students from speaking up in class. As a Chinese colleague pointed out when this Western researcher from Canada first arrived in Hong Kong in 1998:

*The teacher functions as the 'sage on the stage' and transmits knowledge directly to the students. Generally, the students are passive recipients of this knowledge and their role is to absorb it and then regurgitate all that they have learned during the final exam (***Personal correspondence, 1998***).*

Contrary to the above colleague's assertion that Chinese learners are passive recipients of knowledge, Flowerdew (1998) argues that the use of group work for Chinese students fits very well into their Confucian worldview. As noted by Nelson (1995), 'Students learn through co-operation, by working for the common good, by supporting each other and by not elevating themselves above others' (p. 9). Similarly, Cortazzi and Jin (1996) report that 'in Chinese society – and in the classroom – the priorities are that each person must be part of a group or community; learning interdependency, co-operation and social awareness are the accepted standards' (p. 178). All the students work for the common good of the group. Flowerdew (1998) concludes that teachers may need to adjust their teaching styles in order to accommodate group work as a methodological tool on two accounts:

...either because it exploits the Confucian value of co-operation, which would seem to foster a style conducive to learning; or because it can be used to counterbalance the Confucian concepts of 'face' and 'self-effacement,' which could be considered as aspects which impair the learning process (p. 327).

Students may use a surface approach to learning when there is a great amount of material to be learned and this material will be tested on an examination. This type of approach is associated with time pressures, examination stress, and the use of test items that emphasize low-level cognitive outcomes. On the other hand, students are likely to develop a deep approach if they are encouraged to interact with other students, do task-based learning, and if the assessment requires them to understand the principles rather than reproduce facts and figures. Teachers at the tertiary level can modify their teaching situation by taking into account the learning approaches of students, suggest Gow *et al.* (1996). For example, the teacher can help students to change the learning approach by changing the assessment method. This may change the students' motivation, which affects the outcome, which affects the teacher's perception of the students' performance and of course the students' self-perception.

It is noteworthy to clarify at this point that in Hong Kong English is characterized as a second language (L2): the *lingua franca* of commerce, industry, technology, medicine and education (Gow *et al.* 1996). Although Hong Kong students start learning English as a second language from Primary 1 and all the way through Secondary 7 (a total of 13 years of schooling in English as a 2nd. language), their command of the language is problematic. As Gow *et al.* (1996) report, 'the majority of students do not have sufficient competence in English to learn the subjects' content through this second language' (p.117). This problem is compounded when the students enter university. Pennington *et al.* (1992) demonstrated that tertiary level students used English predominantly within the context of education. The above situation, as elaborated by Gow *et al.* (1996) and Pennington *et al.* (1992), is clearly reflected at this author's university. In many departments, university teachers use a combination of English and Cantonese in their lectures and they encourage their students to read the course textbooks, articles, scientific journals and technical reports in English. Cantonese is the predominant language used in lectures (combined with a large number of English terms that have no equivalents in Chinese). Thus students can attend lectures in their mother tongue (Cantonese), but need to use English to supplement their learning. This creates the situation where students listen to lectures in Cantonese, discuss the lecture material in Cantonese, then read texts in English and write term papers in English.

Gow *et al.* (1991) observed that this lack of competence in English affects how students learn. They are more likely to use learning strategies and approaches to learning that suit the assigned readings. For example, if students have language problems, they are more likely to decipher small sections of a text (the rhetorical aspects) and are not able to comprehend the underlying assumptions and meanings of the text. Students with low ability in English are more likely to use a surface approach to learning. Students with a higher English-language ability, according to Gow *et al.* (1996), may employ a deep approach to learning, although they 'may not do so if they have not learnt an appropriate orientation or have been discouraged to do so by the learning environment' (p. 118). Other studies by Watkins *et al.* (1991) confirm Gow's findings. Students with a lower ability in English resorted to rote learning (surface approach to learning) and relied much less on deeper approaches to learning. As Gow *et al.* (1996) point out, 'If students' approaches are impaired by their inadequate English, it is logical to expect a shift in their approach as the language of instruction changes' (p. 118).

Although the learning style of Chinese-speaking students coming from a CHC background, as outlined above, is a crucial element to consider in the learning process, there is one other issue that needs to be addressed here: the 'paradox of the Chinese learner.' According to Biggs (1996), this paradox is premised upon the fact that CHC students are 'prone to use rote-based, low level, cognitive strategies, both in their own cultures (Hong Kong), and overseas. Yet CHC students achieve at considerably higher levels than their Western counterparts' (p. 46 - 48). The resolution to this paradox lies in the fact 'that highly adaptive modes of learning emerge from CHC classrooms' (*Ibid*, p.50). As discussed previously, Biggs (1996) noted that on the issue of approaches to learning, Chinese students, *particularly those educated in educational contexts such as Hong Kong, are highly cue conscious. That is, they are very achievement oriented and alert to cues from their lecturers and from the assessment system itself as to what is needed to succeed in a particular course (Ibid, p.282).*

Thus, to the Western observer, the assumption that the classroom context and the students' behaviour. (i.e. Chinese students seen as passive learners, unquestioning of authority and compliant) should result in low academic achievement, as argued by Biggs (1996), is simply wrong.

According to Biggs (1994), 'much of the anecdotal evidence concerning Asian students and the apparent paradox of research results about them is based on misperceptions' (p.122). A learning approach that is a combination of memorization and understanding is clearly not a surface approach: it is not memorization in the classical definition of rote-learning. This approach is very different from a surface approach where students have no interest in discovering the underlying meaning (Gow *et al.* 1996). Thus the paradox of the Chinese learner, as Gow *et al.* (1996) posit, could be resolved by the finding 'that the learning approaches of Chinese students cannot be adequately described by the deep/surface dichotomy' (p.123). More research is needed in order to claim with greater certainty that 'memorization could co-exist with understanding in Chinese students' (*Ibid,* p.123). Kennedy (2002) posits that 'The picture that often emerges from the research literature on Chinese learners is a caricature of rote-learning, memorization and passivity' (p. 1). He takes issue with the stereotype.

Hofstede and Bond (1984) and Hofstede (1980), characterize (Hong Kong) Chinese culture as low on individualism and high on collectivism. They identified the following features: power/distance ratios were amongst the highest of all 40 countries surveyed; (stereo-typical) masculinity was found to be 'medium' and; there was weak uncertainty avoidance (i.e. the degree of risk tolerance). Similarly in a study by Trompanaars (1993), it was shown that in Hong Kong Chinese culture, a strong sense of belonging to a social group, a preference for working together in groups to solve problems and a high level of collectivism were present. How does the above relate to the learning styles of Chinese students, and more importantly, what are the cultural roots of 'Chinese' learning?

'Confucian values' are assumed to be the foundation of the learning styles adopted by Chinese learners and 'Confucian heritage' determines Chinese students' apparent reluctance to express and voice their opinions in class. In a study by Murphy (1987) it is suggested that the reason 'Hong Kong students display an almost unquestioning acceptance of the knowledge of the teacher . . . may be a transfer of the Confucian ethic of filial piety, coupled with an emphasis on strictness of discipline and proper behaviour' (p. 43). A student should be modest and self-effacing, not waste other students' time in class by expressing independent opinions and judgments. This is considered selfish and egotistical. In addition, such direct challenges are disrespectful to teachers and may cause them to lose face (Hwang 1987; Chang and Holt 1994). These socio-cultural attitudes reinforce passive compliant roles in the classroom and promote conformity (Watkins and Biggs, 1996). Students are taught not to speak out, not to criticize and question, and not to be the first one to stand up and ask a question for fear of being wrong and consequently losing face (Tsui, 1996). For many foreigners (i.e. non-Chinese) this particular learning style is alien to their culture. In Western countries, students are expected to be more individualistic (as opposed to collectivist) in orientation, and they are encouraged to think critically, form judgments, to formulate questions in order to clarify information, and to be able to communicate both with their peers and teachers (Tang, 1996). Research by Liu and Littlewood (1997) suggest that Hong Kong Chinese L2 learners, who show a certain reticence in class, is *less a question of modesty and face than one of competence [in English] and [lack of] confidence because they are unaccustomed to participative modes of learning. However, students can be guided towards greater autonomy if teachers make explicit their expectations and perceptions, get students to brainstorm ideas and clarify concepts in small groups. Students also have to become familiar with the strategies needed for successful communication in English, such as turn-taking, asking for clarification, giving non-verbal feed-back, etc. (p. 378).*

Kennedy (2002) concludes by stating that 'socio-cultural insights and an understanding of students' previous learning experiences can undoubtedly help L2 teachers to develop more culturally sensitive pedagogies' (p. 13) and, further on he notes that there is always the danger of overgeneralization in that

Chinese learning styles are far more subtle and complex than they are often made out to be. Common assumptions such as the notion that memorization and understanding are mutually exclusive categories may be in need of reappraisal' (Ibid. p. 13).

Furthermore, Kennedy urges researchers to re-think the current paradigm on the Chinese learner (and the Chinese learning style) 'and to go beyond the self-fulfilling prophecies and Confucian confusion' (*Ibid.* p 13). Finally, Scollon and Scollon (1995) writing about cultures state that, 'Cultures do not talk to each other individuals do. In that sense, all communication is interpersonal communication and can never be intercultural communication.' (p.125).

CASE DESCRIPTION

As mentioned previously, an NVivo data analysis was used to examine, in detail, the Chinese learner and how this particular learner's English linguistic features are reflected in the messages posted on the CUForum. NVivo V.1.3 was used to analyse the complete record of the 513 pages of transcript of student messages that were posted on the CUForum. These messages (a total of 1224 messages) were collected throughout the whole 13-week semester between Sept. – Dec. 2004 and were archived on the CUForum course website. The complete NVivo analysis, including the manual coding, was done between July and September 2006.

As noted above, NVivo was used to analyse the transcript, but the first step consisted of manually coding the 513 pages of data. Just reading the information was in itself a daunting task; the manual coding of all the messages was certainly a challenge. The raw text was manually coded by theme and common themes were identified. These themes included *cognitive* (including *linguistic*), *affective* and *social* perspectives, and in that regard, keywords and phrases were identified for their frequency of occurrence in the transcript. The following sample below demonstrates the manual coding that was done.

Node: /Cognitive/Buss Eng/Expectation

The manual coding for the above Node was done with the aim of identifying the key words (and phrases) that occurred with any reasonable frequency in the CUForum data (1. Cognitive, 2. Buss Eng., 3. Expectation).

1. Cognitive

This list includes all the words (and phrases) that are related to the term Cognitive:

- Improved
- Improve
- Improving
- Think
- To know
- Learn

- Training
- Skills
- Knowledge
- Better equip myself
- Brush up
- Present
- Re-boost
- Raise my confidence
- Enhance
- To be confident
- To apply

2. Buss Eng.

The words (and phrases) related to Business English are:

- CV
- Resume
- Interview skills
- Job
- Job interview
- Business
- Career
- Speak English
- Written English
- Cover letter
- Job searching
- Interviewing techniques
- Listening
- Writing
- Communication
- Presentation skills
- Employee
- Manager
- Business letter
- Workplace
- Future career
- Language skills
- Business environment
- Impromptu speech
- Interviewer

3. Expectation

The words (and phrases) related to Expectation are:

- I hope
- I see this course as...
- I would like to...
- I strongly hope...
- I want to know...
- I wish that I can...
- I wish I could
- Hopefully able to...
- What I hope to accomplish...
- I want to be...
- The foremost goal I'd like to accomplish...
- I sincerely hope...
- I would really treat this course as a valuable time...
- What I wanna get from this class...
- I rely on this course to give me...

The sample above demonstrates how the manual coding was done. It is not possible to outline the complete manual coding here since space does not permit it.

Once the coding was done manually, the complete data was analyzed using NVivo V. 1.3. The raw text was manually coded by theme and common themes were identified. These themes included *cognitive* (including *linguistic*), *affective* and *social* and perspectives, and in that regard, keywords and phrases were identified for their frequency of occurrence in the transcript.

According to the summary, the results for the three key features of analysis (*cognitive,* [including *linguistic*]. *affective* and *social*) are listed by the number of messages that each of the sub-themes generated.

The Cognitive theme was sub-divided into:

1) Buss Eng
   a) Expectation
   b) Impromptu speech
   c) Resume
   d) Job interview
2) Linguistics
3) GE (General Education)

4) Week starters
5) Presentation
6) Toastmaster

The Affective theme was sub-divided into:

1) Personal feelings
2) Feelings_Other people
3) Feelings E3110*
4) Responses to humourous articles

The last feature of analysis, namely the Social theme consisted of:

1) Making friends
    a) Special visitor – Kathleen
    b) Personal info
2) Encouragement
3) Caring
4) Entertainment
5) Cultural Issues
6) Asking for help
7) Class-related matters
8) Exchanging news
9) School life

In order to gain a better understanding of the number of messages that were generated by the aforementioned themes, Table 1, Table 2 and Table 3 show the frequency of messages for the Cognitive, Affective and Social themes.

CURRENT CHALLENGES

Although the title of the course used in this study (ELT 3110 – Business Communications) clearly states that the central focus of this course is on 'business communications,' one would expect that the majority of the messages posted on the CUForum would deal with business-related topics. The results of the analysis of the CUForum data, however, tell another story. Also, it was stated in the beginning that *interactivity* is the key to online learning, so the question that arises is fundamental to this investigation, namely

What is the nature of this interaction?

To provide a satisfactory answer to this question, one only needs to look at the numbers. There is unequivocal evidence from the CUForum analysis that the largest number of interactions generated focus on, 1) the *Social* aspect of learning (a total of 688 messages were generated), followed by, 2) the *Affective* aspect of learning (a total of 494 messages were generated), and finally, 3) the *Cognitive* aspect of learning (a total of 162 messages were generated). The percentages for the above translate into: 51.19% (Social), 36.76% (Affective) and 12.05% (Cognitive). Participants in this CUForum study focused on a combination of social and affective factors to help heir learning and the cognitive factor was relegated to a distant third place on their learning horizons. Is online interactivity more suited for this purpose? What is the learning that occurs on the CUForum? And, more importantly, can one assume that students learn best when they are having 'fun' in a social setting? Do students 'learn' more from sites like Facebook, YouTube, MySpace, Skype, ICQ and

Table 1. Cognitive

Cognitive	# of messages generated
1) Buss Eng	
a) Expectation	25
b) Impromptu speech	23
c) Resume	11
d) Job interview	23
2) Linguistics	18
3) GE	21
4) Week starters	14
5) Presentation	3
6) Toastmaster	24
Total number of messages generated	162

Table 2. Affective

Affective	# of messages generated
1) Personal feelings	76
2) Feelings_ Other people	72
3) Feelings E3110*	52
Total number of messages generated	200
4) Responses to humourous articles	
4.1) Get caught sleeping	5
4.2) A little math	1
4.3) Computer gender	2
4.4) Virus alert	1
4.5) Tech challenged	3
4.6) A man & a dog	1
4.7) Do u know how to install	2
4.8) Gwai lo shopping in HK	3
4.9) Why English is so hard to learn	3
4.10) English gone awry	2
4.11) A feel good website	4
4.12) Fun to be in school	2
4.13) aoccdrnig to a rscheearch	4
4.14) Things that movies taught you	6
4.15) Desirable qualities of job seekers	4
4.16) Interesting	6
4.17) A perfect Sunday school	2
4.18) A break up letter from a boy	15
4.19) What is marketing	14
4.20) Female vs. Male	10
4.21) Who is bigger	12
4.22) Animal voice	1
4.23) The Bear story	3
4.24) Friendship	4
4.25) Interesting conversations	7
4.26) The Meatrix	4
4.27) Pay it forward	1
4.28) 25 ways to tell the semester is...	14
4.29) Vocab test	3
4.30) Halloween banquet fotos	13
4.31) Engrish in Japan	7
4.32) True globalization	2
4.33) Oxymorons	8

Table 2. continued

Affective	# of messages generated
4.34) How Japanese learn English	6
4.35) Romantic, la	0
4.36) The latest Oxford definitions	8
4.37) Crazy definitions	4
4.38) Thoughts & beliefs	2
4.39) A poem	3
4.40) A Canadian poem	8
4.41) Words to live by	7
4.42) Alice & Claudia the cow	1
4.43) Being twenty something	4
4.44) Test for dementia	3
4.45) Icon story	6
4.46) Christmas Carol	3
4.47) Try to read it out loud	8
4.48) An inspirational message	2
4.49) How business is done	2
Total for messages 1 – 49	236
5) Humourous shares	58
Total for messages 4.1 – 4.49 + Humourous shares	294
Total number of messages Themes # [1-3] + [4.1 - 4.49] + [5]	494

(*Author's note:* *E3110 refers to the course code for the business communications course - i.e. ELT 3110)

MSN? There are certainly no immediate answers to these questions and further research needs to be done in this area.

CONCLUSION AND IMPLICATIONS

To conclude, this study offered a comprehensive analysis of the CUForum dataset by attempting to answer the research question(s) listed earlier and re-printed below

1. What is the nature of *interactivity* in an online programme of business communication in English?

Table 3. Social

Social	# of messages generated
1) Making friends	94
a) Special visitor – Kathleen	19
b) Personal info	25
2) Encouragement	35
3) Caring	24
4) Entertainment	101
5) Cultural issues	22
6) Asking for help	88
7) Class-related matters	33
8) Exchanging news	150
9) School life	97
Total number of messages generated	688

2. How does *interactivity,* in an online business-writing programme, enhance *Chinese-speaking adult students'* learning (specifically, business communication in English) from *cognitive* (including *linguistic*), *affective* and *social* perspectives?

The author of this investigation performed a comprehensive analysis of the *cognitive* (including *linguistic*), *affective* and *social* factors that determine how Chinese-speaking students interact and learn online. This was done by using, 1) an NVivo V.1.3 analysis [1224 messages posted on the CUForum between Sept. – Dec. 2004].

Is the CUForum a platform for learning (*Cognitive*), having 'fun' (*Social and Affective*) or both? For the participants on the CUForum, the answer to this question lies in 'both' – fun and learning. However, for this group of people termed the global 'net generation,' their learning approaches tend to lean more towards the 'fun' side of learning, which are characterized as integrated and multi-faceted, constructivist and chaotic. Results from this study show that for this 'net generation,' learning is often incidental and a sense of 'fun' is of paramount importance. Law (2004) in a study of the CUForum confirmed these findings. He stated that, 'In CU forum, least students post a new topic of this category. The frequency of have it as a new post is 1 in 10.' (p.3) He does state further that language-related topics which are 'fun' do generate more replies. He cites the following: 'because it [fun with language] shows how we play games using English, though it is not related to Business English.' Another example of 'fun' with language is a game with abbreviations.

This topic is about the abbreviations that people often use and let students guess what they mean. This topic get more than 20 replies, which means quite a large number when compare to others. (p. 5)

In terms of quality, the reply is short in average. They showed they can learn from the post and some shared their knowledge about this field with others. May be due to the nature of this topic, the content of replies are quite restricted and they do not write much on their replies. (p. 8)

However, this type of message, a game, invites them to participate and would encourage more students to read and post reply. It is good to keep a style of interactive game in language postings. Otherwise, students may feel bored. (Law, 2004:7)

The use of English (the *Cognitive* feature in the study) is another area where participants may have a concern. Hong Kong students, as Law (2004) asserts, are particularly sensitive to their use of English (a second language for the majority of them). He states,

I suggest that Hong Kong students are afraid of making mistakes, both in content and grammar. Another reason is that Hong Kong students are fed up with academic stuff. They have to spend most of their time on projects (Business students), assignments (Other faculties' students) and tests (All students). They tend not to touch academic stuff anymore (p. 10).

A possible solution for this problem is to create a norm that mistakes are welcome. Positive incentives are given to students who have try to post quality messages. The incentives do not necessarily have to be grades or gifts. Message posted by instructors that show their appreciation of the student's effort would certainly help. Private messages that enumerate a student's mistakes can be sent to the student directly and anonymously. The instructor should also encourage students to share their mistakes with other students since many students make similar errors in grammar. This is possible because most students are used to correcting their own mistakes by either checking them in their electronic dictionaries, or by sending their questions about grammar to their friends via MSN. (p. 4 – 5)

Personal, as opposed to academic topics, were also found to generate a large number of responses, according to Law (2004)

Among all topics on the CUforum, personal topics are most welcomed. About 6 to 7 in 10 are personal topics. Also, students are quite active in these kinds of topics. Let me show an example which is a personal topic with many replies. (p.12)

Law's findings, as well as the results of this study, appear to contradict what numerous authors have argued, namely that online learning, "is simply another, albeit sophisticated, medium for doing what we have always done" (Stephenson, 2001, p. 219). However, there is an indication that:

A re-balancing of the range of pedagogies in use is slowly taking place as more people begin to exploit the full range of facilities that the medium can offer. The re-balancing being stimulated by online learning is towards giving learners greater responsibility for managing their own learning. (ibid p. 219).

Online learning, however, is much more than an electronic blackboard and a platform for exchanging documents. This research has demonstrated that,

- Chinese-speaking learners, specifically those Ho (1991) refers to as coming from 'Confucian-heritage' cultures (or its abbreviation 'CHC'), could be analysed from cultural, psychological and contextual perspectives. Specifically this research has shown that *cognitive* (including *linguistic*), *affective* and *social* features of online learning are an appropriate way to understand this particular learner. This was one of the principal aims of this study, and the results of the aforementioned analysis have helped in, and offered a clearer perspective in, gaining a greater understanding of the nature of this learner.
- The analysis of Chinese students' online learning was done by the use of triangulation, i.e. the analysis was performed through a multilayered examination by considering: 1) Focus groups, 2) End-of-term Questionnaires, 3) NVivo V.1.3 analysis [1224 messages posted on the CUForum between Sept. – Dec. 2004], and, 4) Participant Observer techniques. Thus, it is evident that the research here was not concerned with a single level of analysis. The online learning platform (i.e. the CUForum), which formed an integral part of this investigation, was considered as a complete system and was analysed from that perspective.
- The application of Anderson's theory of online learning is a step in a new direction, and this was done in this study. This research introduced the theory of online learning as it pertains specifically to Chinese learners.
- By using the theory of online learning as a foundation for the analysis of the data,

this author explored a multitheoretical and multilevel examination – MTML (this framework is adapted from Monge and Carpenter 2003) by the use of quantitative, qualitative and ethnographic approaches to empirically evaluate the concepts and theoretical framework presented here.

- Finally, the uniqueness of this research lies in the fact that it contributed to an understanding of how Chinese students, using English as a medium of instruction, learn online. Specifically, this study elaborated on the nature of the Chinese student (coming from the Confucian-heritage culture – CHC), and the importance of the concept of *interactivity* in the online/face-to-face learning process. All of the abovementioned were explored through the lens of the online theory framework.

Technical and pedagogical innovations of online learning platforms (Moodle, WebCt, the CUForum, etc.), as well as the World Wide Web and the Internet, keep changing and developing at a breakneck speed. By way of illustration, a mere six years ago (2003), there was no such thing as YouTube, MySpace, Facebook, Skype (a software programme that lets users make phone calls all over the globe for free), Podcasting, Video-on-demand and Google Earth, where, one can, through the use of satellite technology combined with the GPS [Global Positioning System], zoom in on a three dimensional building virtually anywhere on Earth. The CUForum, the online learning under study, also keeps pace with these new technological innovations. The following are but a few practical suggestions as to how the CUForum could enhance learning.

- *Higher Bandwidth.* With the availability of Podcasting and Video-on-Demand, the CUForum could be a much more interactive and visual medium. Short video clips could be downloaded and viewed face-to-face in a classroom as well as online 24 hours a day 7 days per week (both synchronously and asynchronously).
- *Language Help.* A built-in edictionary (specifically, Chinese-English-Chinese) to facilitate the processing of textual and media (including audio and video) materials. The Spellchecker and Track Changes functions of MSWord could be automatically triggered every time a student writes a text.
- *Information Access.* Access to a full database of varied examples of previous CUForum learners' assignments, achievements, and activities. This could be extended to access to other university databases which may not be necessarily limited to Hong Kong, but universities worldwide.
- *Webcams and Videos.* Mechanisms to promote more extensive, 'real time' interaction between learners and learners, learners and instructors, and learners and experts/non-experts. This could be accomplished by installing Webcams to record live conversations between participants, upload the videos to the CUForum and archive them for future use.
- *Learning Support.* A learning support environment could be created, which is fully integrated into the CUForum providing easy access to a multi-lingual online support for students, tutors, supervisors and experts from around the world.

It should be noted that this is only a partial list of features that an online learning platform should include. The research that could be done in this area should keep computer experts, curriculum designers, language experts, online practitioners and a host of other specialists in business for years to come.

Finally, perhaps the last word should go to Tim Berners-Lee (Now, Sir Tim Berners-Lee employed at MIT), when he was a researcher in the CERN labs in Switzerland. He could never have envisioned that his simple invention of two computers 'talking to each other' (via HTML, HTTP and URLs) would not only revolutionize local, national and global communications on a massive scale, but would also drastically change the way people live, learn, do business and interact with each other. Tim Berners-Lee stated that he was brought up with the belief that, there have always been things that people were good at, and things computers have been good at and little overlap between the two. This question, from a visionary, inventor and an expert in the field, is certainly worthwhile pursuing.

Outcomes matter, but so do interactions. Bogg, J. and Robert Geyer (http://www.liv.ac.uk/ccr/index.htm)

REFERENCES

Anderson, T. (2003). Modes of interaction in distance education: Recent developments and research questions. In M. Moore & G. Anderson (Eds.), *Handbook of distance education* (pp. 129-144). Mahwah, NJ: Erlbaum

Anderson, T. (2004). Toward a theory of online learning. *Theory and practice of online learning*. Retrieved August 21, 2007 from http://cde.athabascau.ca

Anderson, T., & Elloumi, F. (Eds.). (2004). *Theory and Practice of Online Learning*. Edmonton, Canada: Athabasca University Press. http://www.athabascau.ca (Accessed on May 15, 2005).

Berners- Lee. T. (2007, April). Interview with Tim Berners-Lee. *Business Week*. Retrieved August 29, 2007 from http://www.w3.org/2001/sw/

Berners-Lee, T. (2007). Time Berners-Lee on the Semantic Web. *Video on MIT Technology Review*, March 2007. Retrieved Aug. 29, 2007 from http://www.w3.org/2001/sw/

Berners-Lee, T. (n.d.). Retrieved September 8, 2003 from http://www.w3.org/People/Berners-Lee/Short History.html

Biggs, J. B. (1987). *Student Approaches to Learning and Studying*. Melbourne, Australia: Australian Council for Educational Research.

Biggs, J. B. (1993, November). The quality of teaching and learning in an expanding tertiary system. *New Horizons (Baltimore, Md.)*, *34*, 102–104.

Biggs, J. B. (1994). What are effective schools? Lessons from East and West. *Australian Educational Researcher*, *21*, 19–39.

Biggs, J. B. (1996a). Approaches to Learning of Asian Students: A Multiple Paradox. In J. Pandey (Ed.), *Asian Contributions to Cross-Cultural Psychology,* (pp. 18 – 199). New Delhi: Sage Publications.

Biggs, J. B. (1999). *Teaching for quality learning at university*. Milton Keynes, UK: Open University Press.

Biggs, J. B., & Telfer, R. (1987). *The Process of Learning* (2nd. Edition). Sydney: Prentice Hall of Australia.

Bond, M. H. (1991). *Beyond the Chinese Face: Insights from Psychology.* Hong Kong: Oxford University Press.

Bond, M. H. (Ed.). (1996). *The Handbook of Chinese Psychology*. Hong Kong: Oxford University Press.

Bonk, C. J., & Reynolds, T. H. (1997). Learner-centred Web instruction for higher-order thinking, teamwork and apprenticeship. In B. H. Khan, (Ed.). *Web-based instruction*. (pp. 167 – 178). Englewood Cliffs, NJ: Educational technology Publications.

Bransford, J., Brown, A., & Cocking, R. (1999). *How people learn: Brain, mind, experience and school*. The National Academy of Sciences. Retrieved Aug. 28, 2007 from http://www.nap.edu/html/howpeoplel

Chandler, D. (1994). *The Transmission Model of Communication*. Retrieved May 24, 2006 from http://www.aber.ac.uk/media/Documents/short/trans.html#A

Chang, H., & Holt, R. (1994). A Chinese perspective on face as inter-relational concern. In S. Ting-Toomey (ed.), *The Challenge of Facework* (pp. 95–132). Albany, NY: State University of New York Press.

Cooper, J., & Mueck, R. (1992). Student involvement in learning: Cooperative learning and college instruction. In Goodsell, A. Maher, M. & Tinto, V. (Eds.)

Cortazzi, M., & Jin, L. (1996). Cultures of learning: language classrooms in China. In H. Coleman (Ed.), *Society and the Language Classroom*. Cambridge, UK: Cambridge University Press.

Feenberg, A. (1989). The written world: On the theory and practice of computer conferencing. In R. Mason & A. Kaye (Eds.), *Mindweave: Communication, computers, and distance education*. Toronto: Pergamon Press.

Flowerdew, J., & Miller, L. (1996). Lectures in a second language: Notes towards a cultural grammar. *English for Specific Purposes, 15*(2), 121–140. doi:10.1016/0889-4906(96)00001-4

Flowerdew, L. (1998). A cultural perspective on group work. *ELT Journal, 52*(4), 323–329. doi:10.1093/elt/52.4.323

Flowerdew, L. (1998a). A cultural perspective on groupwork. *ELT Journal, 52*(4). doi:10.1093/elt/52.4.323

Flowerdew, L. (n.d.). *Methodological Issues in Teaching English in Hong Kong*. Retrieved May 27, 2004 from http://ettu18.edu.polyu.edu.hk/abstarcts/10.html

Fung, Y. H. (2004). Collaborative online learning: interaction patterns and limiting factors. *Open Learning: the Journal of Open and Distance Learning, 19*(2), 135 – 149. Retrieved Oct. 13, 2007 from http://dx.doi.org/10.1080/0268051042000224743

Garrison, D. R., & Shale, D. (1990). A new framework and perspective. In D. R. Garrison & D. Shale (Eds.), *Education at a distance: From issues to practice* (pp. 123-133). Malabar, FL: Robert E. Krieger.

Gow, L. Balla, J. Kember, D. & Hau, K.T. (1996). The Learning Approaches of Chinese People: A Function of Socialization Processes and the Context of Learning. In M.H. Bond, (Ed.), *The Handbook of Chinese Psychology*. Hong Kong, Oxford University Press.

Harasim, L. (1989). Online education: A new domain. In R. Mason & A. Kaye (Eds.) *Mindweave: Communication, computers, and distance education*. Toronto: Pergamon Press.

Harasim, L., Hiltz, S., Teles, L., & Turoff, M. (1995). *Learning networks: A field guide to teaching and learning online*. London: MIT Press.

Hine, C. (200). *Virtual Ethnography*. London: Sage.

Ho, D. Y. F. (1991). *Cognitive socialization in Confucian heritage cultures.* Paper presented to Workshop on Continuities and Discontinuities in the Cognitive Socialization of Minority Children, US Department of Health and Human Services, Washington, DC, June 29 – July 2.

Ho, D. Y. F. (1996). Filial Piety and Its Psychological Consequences. In M.H. Bond, (Ed.) *The Handbook of Chinese Psychology*. Hong Kong: Oxford University Press.

Hofstede, G. (1980). *Culture's Consequences: international differences in work-related values*. Newbury Park, CA: Sage.

Hofstede, G., & Bond, M. (1984). Hofstede's culture dimensions: an independent validation using Rokeach's value survey. *Journal of Cross-Cultural Psychology, 15*, 417–433. doi:10.1177/0022002184015004003

Huang, S. J., & Liu, H. F. (2000). Communicative Language Teaching in a Multimedia Language Lab. *The Internet TESL Journal, 6*(2), February 2000. Retrieved February 19, 2005 from http://iteslj.org/Techniques/Huang-CompLab.htm

Hwang, K. (1987). Face and favour: the Chinese power game. *Journal of Sociology (Melbourne, Vic.), 94*(2), 944–974.

Jakubowicz, P. (2006). *Can Online Learning be Fun?* [Unpublished paper]. Paper presented at the 4th Asia TEFL Conference in Fukuoka, Japan, Aug. 18 – 21.

Johnson, D. W. Johnson, R.T. & Smith, K.A. (1998). *Active Learning: Cooperation in the college classroom*. Interaction Book Co.

Jonassen, D. H. (1991). Evaluating constructivistic learning. *Educational Technology, 31*(10), 28–33.

Kekkonen-Moneta, S. & Moneta, G.B. (2001). E-Learning in Hong Kong. *British Journal of Educational Technology.*

Kennedy, P. (2002). Learning cultures and learning styles: myth-understandings about adult (Hong Kong) Chinese learners. *International Journal of Lifelong Education, 21*(5), 430–445. doi:10.1080/02601370210156745

Kozma, R. B. (2001). Counterpoint theory of 'learning with media.' In R. E. Clark (Ed.). *Learning from media: Arguments, analysis and evidence*, (pp. 137 – 178). Greenwich, CT: Information Age Publishing Inc.

Lai, E. (1993). Cooperative learning in a Geography class. In J. B. Biggs, & D.A. Watkins, (Eds.), *Learning and teaching in Hong Kong: What is and what might be*. Hong Kong: Educational Papers 17, Faculty of Education, The University of Hong Kong.

Law, K. K. M. (2004). *Research on finding effective ways to enhance the interest of using CU forum of Chinese speaking students in CUHK.* (Unpublished research report – The Chinese University of Hong Kong).

Lipman, M. (1991). *Thinking in education.* Cambridge, UK: Cambridge University Press.

Liu, N., & Littlewood, W. (1997). Why do many students appear reluctant to participate in classroom learning discourse? *System, 25*(3), 371–384. doi:10.1016/S0346-251X(97)00029-8

Mason, J., & Hart, G. (1997). *Effective use of asynchronous virtual learning communities*. Retrieved Aug. 1, 2007 from http://arch.usyd.edu.au/kcdc/conferences/VC97/papers/mason.html

McConnell, D. (2000). *Implementing Computer Supported Cooperative Learning,* (2nd Ed). London: Kogan Page.

McPeck, J. (1990). *Teaching critical thinking.* New York: Routledge.

Merryfield, M. (2003). "Like a Veil: Cross-cultural Experiential Learning Online." *Contemporary Issues in Technology and Teacher Education.*

Monge, P. R., & Contractor, N. S. (2003). *Theories of Communication Networks*. Oxford, UK: Oxford University Press. Retrieved Aug. 21, 2007 from http://hyperion.math.upatras.gr/commorg/nosh/HOCNets.html

Morrison, K. R. B. (2006). *Promoting the Conditions for Effective Learning in an e-Learning Environment*.

Murphy, D. (1987). Offshore education: a Hong Kong perspective. *Australian Universities Review, 30*(2), 43–44.

Napierkowski, H. (2001). *Collaborative learning and sense of audience in two computer-mediated discourse communities*. Paper presented at the 52nd Annual Meeting of the Conference on College Composition and Communication, Denver, CO, March 14-17.

Nardi, B. (Ed.). (1996). *Context and Consciousness: Activity Theory and Human-Computer Interaction*. Cambridge, MA: MIT Press.

Nardi, B. (1996). Studying context: A comparison of activity theory, situated action models, and distributed cognition. *Context and Consciousness.*

Nelson, G. (1995). Cultural differences in learning styles. In J. Reid (Ed.), *Learning Styles in the ESL/EFL Classroom.* Boston: Heinle and Heinle.

Notess, G. (2002). *The Blog realm: News sources, searching with Daypop, and content management*. Retrieved Aug. 30, 2007 from http://www.onlinemag.net/sep02/OnTheNet.htm

Nunan, D. (1991). Communicative tasks and the language curriculum. *TESOL Quarterly*, *25*(2), 279–295. doi:10.2307/3587464

Nunan, D. (2004). *Task-based Language Teaching*. Cambridge, UK: Cambridge University Press.

Nunan, D. (2005) Important Tasks of English Education: Asia-wide and Beyond. *The Asian EFL Journal, 7*(3). Retrieved June 14, 2006 from http://www.asian-efl-journal.com/September_05_dn.php

Pennington, M. C. Balla, J. Detaramani, C. Poon, A. & Tam, F. (1992). *Towards a model of language choice among Hong Kong tertiary students (Research Report No. 18)*. Hong Kong: Department of English, City Polytechnic of Hong Kong.

Phelps, R., & Ellis, A. (2003). *From page turning to deep learning: A case history of four years of continual development of an ICT course*. Paper presented to E-Learn 2003: World Conference on E-Learning in Corporate, Government, Healthcare and Higher Education, Phoenix, Arizona. Chesapeake, VA: Association for the Advancement of Computing in Education (AACE). Retrieved Dec. 16, 2004 from http://www.aace.org/conf/elearn/

Ring, G., & Mathieux, G. (2002). *The key components of quality learning*. Paper presented at the ASTD Techknowledge 2002 Conference, Las Vegas.

Roskams, T. (1998). Collaborative interactions in networked writing classrooms: The student experience. In *The Japan Conference on English for Specific Purposes Proceedings,* Aizuwakamatsu City, Fukushima, Japan, November 8, 1997.

Scollon, R., & Scollon, S. (1995). *Intercultural Communication.* Oxford, UK: Blackwell.

Sheard, J., & Markham, S. (2005). Web-based learning environments: developing a framework for evaluation. *Assessment & Evaluation in Higher Education, 30*(4), 353–368. doi:10.1080/02602930500099169

Slavin, R. E. (1989-1990). Research on cooperative learning: Consensus and controversy. *Educational Leadership, 47*(4).

Stephenson, J. (2001). *Teaching and Learning Online – Pedagogies for New Technologies*. London: Kogan Page Limited.

Tang, C. (1996). Collaborative learning. In D. Watkins & J. Biggs (eds.), *The Chinese Learner: Cultural, Psychological and Contextual Influences,* (pp. 183–204). Hong Kong: The Comparative Education Research Centre, Faculty of Education, University of Hong Kong.

Thompson, L., & Ku, H.-Y. (2005). Chinese graduate students' experiences and attitudes toward online learning. *Educational Media International*, *42*(1), 33–47. doi:10.1080/09523980500116878

Trompanaars, F. (1993). *Riding the Waves of Culture.* London: Nicholas Brealy Publishing.

Tsui, A. (1996). Reticence and anxiety in second language teaching. In K. Bailey & D. Nunan (Eds.), *Voices from the language classroom,* (pp. 145–167). Cambridge, UK: Cambridge University Press.

Tu, C.H. & McIsaac, M. (2002). The Relationship of Social Presence and Interaction in Online Classes. *American Journal of Distance Education.*

Vygotsky, L. S. (1978). *Mind in Society: The development of higher psychological processes*. Cambridge, MA: MIT Press.

Wagner, E. D. (1994). In support of a functional definition of interaction. *American Journal of Distance Education*, *8*(2), 6–26. doi:10.1080/08923649409526852

Walther, J. B. (1996). Computer-mediated communication: Impersonal, interpersonal, and hyperpersonal interaction. *Communication Research*, *23*(1), 3–43. doi:10.1177/009365096023001001

Watkins, D. A., & Biggs, J. B. (1996). *The Chinese Learner: Cultural, Psychological and Contextual Influences*. Comparative Education Research Centre (CERC) Faculty of Education, University of Hong Kong and The Australian Council for Educational Research Ltd. (ACER).

Watkins, D. A., & Biggs, J. B. (2001). *Teaching the Chinese Learner: Psychological and Pedagogical Perspectives*. Comparative Education Research Centre (CERC) and The Australian Council for Educational Research Ltd. (ACER).

Wenger, E. (1998). *Communities of Practise: Learning, Meaning, and Identity*. Cambridge, UK: Cambridge University Press.

Wenger, E. (n.d.). *Communities of Practice – a brief introduction.* Retrieved March 20, 2006, from http://www.ewenger.com/theory/index.htm

Chapter 18
Using Activity Theory to Guide E-Learning Initiatives

Neal Shambaugh
West Virginia University, USA

EXECUTIVE SUMMARY

This case documents how activity theory can be used as a tool to help educators understand the issues behind deploying online learning programs. Faculty members in higher education are accustomed to teaching online, but are new to the development of online academic programs. This case chapter provides a background to the academic setting and a discussion of activity theory. The specific context of an academic department is described, followed by how activity theory was used to represent the overlapping goals of faculty, students, and administrators, and to understand the contextual issues of roles, community of practice, and division of labor to reach the desired goal, which was to implement their academic programs online. Guidelines for using activity theory are provided.

BACKGROUND

Organizational Issues

Higher education institutions historically have been slow to adjust to changing pressures and environments. Administrators now push for E-Learning initiatives, which increase student enrollments through online or blended learning courses. E-Learning, of course, can be used in a face-to-face environment or in blended instruction. However, E-Learning is defined in this case as the use of digital and networked technologies for online instructional programs.

Faculty who teach courses, conduct research, and contribute to service activities are increasingly called on to develop plans to market, recruit, and retain students in new online programs, tasks that are new to them. In addition to taking considerable time to design, faculty groups work with little organizational experience to implement and manage E-Learning initiatives. Traditional models of **curriculum development** take too long and are linear in nature, ill-suited to the needs of institutions to

DOI: 10.4018/978-1-60566-942-7.ch018

make decisions faster than they are accustomed to. Faculty members typically discuss program objectives, program features, courses, and assessment, while administrators deal with the issues of student recruitment, course staffing, and program coordination and evaluation. E-Learning initiatives, however, require academic departments to address issues of both curriculum and organization simultaneously.

To involve faculty, students, and administrators in making faster and more responsive curricular and organizational decisions, academic institutions must understand the complex contexts surrounding fast-changing market conditions. This case documents the use of **activity theory** as a tool to help faculty members of an academic department understand the issues of developing their online academic programs and making decisions. Viewing E-Learning initiatives through "activity" acknowledges the different constituencies that have a stake in such programs, including faculty, students, and administrators. An activity perspective acknowledges the context of both curriculum (what is to be taught) and organizational needs (how to implement and manage).

Activity theory has been used to analyze educational settings ranging from computer-based training to better understand the workplace in which the training was used (Pang & Hung, 2001), as well to acknowledge teachers' beliefs about teaching and the power issues between teachers and administrators in public schools (Robertson, 2008). Activity theory has been used in higher education strategic E-Learning initiatives (Salomon, 2005) and to look specifically at asynchronous learning networks (Li & Bratt, 2004).

Activity Theory as a Tool

Activity theory is a socio-cultural perspective on understanding the interconnections of people, organizational rules and culture, and tools, all directed to some outcome or goal (Bertelsen & Bodker, 2003; Cole & Engeström, 1993). To achieve E-Learning in higher education involves different human constituencies including faculty, students, administrators, and the influence of social and cultural norms, values, language, and tools on these humans (Vygotsky, 1978). Activity theory is represented in Figure 1 as a collective image of several components or nodes (Engeström, 1987).

The top triangle in the visual involves users, goals, and tools. The goal node in the activity system visual itemizes how human activity is directed. These goals are accomplished with the help of tools, such as a learning management system in E-Learning. The bottom half of the triangle identifies three categories of contextual issues that involve all human activity. The community of practice is made up of individuals and groups who share the same goals and have developed specific ways of working. The division of labor node refers to the roles and tasks of the community members and a division of responsibility and control. Rules and norms refer to the explicit and implicit regulations, norms, and conventions that constrain actions and interactions within this activity system.

On a pragmatic level, activity theory can be used as a tool of **analysis** to examine the interconnected activity of faculty, students, and administrators. Analysis is a distinct feature of human-computer interaction design where systems analysis, the gathering of data on the information needs of a unit (e.g., data flow diagrams, entity-relationship diagrams), leads to a system design which meets those needs. In education instructional design includes analysis as a front-end activity in a cyclical systematic process, which proceeds to design, implementation, and program evaluation/revision. Instructional design taps many analysis methods, including task analysis, content analysis, instructional needs analysis, and knowledge elicitation techniques, features which attempt to identify what is to be learned before the actual performance (Jonassen, Tessmer, & Hannum, 1999). Activity-based approaches assume that

Figure 1. Visualizing activity system components

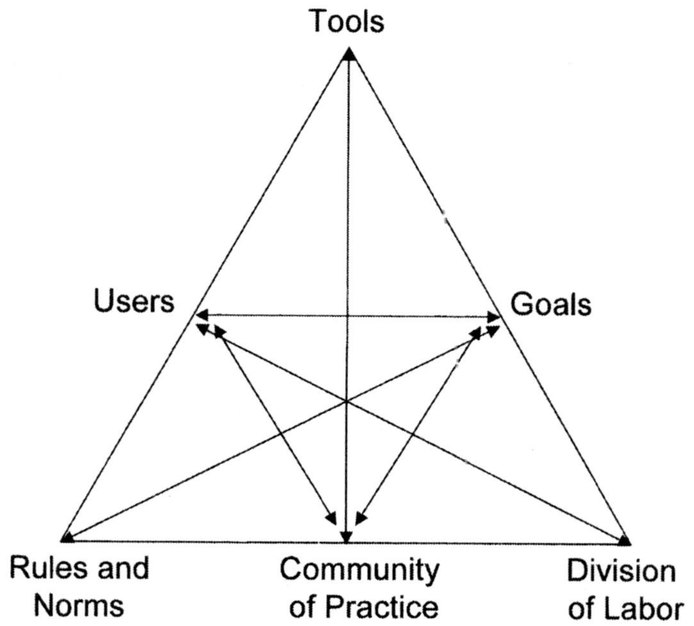

knowledge and performance are connected, and that activity structures serve to help understand how humans organize to accomplish specific goals. Thus, activity theory is related to other learning theories including situated cognition, distributed cognition, and constructivism.

The overall advantage to activity theory as an analysis tool is that the model raises awareness of the players to contextual and historical factors that comprise human activity. The different players can then discuss the potential impacts of this context on reaching the goals. Activity theory is not just a front-end analysis tool, as needs assessment is in **instructional design**, but a "heuristic aid" (Gay & Hembrooke, 2004) for the players to continually evaluate implementation and make revisions, but also to keep the needs and concerns of the interconnecting players in front of everyone.

The cultural-historical approach discussed here has focused on 'higher psychological functions,' Cole (1988) but Griffin & Cole (1984) pointed out the insensitivity of activity theory towards cultural diversity. Now questions of diversity and dialogue have become increasingly serious challenges to advance the usefulness of activity theory in organizations. One example, are the power relationships between public school teachers and administrators (Robertson, 20008). Future uses of activity theory need to develop conceptual tools to understand dialogue, multiple perspectives and voices, and networks of interacting activity systems.

SETTING THE STAGE

Administrative

The department is one of four academic departments in a college of education. The college is part of a land-grant university, which in the United States means that the university offers a comprehensive range of degrees, conducts doctoral level research, and provides service to its state. The department consists of 15 faculty members, each of which serves one or more program areas. A total of four program areas exist consisting of child development, educational psychology, research methods, and instructional technology. The child development program offers an undergraduate degree (B.S.) and a master's (M.A.), while educational psychology and research methods offer master's degree (M.A), and the instructional technology program awards both a master's (M.A.) and a doctoral degree (Ed.D).

Within the past ten years the department has reorganized twice, retitled itself, adding a child development program from another college, and losing a program to another department in the college. In general the department serves graduate students, but with the addition of a new program a significant number of undergraduate students have improved the department's overall student numbers. The department is supervised by a department chair, who reports to the college dean. Department issues are discussed at twice-monthly leadership team meetings, composed of the dean, associate deans (3), department chairs (6), and center directors (3). Within the department one of the faculty members coordinates each of the four programs. A small stipend is paid to each program coordinator. Funds are allocated to each department based on the student numbers across a fall and spring semester academic year, as well as any summer enrollments.

Three competent office staff, each with 20+ years of experience, handle the financial details of each program, including purchasing of materials and supplies and travel reimbursement, as well as providing program support dealing with faculty teaching, student queries, and student records.

Faculty

Faculty members teach courses to serve their program areas, but may also teach courses in a dual-degree five year teacher education program. All faculty members have a doctorate degree. Adjunct faculty members are sometimes used for summer courses. Of the 15 faculty, 13 are tenured or on a tenure-track. These 13 have written expectations for high quality teaching and research dissemination, with moderate expectations in the service category. The other two faculty members are clinical appointments, meaning that their primary activity is teaching. Each of these two faculty members teaches 4 courses, while the other 13 teach between 2 and 3 courses, depending on their other responsibilities.

As this department primarily serves graduate students, its focus is graduate level teaching and conducting research. Tenure-track expectations place a considerable pressure on new faculty members on their first 6 years, as they are expected to publish regularly to achieve tenure, while also receiving good-to-excellent ratings from students on their teaching. Of the 13 faculty members, 8 have tenure at the associate professor or professor level with 5 faculty members at the assistant professor level. Additional pressure for all faculty is the amount of graduate advising that must occur. Faculty members overage over 12 chair duties and sit on 50+ committees. Research methods faculty who teach research methods courses sit on 50+ master's and doctoral committees, as they are likely the methodologist on the committee. Advising in the child development program area is principally undergraduate and advising students on courses and graduation requirements. The one faculty member who teaches the online courses in child development also performs many advising duties at a distance with online students.

Students

Traditional age students are enrolled in the child development undergraduate program. Each of these students is advised by a child development faculty member or an office staff person. A wide range of ages characterizes the students enrolled in the graduate programs. Student enrollments in the graduate programs consist of 30% international students. Advising for the graduate programs involves a faculty member who is assigned to be the student advisor. At the master's level a program of study with a coursework option specifies courses and requires one faculty member signature. Students who choose a thesis option submits a program of study specifying courses and requires three faculty members to attest to the thesis through a written document and a defense. The instructional technology doctoral program requires a committee of five faculty members, a program of study, a year residency requirement, a candidacy examination, and a dissertation document with defense. Online courses are usually fully enrolled (caps of 20-30 students).

Technological Infrastructure for E-Learning

The university's administrative and educational technology needs is supported by an office for **information technology**. This support group provides web hosting space on its server, including space for departments and student organizations, online courses, and personal web hosting for faculty, staff, and students. Wireless network access is provided, both encrypted for staff and students, and non-encrypted for guests.

Distance learning opportunities for students across all programs is marketed by an office of distance learning. However, E-Learning courses can be developed and delivered by any academic program unit. Instructional support is coordinated by an instructional technology group and consists of course development, media development, including blogs/wikis, multimedia and streaming media.

Instructional technology support for in-class use within the college is supported in two ways: A university unit for classroom technology supports eight classrooms in the college building, which are equipped with a PC (with DVD playback capabilities), data video projector, document camera, auxiliary input panel, and touchpanel control system. Additional "optional" equipment such as microphones, personal response system clickers (PRS), Macintosh computer, VCR, and dual projection screens are available depending on the room. Second, the college's computer lab manages a full floor of 5 classrooms, 2 labs, and public computing space, as well as computing consulting, mobile computing, reserve materials, and audiovisual support. The college lab also sponsors periodic workshops, software guides, and an end-of-the-year professional development event.

Online delivery is coordinated through the use of a **learning management system** (LMS), which is integrated with the university's administrative system so that faculty and students are automatically registered for their online courses as they are for face-to-face classes. Faculty use the LMS to develop online courses using the standard features. These features include posting of materials and links, blogs/wikis, web chats, synchronous audio, as well as learning outcomes, assessment and grade data. Instant messaging is also available for each course. All faculty and students are given email addresses for communication within the LMS and outside using other mail clients. Within three years (Fall, 2005 through Fall, 2008) the number of sections grew from 1454 to 5994.

E-Learning History

The child development program offers a master's degree through online delivery. A non-tenure track faculty member coordinates the online courses and teaches full time in the department. A total of 5 faculty members teach in this program. A second program, instructional technology with 4 faculty members, delivers its master's courses face-to-face (F2F) or online, either through **synchronous** (e.g., video) or **asynchronous** modes, depending on the faculty member and the individual student. The goal is to enable any student to take a course online or F2F for the instructional technology master's degree, however, the master's program is not marketed as an online program. This online teaching decision was made recently as a first step in offering an official online master's instructional technology program.

Administrative pressures over the past 10 years have stressed the need for student numbers, a common issue at most academic institutions. Faculty members have been under this pressure to keep course enrollment high, although the majority of the programs in this department are graduate programs, and issues of program quality have arisen. Student enrollment targets and a student recruitment process have been adopted. Curriculum issues and online program features, questions of quality control for these programs, have not been discussed at the department level.

During periodic meetings of the department issues of online courses come up for discussion but no organized discussion, design, or professional development has occurred with overall program goals in mind. **Professional development** in the college has focused on individual faculty members' needs to understand online tools and how they might be used in their courses. There was, however, a need to discuss program issues regarding online programs and delivery.

Department Management Practices

Decision-making in this academic department is solicited and managed in a top-down mode typical of academic institutions: namely, dean, leadership team, and department chairs. Faculty input and decisions occur through standing committees and a faculty constitution. Within this department faculty have a significant voice in all matters. The downside is that there are many academic matters to contend with and periodic meetings have long agendas. Not all issues can be addressed during the academic year; only a minority of the issues can be adequately addressed through department meetings.

The case description provides an overview of **activity theory** as an analysis tool, and demonstrates how activity systems can be depicted for the major E-Learning constituents, which include faculty, students, and administrators. Robertson (2008), meanwhile, labels his three activity systems as organizational, technological, and pedagogic. Subsequently the case describes how the activity system model was used to analyze overlapping goals of the three constituent groups and better understand the context in which each group operated. While this approach is ongoing, the case describes how the system representation has been used to prompt changes in rules, work culture, and roles. The idea of boundary crossing actions was used to prioritize E-Learning decisions for future curriculum and implementation.

CASE DESCRIPTION

E-Learning Activity Systems for Faculty, Students, Administrators

Specific components of a generic activity system model for E-Learning initiatives (see Figure 2) include the curricular degree-granting program, which is the goal of the activity systems. The users include students, faculty members, and administrators and require activity systems of their own to examine their distinct contextual features (to be described below). These users tap tools, such as the **learning management system**, which coordinates student activity in courses, and the administrative system, which coordinates student matriculation through a program. Sometimes these systems are connected, sometimes they are not. This tool category would also include the technical support and instructional design support personnel. The division of labor node identifies "who does what," while the rules-and-norms node identifies the policies and procedures of that department in regard to hiring, promotion, teaching, research, and service activities. The cultural features describe a unique **community of practice** that describe the unique ways of working from faculty members, students, and administrators.

E-Learning requires three activity systems to address the goals of faculty, students, and administrators. Initiatives must take into account

Figure 2. Academic e-learning activity system

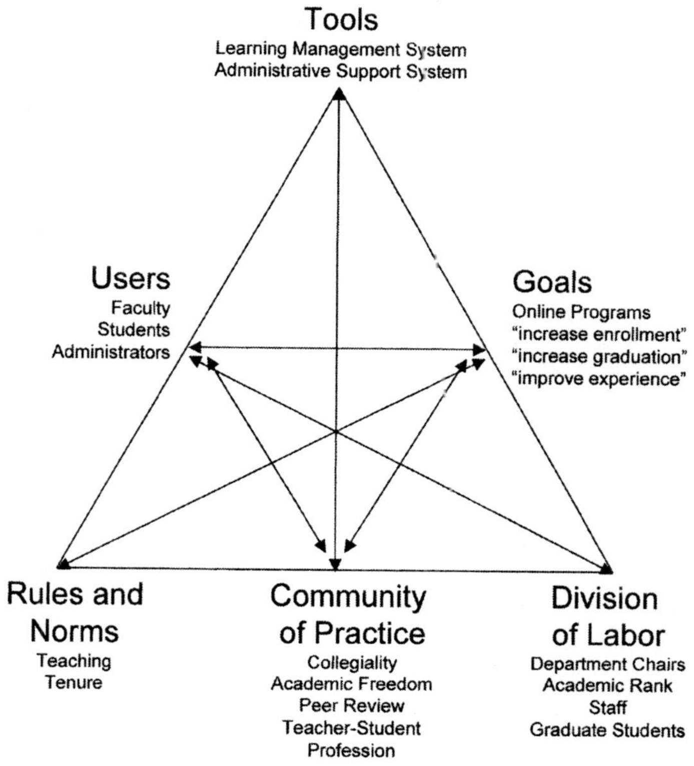

issues of *what to teach* (curriculum) and *how to implement* this curriculum (organizational) within a complex academic context. Both of these issues also occur within different systems of activity for faculty, students, and administrators.

Faculty Activity System

Faculty activity systems can be visualized using an activity system (see Figure 3). The items listed under each node were gathered from faculty meetings in which online program development was discussed. Faculty goals with E-Learning initiatives involve the preparation, delivery, and evaluation of online courses. Faculty use student course performance as an outcome in the advising of a students' program of study and mentored research activities. The rules and norms for faculty can be divided roughly by those seeking tenure and those who have tenure. Tenure-track faculty operate under expectations to publish and achieve good-to-excellent teaching ratings, all within a six-year time frame. Faculty members themselves have expectations for a successful delivery of a course, as well as mediating how students regard their expectations.

The **community of practice** for faculty involves shared decision-making when it comes to programs, courses, and their role. Faculty have varying views of collegiality and academic freedom in regards to their activity. Their view and treatment of staff members is an important community of practice. Faculty also have varying views of their role in developing the skills and attitudes of graduate students. Faculty assume different roles within the college and department, include the new faculty member, the tenured faculty member, and a role of annual peer review

Figure 3. Faculty activity system

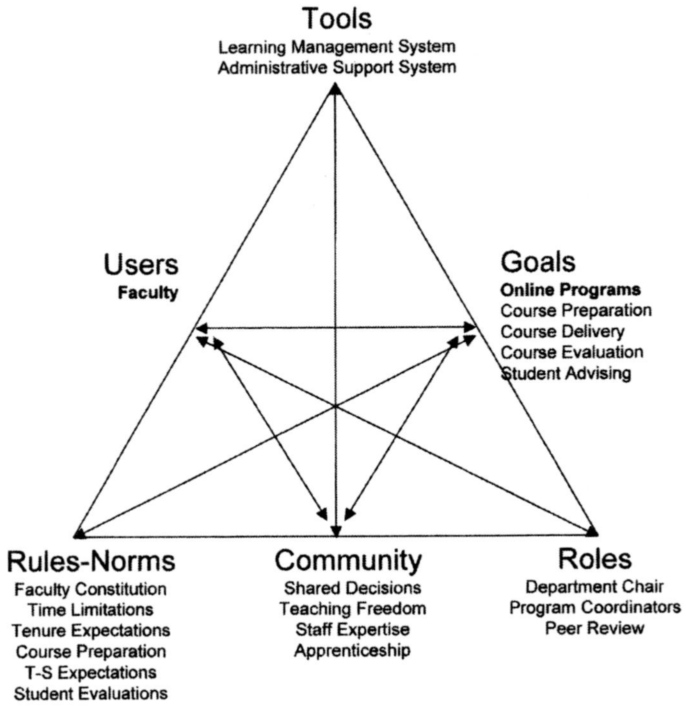

for promotion. Faculty members may also assume different administrative functions.

Student Activity System

Student activity systems can also be visualized (see Figure 4). Students have somewhat different goals than faculty and focus primarily on the course experience and resulting grade. Students are also concerned with completing the program requirements, which may include a thesis or dissertation. In terms of rules and norms, students "navigate" a set of admission and program completion requirements and procedures, address financial needs, and complete semester courses and other expectations. Student **communities of practice** and inherent cultures include temporary communities found in course, whether face-to-face (F2F), online, or in blended course deliveries. Ad hoc study or project groups may form and disband over time. A larger level of community involves the relationships developed between faculty members and their students in coursework, program advising, and mentoring of research projects and a thesis or dissertation. While graduate programs in particular focus on the development of future faculty who may teach or conduct research, the roles for these students become that of course performance, review of faculty teaching, and activity in other program related functions.

Administrator Activity System

The activity of administrators, which is seldom examined, can also be visualized (see Figure 5). Their goals in terms of E-Learning overlap to some extent but focus on student recruitment, enrollment, and graduation. Administrators are also responsible for faculty teaching assignments and overall workload, as well as the quality of their academic programs.

Figure 4. Student activity system

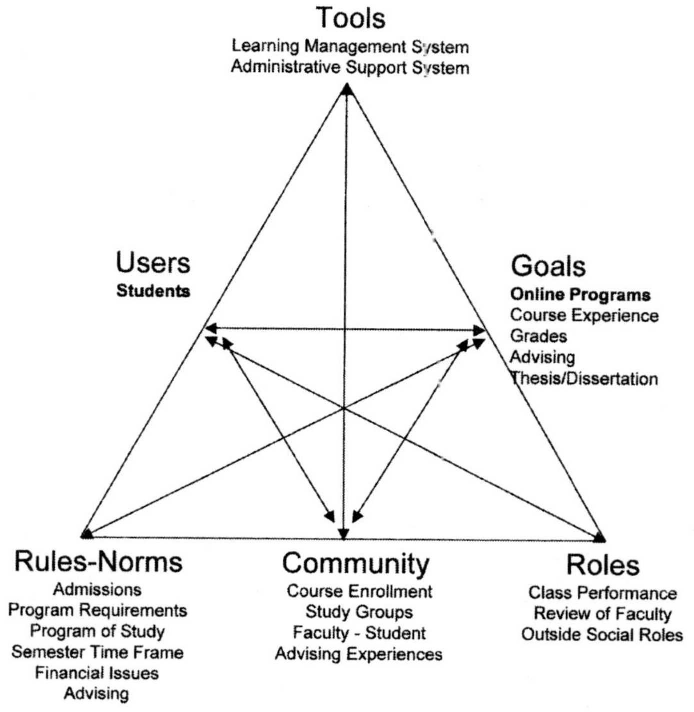

The rules and norms for administrators who are also faculty members, involve the upward (dean's office) and downward (faculty assignments, student admissions and financial support) reporting of goals and results, as well as the enforcement of department policies with staff, students, and faculty members. The activity of administrators is bound by academic policies and college/department norms for shared decision-making. The **communities of practice** experienced by administrators can be found at the leadership team level, which is composed of department chairs, the dean and associate deans, a staff member, and center directors. Each administrator is also bounded by historical department practices and expectations, as well as the different expectations for faculty who are new, tenured, or those without a tenure-track appointment. Students and the administrator form a unique community, mostly procedural, policy-driven, and management-by-exception. Administrators take on different roles and power relations providing guidance for new faculty, the continued success of all faculty members, coordinating program initiatives and teaching, as well as supporting staff members.

Using Activity Systems for E-Learning Analysis and Synthesis

Child development online programs were in place prior to official faculty meetings on E-Learning programs, which occurred during the Fall 2008 semester. In addition, master's courses in the instructional technology program were being offered both face-to-face and synchronously for distance students. Monthly meetings included on their agenda discussions on E-Learning programs, first in a brainstorming session, which allowed faculty to share their past experiences or concerns. These issues were then analyzed using the activity system structure to help organize their concerns.

Figure 5. Administrator activity system

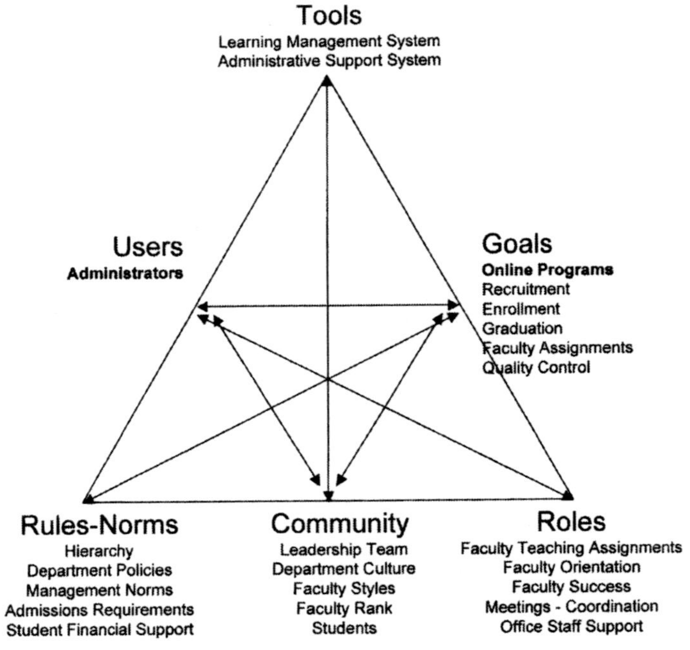

Analysis of Overlapping Goals

Analysis of the activity system components across the three groups provided immediate insight. Breaking out the shared but different aspects of activity between faculty, students, and administrators identified an overlapping set of goals (see Figure 6). This overlap suggests that E-Learning was a mutual concern and that changes to address the needs of one group would impact one or more of the other two groups.

Figure 6 was developed by examining the E-Learning goals for faculty, student, and administrator in Figures 2-4 and looking for the overlaps. A significant overlap of concerns existed with online courses and advising, meaning that any implemented changes will impact all three groups. One area of less concern for faculty and students, but significant for administrators, was in evaluating and improving the quality of the overall academic programs, including E-Learning components. This area of concern becomes an

Figure 6. Shared, non-shared goals for e-learning

Faculty Goals	Student Goals	Administrator Goals
Online course preparation, delivery, and evaluation	Online course expectations, grades, and performance	Faculty assignments
Student advising	Program advising online/F2F Thesis/dissertation mentoring	Student recruitment, enrollment, graduation
		E-Learning program quality control

acute issue with new programs that are completely online. Despite overlapping goals, the immediate concerns of faculty, students, and administrators can be regarded as "divided terrain" where these individuals do not always talk or work with each other. The activity system suggested the possibility of analyzing a multitude of relations within the triangular structure of activity, including shared goals and differences in roles, and ways of working together. However, the essential task was always to grasp the systemic whole, not just separate connections.

Synthesis of Decisions for E-Learning

The visual activity representation provided both analysis and synthesis opportunities. **Synthesis**, in terms of decision-making, can be facilitated by discussing how the use of E-Learning tools influences or mediates the goals of the three groups and what changes in rules-norms, communities of practice, and roles may be needed. Synthesis, in terms of decision-making, can be facilitated by discussing how the use of E-Learning tools influences or mediates the goals of the three groups and what changes in rules-norms, communities of practice, and roles may be needed. Based on Figure 7, online courses and E-Learning programs require what Engeström (2002) calls **boundary-crossing** actions. Boundary-crossing actions are two-way collaborative interactions requiring both renegotiation and reorganization decisions.

Figure 6 depicts what the department has identified as such opportunities between Faculty – Administrators, Faculty – Students, and Administrators – Students. Benefits of identifying such opportunities forces joint responsibility for E-Learning courses and programs on all three groups, that program development in E-Learning,

Figure 7. E-learning boundary crossing actions

Existing Faculty Actions Faculty Constitution Time Limitations Tenure Expectations Course Preparation Teacher-Student Expectations Student Evaluations	*← Boundary Crossing Actions →* *College provides support for graduate programs and E-Learning* *Department support of E-Learning program development* *Department support for online course development* *Department documentation of a faculty member's online activities for tenure* *Faculty collaborate on program features, goals, curriculum scope and sequence, student assessment, student advising, program evaluation* *Faculty dissemination of online teaching best practices*	**Existing Administrator Actions** Hierarchical Reporting Department Policies Management Norms Admissions Requirements Student Financial Support
Boundary Crossing Actions *Instructor provides clear specifics on student performance in an online course* *Instructor provides prompt, consistent, and constructive course feedback and student advising* *Instructor provides flexibility in course delivery features based on students' technological capacity.* *Instructor identifies student differences, including international students, and makes adjustments in course implementation.*	**Existing Student Actions** Admissions Program Requirements Program of Study Semester Time Frame Financial Issues Advising	*Boundary Crossing Actions* *Department provides clear specifics on online program requirements* *Department provides clear policies and availability of financial resources* *Department examines what a contact-hour means in an online course* *Department solicits data on improving course and program features* *Department identifies program evaluation details*

given its uniqueness and newness, requires an ongoing, iterative, and collaborative set of practices quite different from traditional academic courses and programs. The one boundary crossing action on collaborative discussion of program features necessitates continual involvement of faculty, students, and administrator across many issues. Thus, program development is not a linear history of **curriculum development**, followed by administrator implementation, as in traditional academic programs. E-Learning requires a dynamic involvement across multiple issues, including program purpose, marketing, program features, assessment, and matriculation of students through the program.

One specific conflict identified from these **boundary-crossing** activities involves the redefinition of what constitutes a faculty-student contact hour. Traditionally, courses have required 45 contact hours (3 hours/week X 15 weeks). Online courses, which operate in an **asynchronous** delivery mode, are at odds with a **synchronous concept**. Rather than defining contact as "seat-time," student performance is being discussed. Rather than a unit-of-delivery, such as contact hours, specific types of student performance for each course is specified, such as papers, designs, critiques, or chats. Thus, contact hours are specified in terms of a unit of task performance.

CURRENT CHALLENGES FACING THE ORGANIZATION

Short-Term Needs

An important administrative shift from the university has been an increased priority on graduate programs. Such a shift signals college administrators to re-visit existing graduate programs and identify opportunities for student recruitment and support, in terms of finances and advising quality.

Faculty are being asked to invest additional time in the design of online courses and programs. Department faculty, however, are still concerned with online course development, rather than online program development. Finding time to discuss these issues is problematic in any academic setting. Agendas that are specific, limited in scope, and achievable provide have been useful to establish the norms and community for this activity.

Staffing of courses for instructors is an issue, given existing faculty workloads and responsibilities to conduct research and service. Some of the department's courses, particularly in research courses, must look across program areas to staff courses. This flexibility enables faculty members in the department to concentrate on course development or research projects, or E-Learning development.

How learning will be assessed in E-Learning programs is an ongoing issue that is currently addressed on by individual online programs in the department. The undergraduate child development program uses the current learning management system to provide security on individual student work, while graduate programs focus primarily on designed artifacts that are open for critiques from class-enrolled students. Online portfolio products are being tested to house these artifacts. Matriculation issues, documenting that students are completing program requirements, still rest on prior form-based mechanisms, rather than an online system that might be used in business training systems.

Long-Term Needs

For faculty E-Learning initiatives must figure into faculty members' teaching and research agenda, and that adequate professional development and time be accorded. Particular attention needs to be given to tenure-track faculty who have online teaching expectations, which provides some "success" issues in terms of student evaluations and course preparation time. Guidelines for peer review need to be re-examined in terms of course design and student evaluations.

Students have higher expectations for online courses than traditional F2F courses, resulting in a significant increase in online faculty attention. Program and course clarity in terms of graduation and performance need to be clearly specified. Online training might also be needed for some students who do not fare well in non-F2F settings.

For administrators, E-Learning provides opportunities for significant student enrollment increases, but such initiatives carry with them the need for quality control and ongoing development. Program area coordinators will require a significant amount of time to coordinate E-Learning programs. Departmental meetings, with a long list of other issues, cannot address E-Learning in these settings. An ad hoc work group has formed so that a small group of faculty members can directly tackle E-Learning issues across the academic year. Working groups may providing a more effective organizational structure to identify work priorities and report out to the department results from small cycles of work-decisions.

GUIDELINES FOR USING ACTIVITY THEORY

Brief Constituents on the Purpose of Activity System

Higher education administrators press faculty members for online delivery of courses mostly out of a need to increase student-paying enrollments. Years ago, online teaching experience was a rarity, now it is looked for in new faculty. Developing online programs, however, is a new experience for many faculty members. Activity theory provides a tool to help faculty members unpack the complexity of delivering online programs. The approach provides a working structure to organizing the discussion of developing E-Learning programs. Another value to the approach is that the entire process of E-Learning development is occurring systematically, useful for program evaluation documentation, but also to disseminate what they learn in publications and conferences.

Specify Procedures

What was learned from the use of **activity theory** as an analysis tool was the need for a set of procedures. This case describes a three-step process, including **analysis**, **synthesis**, and **boundary crossing** activities as its procedure. This was done to keep the procedure simple for the faculty meetings. The activity theory structure very much assisted in making sense out of faculty discussions.

Jonassen, Tessmer, and Hannum (1999) suggest six steps. Step 1 clarifies the purpose for the activity system. Step 2 analyzes the activity system by defining the subject, communities, and goals. Step 2 provides a big picture look at the overall initiative. Here faculty decide on such questions as what distinguishes a program from other programs, and what skills and competencies students learn from the program. Step 3 specifies analyzing all of the activities that involve the participants. Here purposes to support the goals are re-examined, action steps are determined, and any design or development work undertaken. Step 4 examines the role of the tools, specifically, the Learning Management System, portfolio software, and synchronous or asynchronous E-Learning tools, on student learning, faculty, teaching, and administrative management. Step 5 addresses the internal and external contextual features. Internal features can include faculty workload concerns, getting along with other members, experiences in teaching and online instruction. External features of the context can include faculty reward structures, external support from administrators, the sharing of tasks, and expectations for performance. Step 6 prompts one to examine what is occurring and progress towards the intended goals, as well as any changes in relationships and understandings of the overall process.

Collaborate Continually and Frequently

E-Learning programs, given their dynamic nature, require a continual collaborative involvement of all constituents. Activity systems visually depict different features of faculty, students, and administrators. The various nodes along the triangle raise an awareness of different rules/norms, community, and roles for these three constituent groups. Comparing goals depicts a possible overlap and potential for a change in existing practices and boundary crossing actions to take place. Conflict will arise, but these conflicts will provide opportunities for discussion and clarity or change. The overall value of activity theory is that the contextual details inherent in three activity systems come into view.

E-Learning programs require a joint involvement of time. Any initiative that asks more time of faculty members will undergo careful scrutiny. The design of E-Learning programs requires significant time to develop, but an ongoing commitment is necessary in their implementation and revision. One way to address the time investment is to identify a working group of faculty members who share joint interest in developing E-Learning programs. The key is to frequently report back to the large faculty group and to solicit input and delegate action across the faculty group.

Share in the Decision-Making Process

While faculty are accustomed to developing and revising their own courses, E-Learning programs require more collaborative development to ensure that they are successful. The definition of what constitutes "success" must be clearly spelled out in advance and subsequently evaluated on a continuing basis. Student assessment data will need to become a regular activity which goes beyond the assignment of grades to data used by the program to partially measure success. Program features will need constant revision, and initial assumptions will need to be scrutinized by all three groups, as the overlapping goals of the three groups determined from the activity systems approach, necessitate this attention. Thus, **curriculum development**, and organizational development, which implements the designs, merges over time. The role of the administrator evolves to facilitate collaboration of the constituent groups and keep the overlapping goals and issues in front of everyone.

Manage the Complexity of Information and Contexts

One disadvantage of any context-based approach is that it will generate a lot of information. The activity system nodes and labels under those nodes help to keep the major issues in front of everyone. The management of that information will need to be facilitated by a working group of faculty members. Summary tables serve to communicate periodic working group activity to other members of the faculty group.

Context complexity will emerge from faculty discussions, a complexity beyond their own program or department. Micro and increasingly broader macro contexts at play have their own pace and rate (Boer, van Baalen, & Kumar, 2002), but this understanding prompts members of the faculty to attend to some issues more than others. For example, it may be needed to spend some time working with other groups who have a stake or a say in your E-Learning program. With E-Learning the broader contexts can include college and university agendas, including university-wide instructional support groups. Even with large-scale E-Learning initiatives, de-centralizing the discussion enables faculty in academic programs to maker their own decisions (Sharpe, Benfield, & Francis, 2006). Attention may need to be applied to new faculty who have tenure-track pressures. These groups are at a different stage of stability. Priorities can then be determined and action steps be assigned with deadlines.

REFERENCES

Bertelsen, O. W., & Bodker, S. (2003) Activity theory. In J. M. Carroll (Ed.), *HCI models theories, and frameworks: Toward a multidisciplinary science*, (pp. 292-324). San Francisco: Morgan Kaufmann.

Boer, N., Van Baalen, P., & Kumar, K. (2002). An activity theory approach for studying the dynamics of knowledge sharing. In *Proceedings of the Thirty-fifth Annual Hawaii Internatioinal Conference on System Sciences (HICSS 35)*, (pp. 90-93). Los Alamitos, CA: IEEE Computer Society Press.

Cole, M. (1988). Cross-cultural research in the sociohistorical tradition. *Human Development*, *31*, 137–151.

Cole, M., & Engeström, Y. (1993). A cultural-historical approach to distributed cognition. In G. Salomon (Ed.), *Distributed cognitions: Psychological and educational considerations*, (pp. 1-46). New York: Cambridge University Press.

Engeström, Y. (1987). *Learning by expanding: An activity-theoretical approach to developmental research*. Helsinki, Finland: Orienta-Konsultit.

Engeström, Y. (2002). *The horizontal dimension of expansive learning: Weaving a texture of cognitive trails in the terrain of health care in Helsinki*. New Challenges to Research on Learning, Helsinki: University of Helsinki.

Gay, G., & Hembrooke, H. (2004). *Activity-centered design: An ecological approach to designing smart tools and usable systems*. Cambridge, MA: MIT Press.

Griffin, P., & Cole, M. (1984). Current activity for the future: The zo-ped. In B. Rogoff & J. V. Wertsch (Eds.), *Children's learning in the zone of proximal development*. San Francisco: Jossey-Bass.

Jonassen, D. H., Tessmer, M., & Hannum, W. H. (1999). *Task analysis methods for instructional design*. Mahwah, NJ: Erlbaum.

Li, J. Z., & Bratt, S. E. (2004). *Activity as tool for analyzing asynchronous learning networks*. *LNCS*, *3143*, 19–26.

Pang, P. M. N., & Hung, D. W. L. (2001). Activity theory as a framework for analyzing CBT and E-Learning environments. *Educational Technology*, *41*(4), 36–42.

Robertson, I. (2008). Sustainable e-learning, activity theory and professional development. In *Hello! Where are you in the landscape of educational technology? Proceedings ascilite Melbourne 2008*, (pp. 819-826). Retrieved from http://www.ascilite.org.au/conferences/melbourne08/procs/robertson.pdf

Salmon, G. (2005). Flying not flapping: A strategic framework for pedagogic innovation in higher education institutions. *ALT-J Research in learning technology*, *13*(3), 201-218.

Sharpe, R., Benfield, G., & Francis, R. (2006). Implementing a university e-learning strategy: Levers for change within academic schools. *ALT-J Research in learning technology*, *14*(2), 135-151.

Vygotsky, L. S. (1978). *Mind in society: the development of higher psychological processes*. Cambridge, UK: Harvard University Press.

ADDITIONAL READING

Dede, C. (2006). *Online professional development for teachers: Emerging models and methods*. Cambridge, MA: Harvard Education Press.

Engeström, Y., Miettinen, R., & Punamäki, R. (1999). *Perspectives on activity theory*. New York: Cambridge University Press.

Haertel, G. D., & Means, B. (2003). *Evaluating educational technology: Effective research designs for improving learning*. New York: Teachers College Press.

Kaptelinin, V., & Nardi, B. A. (2006). *Acting with technology: Activity theory and interaction design*. Cambridge, MA: MIT Press.

Mishra, P., Koehler, M. J., & Zhao, Y. (2007). *Faculty development by design: Integrating technology in higher education*. Charlotte, NC: Information Age Publishing.

Salomon, G. (1993). *Distributed cognitions: Psychological and educational considerations*. New York: Cambridge University Press.

Shulman, L. S. (2004). *Teaching as community property: Essays on higher education*. San Francisco, CA: Jossey-Bass.

Tierney, W. G. (2001). *Faculty work in schools of education: Rethinking roles and rewards for the twenty-first century*. Albany, NY: SUNY Press.

Van Oers, B., Elbers, E., van der Veer, R., & Wardekker, W. (2008). *The transformation of learning: Perspectives from activity theory*. New York: Cambridge University Press.

Wenger, E. (1998). *Communities of practice: learning, meaning, and identity*. New York: Cambridge University Press.

Chapter 19
Addressing Online Student Learning Environments and Socialization Through Developmental Research

Ruth Gannon Cook
DePaul University, USA

Caroline M. Crawford
University of Houston – Clear Lake, USA

EXECUTIVE SUMMARY

The chapter looks at the online learners in the course to distinguish whether interactivity and an online community was established. This case study also considers the shift that took place in the learners' focus from simply participating in an online course to reframing their understanding of the course content and whether this holistic approach reflects both the students' and instructor's learning objectives and anticipated outcomes. Design, development and implementation of online learning environments have predominated distance education research over the past fifteen years. Since 2006, dynamic communities of learning have begun to emerge that encompass a more expansive learning environment, addressing the needs of adult learners and their sociocultural environments as well as content materials. This study employs developmental research to examine online learners engaged within a dynamic learning community and provides detailed feedback on the strengths and potential weaknesses of the online course employed in the study.

BACKGROUND

The university that provided the course and sample student population was a small public university located in the southwestern area of the United States. The course was titled *INST 6137: Technology and eLearning* and the research was conducted during the fall 2008 semester, from August 2008 through December 2008.

At the beginning of the course, there were 13 subjects enrolled and, due to different personal reasons, four dropped the course within the first week, and nine students went on to successfully complete the

DOI: 10.4018/978-1-60566-942-7.ch019

course. Three of the subjects were male, and six of the subjects were female. Concerning the subjects' ethnicity, one subject was African American, three Mexican-American, one American Indian, and five Caucasians. These students had several aspects in common; they

- were Instructional Technology degree-seeking graduate students;
- had successfully completed a minimum of 24 graduate-level course credits (8 courses) within the graduate degree program of study; and,
- all had successfully completed the majority of their graduate-level courses within the online learning environment.

The online course was designed with innovative course features, such as embedded metaphorical representations and innovative classical music underlays, to enhance the course atmosphere, support the learner's developing conceptual frameworks, and engage learners in multi-sensory stimuli. Multimedia components were integrated with graphics, audio, video, interactive gaming, and web conferencing to help reframe basic online text-based content. There were also some activities designed to let students experiment with *Moodle* (Moodle.org, 2008) learning community features.

The researcher who was embedded in the course had been designing, developing and teaching online courses for at least ten years. In addition, the researcher had developed collaborative research partnerships with colleagues, such as the second researcher, who were interested in affective elements, like icons, metaphors and narratives, which seemed to mediate recognizable symbols with new materials and enhance students' conceptual frameworks of understanding. The study of affective elements has recently become a primary consideration within online learning environments (Ahrlberg, 2008; Crawford & Gannon-Cook, 2008; Doyle, Radzicki, & Trees, 2008; Gannon-Cook & Crawford, 2004, 2006; Pink, 2005). The researchers sought to explore whether these representations could have any documentable impact on students' learning in the course.

SETTING THE STAGE

Much research on online instructional design has focused primarily on multimedia, assessment, and text, but researchers are now beginning to look more at topics, such as, what actively engages students, how can online students and instructors interact more effectively, and what makes an online class become a learning community? Dynamic learning communities encompass more comprehensive, multi-layered learning environments which include adult learner and sociocultural perspectives. This study conducts developmental research in a case study of online graduate students at a public university in the southwestern United States. The research provides detailed feedback on the strengths and weaknesses of dynamic learning environments, as focused upon the course design, content, and learners in online courses in holistic learning and practice communities. It also looks at the student reflections and evaluations that reveal their thoughts about this type of dynamic learning experience compared with prior online learning experiences.

The study looks at whether one of the reasons for the deep level of student comprehension (and retention) may be due to the embracing architecture of Web 2.0 technologies that is focused upon social engagement and the community of learning that help scaffold the course content on prior learning (Vygotsky, 1935, 1962, 1978, 1981; Wertsch, 1985). One more factor reviewed whether the inclusion of semiotic tools could have an influence on students' receptivity to the content conveyed by new technologies (Chomsky, 2004; Cook, 1985; Cobley, Jansz, 2003; Gannon-Cook & Crawford, 2006; Hamilton, 1969; McLuhan,

1964 a, b; McLuhan & Fiore, 1967; Rothstein, 1995; Schlain, 1998).

Developmental research is often limited to one case study, such as this one, and hence, is not generalizable; but, the findings of the study can still provide important insights and recommendations based on certain areas of design, development, and effective implementation of online learning. These environments nurture deeper levels of understanding that facilitate students' abilities to reorganize and assimilate the information into novel and enhanced understanding. The authors utilize developmental research to explore this deeper level of understanding in a dynamic online learning community.

A Review of Literature

The Shift from the Information Age to the Conceptual Age

The term Information Age evolved in the early 1990s and "as our students enter the workforce, the ability to deal with complex and often ambiguous information will be more important than simply knowing a lot of facts or having an accumulation of knowledge" (Frand, 2000, p. 17). This characterization of time is based on the widespread proliferation of information and communication technologies and the capabilities that those technologies provide to overcome the barriers imposed on communications by time, distance, and location. Advocates of the concept of the Information Age maintain that humans "have embarked on a journey in which information and communications become the dominant forces in defining and shaping human actions, interactions, activities, and institutions" (Alberts & Papp, 1997, p. 13). The shift from the Information Age towards the Conceptual Age (Brooks, 1999; Pink, 2005) has focused upon the important shift from mere delivery of information via the Internet towards the ability to organize, apply, analyze, synthesize and evaluate information on a more integrative level (Bloom, 1984; Bloom, Englhart, Furst, Hill & Krathwohl, 1956). The driving force has turned out not to be globalization, but the skills revolution that thrusts humans into a more demanding Conceptual Age in which cognitive strength is an essential.

This is happening in localized and globalized sectors.... The globalization paradigm emphasizes the fact that information can now travel 15,000 miles in an instant. But the most important part of information's journey is the last few inches — the space between a person's eyes or ears and the various regions of the brain. (Brooks, 2008a, ¶ 10-11)

Twenty-First Century education could fall short if it only provides learners with some new content and technology skills. The education of humans in a global economy will be profoundly affected by technology, with competition for jobs occurring throughout the world. For education to become more effective, without the current high dropout rates (some racial and ethnic group dropout rates as high as 22.1% in 2006 [U.S. Department of Education, National Center for Education Statistics, 2008, ¶ 3]), there must be more ways to reach students and get them excited about learning. Research has indicated that social engagement, establishing communities of learning, and the creation of a supportive environment that embraces diverse cultures and demonstrates respect for knowledge (whether online or on-ground) are keys to students' successful attainment of their learning goals (Bonk & Zhang, 2008; Bonk & Graham, 2006; Bonk & King, 1998; Burnham & Walden, 1997; Crawford, 2006; Darling-Hammond, 2006; Deci, Vallerand, Pelletier, & Ryan, 1991; Lazarus, Wainer, & Lipper, 2005; Liu, Magjuka, Lee, 2008; Min, 2008; Vansteenkiste, Simons, Lens, Sheldon, & Deci, 2004; Wertsch, 1985).

Communities of Learning

Communities of learning are environments that successfully engage both the learners and course instructors and encourage higher order thinking skills, such as those defined by Bloom (1984), which are the abilities to access, evaluate, organize, comprehend, apply, analyze, synthesize and evaluate knowledge (Bloom, 1984; University of Victoria Counseling Services, 2003). Learning at the higher order levels is dependent upon having attained prerequisite knowledge, acquired at more basic thinking levels; then affective processes, such as receiving, responding, valuing, organizing, and characterizing, may engage with psychomotor and cognitive skills which can scaffold higher order thinking into application, analysis, synthesis, and evaluation. Ideally, milieus that incorporate these learning sequences provide rich learning environments by encompassing not only knowledge content, but interactive activities that enlist participants on higher, more engaging and personal levels.

CASE DESCRIPTION

The researchers sought to explore reasons for the deep level of student comprehension, whether it could be due to the inclusion of Web 2.0 tools focused upon social engagement, or due to the inclusion of semiotic tools that were strategically embedded to convey a degree of familiarity to learners (Chomsky, 2004; Hamilton, 1969; McLuhan, 1964 a, b; McLuhan & Fiore, 1967; Rothstein, 1995; Schlain, 1998).

Developmental Research

The research utilized developmental research, focusing primarily on both action research and case study qualitative methodologies. Developmental research supports the premise that "research is critical with respect to the models and processes employed by designers and developers" (Richey & Klein, 2007, p.3). Three lines of research inform the design and development knowledge base: psychological and learning theory; instructional theory and teaching-learning; and, communication theory and message design. While all three have implications for design and development, they do not explain the "role of the designer or the context in which design and development takes place" (Richey & Klein, 2007, p.6), so important contributions to the field of instructional design can be made with more developmental research. This research can involve studying the research of others, involve someone performing design and development activities in the study, or have the researcher embedded in action research throughout the study (Richey, 2005; Richey & Klein, 2007).

Action Research

Action research employs field research (Lewin, 1951) where the researcher observes and reports while embedded within the study. It uses "practical application of scientific method or other forms of disciplined inquiry to the process of dealing with everyday problems" (Vockell & Asher, 1995, p.445). Teacher as researcher collects data and provides needed documentation to journal processes of discovery and steps taken to provide problem solutions. It provides valuable insights into curricula and instruction, and serves as an excellent source for important archival data (Avison, Lau, Meyers, & Nielsen, 1999; Argyris, Putnam, & Smith, 1985; Gask, 2005; Nelson, 2004). In this study, the researcher observed and chronicled student responses while embedded in the course as the facilitating instructor; the students were encouraged to journal responses based on their class experiences (Appendix A).

How the Study Was Conducted

The researcher recorded events, interactions, and conferences throughout the Fall 2008 semester (Representative student communications from the course can be found in Appendix A). While embedded in the course, she noted problems as she uncovered them and followed through on a rapid prototype structure to revise and implement the course changes "on the fly" and without significant impact upon the students.

The researcher conducted both formative evaluations throughout the course and a summative evaluation after the completion of the course. The formative evaluations collected included reviews of the course for appropriate graphics, navigation, or missed errors. Also, through each module of the course, analysis of the design and implementation were reviewed and evaluated. In addition, ongoing feedback of the course was requested from the students and formative assessments were encouraged, through all of the course interactivities. Purposeful sampling techniques, such as student feedback and assignment samples, were employed by the researcher so as to ensure information-rich data.

These (qualitative research) techniques, combined with careful non-biased observations, can lead to sound interpretation and inference even from small samples. Such conclusions are often the primary goal of qualitative researchers, rather than aiming to generalize their conclusions to a broad population. (Richey & Klein, 2007, p. 38)

Since formative assessments were done to measure learning, not to allocate grades, the students were willing participants in providing course feedback to the researcher through emails, and course postings. In this manner, the teacher as researcher obtained "a view of both individual and class performances while students learned how well they had done" (Georgetown College Center for Advanced Study of Assessment, n.d., p. 2; Wikipedia, 2009, ¶ 6). The findings are discussed in the results section of this study.

Summative evaluations provide a basis for the teacher to re-visit topics in the course that may need work or, conversely, that provide positive feedback to reinforce instructional units that were helpful to students; further, summative assessments are required for the course which were reviewed and graded by the course instructor. The researcher also requested final reflective feedback from the students, to allow each to offer thoughtful insight into the course experiences and instructor, and self-reflection about her or his personal learning experience.

Factors Addressed During the Study

Factors that could impact the case study related not only to the online design and environment in which the case study occurred, but also to a number of other issues. These issues related to the appropriateness of theory applied to the course, instructor ability and responsiveness, student learning styles and cognitive load, issues of socialization and community, and issues of how to conduct constructive interventions.

Philosophy Used in Course Design

The underlying philosophical belief systems of the course designer, course instructor and course learners directly impact the success of the online learning environment. This is apparent through the style of the online course structure, the interactivities and also through the instructional support aides available to the learner. The course philosophical framework also affects the learners, along with instructional aspects (teaching), learning styles and other factors, such as cognitive load, that impact success within the learning environment.

There are a number of learning theories, with the primary being behavioral (teacher-driven), cognitive (pragmatic), and constructivist (learner-centered) theories, plus an emerging learning theory called connectivism. These learning theories represent the lens through which designers and instructors view teaching/instruction and learning. Everyone has her or his own prior experiences that contribute to their beliefs and understandings; from this, they develop certain understandings and beliefs about the most appropriate way(s) to teach and the most appropriate way(s) to learn. As such, it is important to develop an understanding of one's own underlying philosophies and beliefs related to the three primary realms of learning theory that directly impact this understanding.

Philosophies of learning are articulated through different theoretical belief systems that appropriately work together on the part of different instructional professionals, such as: instructional designers; instructors; and, developers. There are myriad ways that learning philosophies present themselves within the professional arena, such as: the format through which to represent a face-to-face, hybrid or online course; the learning objectives, assignments and assessments that are articulated; the ways in which knowledge is represented within a learning environment; and, the instructional support and expectations within a learning environment.

Instructional Design-Related Issues

The purposes of instructional design for online learning are threefold:

- to meet learning goals and learning objectives;
- to develop an "equal but distinct" learning experience from a traditional classroom experience; and,
- to develop an active, professional community of learners to sustain and enhance the instructional experience within a pseudo real-world environment.

Before considering the technological systems through which online learning environments are housed, the instructional design process must be addressed. Berger and Kam (1996) suggest that "instructional design is the systematic process of translating general principles of learning and instruction into plans for instructional materials and learning" (¶ 1). Of all the potential instructional design models that have been offered over the years, the generic ADDIE (analysis, design, development, implementation, and evaluation) model offers the basic phases through which an instructional designer travels. The constructivist instructional design models (Crawford, 2004; Seels & Glasgow, 1998) are also based upon the ADDIE model components; the synergistic design is nonlinear and recursive. In this study, the online course learning environment is constantly evaluated by both the students and the instructor and the feedback can enhance the design and course structure.

The primary technology components that frame the online course learning environment within this case study are the Learning Course Management System (LCMS) designated as *WebCT* (*Blackboard Inc*., 2008) and Internet accessibility and speed of access issues.

Tools in Course Design

While most instructional designers consider technology, philosophy, and design features, seldom do they consider the use of tools, such as icons, signs, metaphors and targeted graphics. These representations are part of the field of semiotics which is the study of patterned human communication behavior, including auditory/vocabulary, facial expressions, touch (proxemics), signs, and symbols (Webster, 1989). As sociocultural tools, the signs and symbols of semiotics take on enriched meaning, affecting the functions of human consciousness as well as environment. Ultimately, everyday language and discourse come under the scrutiny of this discipline since it becomes a metalinguistic descriptor of ordinary communication (Brooks, 1999; Cook, 1985; Dant, 1991, as cited in Gannon-Cook, 1998). Ordinary language identifies and uses written material and verbiage to communicate and express meaning. It is also uses representations as tools, to construct meaning (Gannon-Cook, 1998). The primary tools of activities, represented in signs and symbols, act as agents for culture and serve as intervening links to consciousness. These mediation tools (Wertsch, 1985) are the structural and genetic central features of mental functioning. Vygotsky believed the "historical study of behavior is not an auxiliary aspect of theoretical study, but rather forms its very base" (Lee, 1985, p.105). The thoughtful inclusion of strategic symbols and metaphors can resonate with students and help them to be more receptive to learning. The student postings in this study supported this premise; they seemed to enjoy the idea that they could help predispose learners through semiotic tools.

Instructor Ability and Responsiveness

As indicated by some of the students' postings, since online students do not see the instructor, they make quick judgments about their instructor's abilities and willingness to help students (Appendix A). The researcher found that instructor ability was reflected in the responsiveness to students' questions and to the organized learning activities and social networking encouraged by the instructor.

Prompt Instructor Interactions and Feedback

What has emerged in the world of text messaging, blogs, *You Tube*, *Twitter* and other social networking communication environments is the shift in social communication expectations. Communication styles have changed dramatically over the last several years, but also changed are student expectations concerning course instructor feedback. Students have come to expect instant messaging and instant feedback. So, within an online learning environment, it is important for the course instructor to maintain daily course interactions. Instead of a face-to-face course that meets once a week, twice a week, or once a day Monday through Friday, an online course learning environment has significantly and inherently different instructor expectations. Managing student expectations, such as specifying 24-48 hour response time during the week, or specifying that no responses will be provided on Sunday, is critical.

Since research supported the need for an online course instructor to maintain quick assignment assessment turn-around and timely feedback, the researcher observed instructor maintenance of communications with students to see whether the levels and quality of guidance helped them feel comfortable in the course. The students' feedback indicated their expectation of specific directions, prompt instructor interactions and feedback, so they would not feel anxious. The researchers understood that the expectations of online instructors could be overwhelming due to the time obligations of online instruction, but also realized that the learners' needs must be supported.

Instructor Monitoring of Learners

Due to the importance of the instructor's role in the guidance and continuous monitoring of the learner's progress, the ability to monitor the learner's progress throughout the online course is vitally important. Students need to feel they are "on the right track" and that they can ask questions that receive timely answers from the instructor. Feeling supported helps to lesson anxiety and cognitive load; it also helps the student feel she or he can scaffold new learning because of the instructor reinforcement. This topic links into the previous discussion related to prompt instructor communications and assignment feedback, and the importance of this factor is reflected in student postings (Appendix A).

Student Learning Styles

While there is little research to support the concept of learning styles, this has not dampened the classroom instructors' nor trainers' beliefs that learning styles are important considerations within the classroom environment. The gut feelings of K-12 professional educators and corporate trainers should not be dismissed. Learning styles can influence how a student learns. The styles are identified as: Visual (spatial); Aural (auditory-musical); Verbal (linguistic); Physical (kinesthetic); Logical (mathematical); Social (interpersonal); and, Solitary (intrapersonal) (Advanogy.org, 2007, ¶4; Gardner, 1993). Currently, it is difficult to meet all of these learning styles through available online technologies, but most can be addressed through concerted efforts to provide linked exercises, auditory stimuli, and interactive activities. The researcher asked students to post how they learned based on these learning styles and then asked those students with auditory or kinesthetic styles how she could help them get the most from the course. She communicated through telephone conversations, audio and video representations of knowledge-related discussions, and included planning related to conferencing considerations, so as to enrich their learning experience in the course.

Student Cognitive Load

Cognitive load issues are important considerations when designing an online course. The learner can easily become overwhelmed with information and, as online learning may be less structured than within a face-to-face classroom, could cause confusion to the learner. Therefore, it is important to structure the online course information to emphasize a simplistic presentation of information that progressively develops a cognitive and conceptual framework of understanding on the part of the learner. The learner must develop a knowledge base before moving on to the next bit of knowledge and learning experience. It takes a learner new to the subject matter (novice understanding of the knowledge) a longer period of time to understand and develop an understanding of the subject; however, a learner with prior knowledge and understanding of the subject matter (towards an expert understanding of the knowledge) will take a shorter period of time to understand and conceptually frame the new information within prior knowledge. "By simultaneously considering the structure of information and the cognitive architecture that allows learners to process that information, cognitive load theorists have been able to generate a unique variety of new and sometimes counterintuitive instructional designs and procedures" (Paas, Renkl & Sweller, 2003, p. 1). As the conceptual framework of understanding is developed and realized, the amount of information that leads towards cognitive load issues lowers. "Then, once expertise is gained the newly crowned expert can reinvest the extra cognitive load into other things" (Wilson, 2008, ¶ 3).

The concepts emphasized in this course were grounded in the belief that information should be "chunked" into workable bits of information so that the learner has the time to conceptually understand the information and move from a learner's working memory towards more long-term information understanding. Their posts indicated a desire to delineate tools which could support the learner's cognitive load issues within the online learning environment and that this should be integral towards active instructional design considerations, particularly for novice learners.

Socialization within the Learning Community

It is an artistic task to successfully develop a community of learning within an online course learning environment. Researchers have been studying this task for over a decade, with interesting opportunities for further consideration and further research. As such, it is important to develop an understanding of the potential for the successful development of an online community of learning. The initial consideration should revolve around defining learning communities, with Buffington (2007) offering some insights:

A basic structural model for learning communities... includes three fundamental elements: domain of knowledge, community of people, and the shared practice that they are developing. The domain creates common ground and a sense of common identity... and gives meaning to their actions. The community creates the social fabric of learning, because it encourages a willingness to share ideas, expose one's ignorance, ask difficult questions, and listen carefully. The practice is a set of frameworks, ideas, tools, information, styles, language, stories, and documents that community members share. (Wenger, McDermott & Snyder, 2002, as described by Buffington, 2007, ¶16)

This socialization is addressed through different forms of interactive activities. Interactive activities are the embodiment of different forms of communication that occur within any instructional framework that engages the conceptual development and understanding of information. These type of interactions may be designed into the learning environment or develop as naturally occurring entities, but the integral aspects associated with the different types of interactive activities delineate the importance of interactions thorough out the learning environment so as to work with the knowledge and conceptually frame the learner's understanding. The following is a framework of interactive activities within a learning environment, all of which were employed within the course used in this study:

- learner-content (Moore, 1989);
- learner-interface (Hillman, Willis & Gunawardena, 1994);
- learner-instructor (Moore, 1989);
- learner-learner (Moore, 1989);:
- learner-self (Crawford, 2001);
- learner-community (Burnham & Walden, 1997);
- instructor-community (Crawford, 2001);
- instructor-content (Crawford, 2001);
- instructor-interface (Crawford, 2001); and,
- instructor-self (Crawford, 2001; 2003).

The researcher noted the interactive activities of the learners' interactions with the content, interface, instructor, other learners, self and community; and also noted the instructor interactions with the learners, content, interface, self, and community. Initially the instructor had some difficulty getting students to participate in the social networking, but as the students became comfortable within the course environment and with the course colleagues it became a regular occurrence.

As one of the first posts in the course, the learners were required to post an introductory discussion board communication, to introduce themselves professionally. This provided a non-threatening virtual space where learners were encouraged to become acquainted and reconnect with each other. The teacher shared relevant stories that were directly linked to the professional background of each learner's post, which not only engaged the learner and developed a level of learner confidence within the course structure, but also framed an opportunity for the learners to make connections with their learner colleagues who held similar backgrounds or professional interests. This type of interactive sharing and support modeled the openness and communication that nurtured a level of trust and connection amongst the learners and the course instructor and, after a few days, began to form a learning community. These types of posts were designed into the course instruction, but the instructor further enhanced the expectations for community building through responses and affirmations regarding the learner's efforts.

The interactivities described in the learning environment and supported the findings of other research (Crawford & Gannon Cook, 2004, 2008), suggested that the learners found learner-learner, learner-instructor, learner-content, and learner-community, to be the most relevant interactive activities to them. Miranda and Saunders (2003) discuss that:

... information possesses radically different meanings for different individuals, based on their biographies and positions in the social setting. The very social setting in which information is encountered contributes to its meaning. Schutz introduces the notion of intersubjectivity in describing the understanding that emerges from shared human experiences. (Schutz, 1967/1997, as presented by Miranda & Saunders, 2003, p. 87)

Intervention Measures

Intervention measures can be useful tools, and support the learner's active engagement in monitoring her/his student progress. In addition to instructor interventions, such as feedback suggestions and course aids, there are interventions that require the learner to take more responsibility for her or his learning. Checklists, calendar layouts of homework and/or assignment submission deadlines and other tools help support the learner when an intervention plan is necessary. Also, task management skills, such as self-regulation, and self-reflection can become habitual practices that perpetuate lifelong learning. Interventions can also include interactive activities designed to get students out of their silos and into collaborative group activities.

Self-Regulation

Holmberg suggests that "A basic general assumption is that real learning is primarily an individual activity and is attained only through an internalizing process" (1995, p. 47). The learner's ability to control their learning environment, as well as themselves as learners remains central to the learning process. Much research has focused on self-regulation and activities to enhance the learner's success within the learning environment (Bandura, 1991; Corno & Mandinach, 1983; Garcia, 1995; McManus, 1995; Pintrich, 1989; Pintrich & De Groot, 1990; Schunk, 1989, 1994; Zimmerman, 1989, 1990, 1994).

Self-regulation refers to "the use of processes that activate and sustain thoughts, behaviors, and affects in order to attain goals" (Schunk & Zimmerman, 1997, as stated by Vockell, n.d., ¶ 5). Self-regulation consists of three components:

- Self-observation. Deliberate attention to specific aspects of one's own behaviors.
- Self-judgment. Comparing one's current progress toward a goal with a standard.
- Self-reaction. Making evaluative responses to judgments of one's own performance. (Vockell, n.d., ¶ 5)

Self-regulation empowers the learner to control her or his learning environment and, therefore, to control learning. Learners need to have information clearly articulated so as to ensure that all expectations are met, particularly within an online course environment.

Within the course utilized in this study, the information was clearly articulated and the Course Deliverable Timeline further articulated the learner's assignment expectations with a one-page format that supported and modeled self-regulatory tools. These type of tools helped the learners better understand the basic course layout and expectations, which then allowed them to focus more upon their learning instead of what was expected of them.

Self Reflection

Self reflection helps the learner to integrate what she or he learned, to find meaning in her or his professional and personal life. A definition of self reflection is offered as:

Self-reflection is the process of examining the impact of personal values, beliefs, styles of communication, and experiences. This process develops a deeper understanding of one's culture, personal and cultural biases, experiences, and beliefs as these may influence future action and learning. (MCH Training Trantee Network, 2007, ¶2)

The modeling that occurs within the online learning environment emphasizes higher order thinking skills through interactivities and discussion board postings which develop a sense of self-awareness within the online learning environment.

Interactive Activities

Collaborative group projects are important to enhance the learner's understanding of the course knowledge, as well as socially frame and reframe information as appropriate. Within the instructional realm of online environments, group projects or collaborative projects may be difficult to design, integrate and implement, because special consideration must be given to time zones, extra time for students to plan meetings for projects, and to countering resistance from students who do not like participating in group projects. The desire for online learning to occur in an "anywhere, anytime" style of environment does not operate well with the synchronous collaborative projects. But if the project is interesting enough, or provides an exciting learning opportunity, most students will make the efforts to work together in groups so as to accomplish the project objectives.

An interactivity that enlisted group participation was the creation of a student "water cooler" site where students could gather and interact with each other without interference from the instructor. While the instructor agreed not to post to the water cooler site, she did inform students that periodically she would just check in to be sure that the postings were ethical and not too personal in nature. As researcher, she found that the water cooler site was very helpful to students who seemed to be a bit introverted or reticent to participate in the more structured course online discussions.

CURRENT CHALLENGES/ PROBLEMS FACING THE ORGANIZATION

Results of the Study

The students in the case study were experienced online learners; through their careers and academic pursuit they had experienced different styles of online course instructors and they had worked with actively engaged online instructors, short-response instructors, and the "where are they?" online instructors. Of the nine students, six remarked about the easy navigation and easy accessibility of information; seven remarked about how they enjoyed learning about the use of semiotic tools and metaphors in their course designs; and, two students remarked positively about the use of new technologies in the course.

In the feedback from the students in the course, the emphasis was also on how online courses could be made more engaging to learners. Alternatives to text-based forms of subject matter communication and alternative forms of communication within the online learning environment seemed to be important to the learners in order to engage them at different levels of understanding.

Although the case study students' comments were framed through their own professional background experiences and interests, the recurrent theme that ran through the students' commentaries was the importance of appropriate course design and development; that priority should be focused upon higher order thinking skills, while ensuring that learners could scaffold their subject matter understanding as they progressed through the course. The students had not only been learners, but active participants in the course learning community; they had contributed to their own and the next learning community's educational evolution.

In the retrospective analysis of the research course journals, course postings and completed assignments, the teacher as researcher gleaned that the course reviewed for this study was thorough, offered strategic graphics and semiotic representations, provided ample audio-video insertions and interactivities, to keep the enrolled students interested and retained in the course. While the student postings supported these findings, it was impossible to tell whether a few more or a few less of any of these components would have provided the same result. What was clear, however, was that the students did expect a high level of interactivity. Perhaps it was because they were graduate students, or because they were technologically experienced. But the researcher realized that, without the varied course interactivities and high participation by the instructor, the results of evaluations and feedback would have been less favorable.

Other findings included the observations that students were somewhat complacent about the technologies, but seemed genuinely interested in the concepts of online narratives and ways that they could give both themselves and their students a voice in their learning and a sense of self online. Even at their sophisticated experience level, the students liked having instructor feedback and communication. While a few years earlier it may have seemed either condescending or over the top to take such personal interest in students' success, the massive exposure to online, iPhone, and Twitter make it increasingly more important for instructors to stay connected with students.

The weaknesses discovered in the study also related to the effects of speed on students in today's global electronic societies; students needed to find their voices amidst all of the noise and distractions in their lives. The researcher came to the realization that while it was important to be responsive and facilitative, enabling students would not be productive. Students needed to take responsibility for their learning and their lives. It also seemed that the inclusion of symbolic representations and narratives did have some impact on the students, in that these artifacts put mental landmarks in place to remind them of something they recognized somewhere in the recesses of their inherent knowledge base. At any rate, these were fairly advanced learners, so it is difficult to tell to what extent these may have affected the learners or whether they would have been as effective with more novice learners.

Also, as evidenced by the student postings (Appendix A), the students liked feedback, and they also liked to offer it. As adult learners, they wanted to feel that they brought their experiences with them into class and that they were not just empty vessels being filled by the instructor. The weakness here was the risk that unmotivated learners could use feedback as an excuse for complaints or worse; but fortunately, none of this occurred in this course during the study. Steps to prevent online student "attitude" were preliminary but also embedded in the Netiquette section of the course. The learning course management system (LCMS) was carefully structured and designed to be easily maneuvered by the learners and the course instructor, to ensure that the subject matter was the primary focus of the learner's cognitive focus; however, there was not one comment from students that directly addressed the learner's ease of use through the course interface set up for this course. Learners seemed to assume that navigation and maneuverability were going to be in place for every course, whether that is the case or not.

Last, while there were significant limitations to the study in size and generalizability, as pointed out in the developmental design research (Richey & Klein, 2007), the observations and findings were rich in terms of reinforcing what worked and did not work in this course. In addition, the potential liability of teacher as researcher impeding the objectivity of the results was lessened by the sheer fact that there was no reason to skew or misrepresent the results. As mentioned in the methodology section, development researchers conduct research to observe and honestly report the findings of their work. That is their strongest motivation, along with making a contribution to the body of research on development design that assists others in their chosen field. While that reasoning may be insufficient for generalizability of the findings, it is sufficient for recommendations for further research in the field of development research. The terms "purposeful", "in context" (Richey & Klein, 2007, p.91), and "affective" (Richey & Klein, 2007, p.92), along with the commitment to integrity, were all key terms in this type of research. The researcher was very aware of the need for all of these factors and endeavored to ensure that all were carefully included in the research undertaken in this study.

Summary

Due to the "silo" effect of online learning environments, the learner's desire to maintain a positive image within an online learning environment (such as being in control and "on top of" the course expectations) and quick instructor response rates proved to be extremely important to the learners in this study, so this must be addressed whenever possible. While the researcher thought that students would notice good course design in the online course interface used in this course, no notice was taken because students expected interface transparency. Interface should be simplistic in nature and usability considerations so as to ensure that the course environment is clear, concise and does not impede the learner's ability to easily locate pertinent information. All information needs to be available within a few clicks of a mouse through one specific access, unlike the natural inclination to make information available in a few locations for ease of learner use; this inclination initially appears to be a good idea but, in reality, merely causes confusion for the learners. The more simple, straightforward and clear the online learning environment, the stronger the learning environment supports the learner.

The artistic endeavors and semiotic, representational, features that frame the online learning environment are focused upon the enhancement of the learner's conceptual framework of understanding of the subject matter. These, while working towards ensuring that all learners within the online course are actively engaged in the course, establish a resonance with the learner that assists receptivity to new learning. If attention is given to responsiveness and interactivity, students also develop a feeling of personal importance along with a sense of being a part of the online course community. Herein resides the meaning-making and structure of the online learning community.

This case study reviewed strengths and weaknesses of an online course as a dynamic learning environment, particularly related to the course design, content, and community of learners. It also looked at the student reflections and feedback as a result of the holistic approach utilized in the course design compared with their prior online learning experiences. In addition, the study also looked at the online learners in the course to observe whether interactivity and an online community were established. In the end, while the findings were not generalizable, it was clear that a shift occurred in the learners' focus, from simply participating in an online course to reframing their understanding of the course content and developing a voice within the online community. In this sense, both the students' and instructor's objectives and anticipated outcomes were met in the study.

REFERENCES

Activeworlds Inc. (2008). *Active Worlds: Home of the 3D Internet, virtual worlds and community chat*. Retrieved on December 4, 2008, from http://www.activeworlds.com/

Advanogy. (2007). *Learning styles*. Retrieved on November 4, 2008 from http://www.learning-styles-online.com/overview/

Ahrlberg, M. (2008). *Practical methods and techniques of knowledge representation in particular those related to concept mapping and mind mapping*. History, theoretical background, software, and comparison table. Paper presented October 6, 2008. Retrieved February 20, 2009 from http://itech1.coe.uga.edu/itforum/paper109/Ahlberg_manuscript.pdf

Alberts, D. S., & Papp, D. S. (1997). *The information age: An anthology on its impact and consequences*. Washington, DC: CCRP Publication Series. Retrieved on November 10, 2008 from http://www.dodccrp.org/files/Alberts_Anthology_I.pdf

Allen, I. E., & Seaman, J. (2007). *Online nation: Five years of growth in online learning*. Needham, MA: Sloan Consortium. Retrieved on December 1, 2008, from http://www.sloan-c.org/publications/survey/pdf/online_nation.pdf

Argyris, C., Putnam, R., & Smith, D. (1985). *Action science: Concepts, methods and skills for research and intervention*. San Francisco, CA: Jossey-Bass.

Avison, D., Lau, F., Meyers, M., & Nielsen, P. A. (1999). Action research: To make action research relevant, researchers should try out their theories with practitioners in situations and real organizations. *Communications of the ACM, 42*, 93–97. doi:10.1145/291469.291479

Bandura, A. (1991). Social cognitive theory of self-regulation. *Organizational Behavior and Human Decision Processes, 50*, 248–287. doi:10.1016/0749-5978(91)90022-L

Berger, D., & Kam, R. (1996). *Training and instructional design*. Retrieved October 24, 2007, from http://www.umich.edu/~ed626/define.html

Blackboard Inc. (2008). *Blackboard: Educate. innovate. everywhere*. Retrieved on December 2, 2008, from http://www.blackboard.com/us/index.bbb

Bloom, B. S. (1984). *Taxonomy of educational objectives*. Boston, MA: Allyn and Bacon.

Bloom, B. S., Englhart, M. D., Furst, E. J., Hill, W. H., & Krathwohl, D. R. (Eds.). (1956). *Taxonomy of educational objectives, the classification of educational goals, Handbook I: Cognitive domain*. New York: Longmans.

Brett, J. (2006). 360 Degree feedback to leaders: Does it relate to changes in employee attitudes? *Group and Organization Management 36*(5). 578-600. Los Angeles: SAGE Publications. Retrieved on December 1, 2008, from http://gom.sagepub.com/cgi/content/abstract/31/5/578

Brooks, D. (2008). The cognitive age. *The New York Times*, May 2, 2008. Retrieved on November 15, 2008, from http://www.nytimes.com/2008/05/02/opinion/02brooks.html?_r=1&em&ex=1209960000&en=1628bc39165590dc&ei=5087%0A

Brooks, D. (2008). Demography is king. *The New York Times*, May 3. Retrieved on November 18, 2008 from http://neuroanthropology.net/2008/05/03/david-brooks-part-two-demography-is-king/

Brooks, F. P. (1999). *The mythical man-month: Essays on software engineering*. Reading, MA: Addison-Wesley.

Buffington, J. (2007). *Learning communities as an instructional model.* Retrieved on January 12, 2008, from http://projects.coe.uga.edu/epltt/index.php?title=Learning_Communities_as_an_Instructional_Model

Burnham, B., & Walden, B. (1997). Interactions in distance education: A report from the other side. In *Proceedings of the Adult Education Research Conference*. Oklahoma State University. Available: http://www.edst.educ.ubc.ca/aerc/1997/97burnham.html

Chomsky, N. (2004). *Language & thought*. Wakefield, RI: Moyer Bell Publishers.

Cobley, P., & Jansz, L. (2003). *Introducing semiotics*. Royston, UK: Totem Books.

Cook, A. (1985). *Changing the signs*. Lincoln, NE: University of Nebraska Press.

Corno, L., & Mandinach, E. B. (1983). The role of cognitive engagement in classroom learning and motivation. *Educational Psychologist, 18*, 88–108.

Crawford, C. M. (2001). Developing webs of significance through communications: Appropriate interactive activities for distributed learning environments. [CWIS]. *Campus-Wide Information Systems Journal, 18*(2), 68–72. doi:10.1108/10650740110386675

Crawford, C. M. (2003). Emerging learning environments: Enhancing the online community. *Academic Exchange Quarterly*, 7(4), 131–135.

Crawford, C. M. (2004). Nonlinear instructional design model: Eternal, synergistic design and development. *British Journal of Educational Technology, 35*(4), 413–420. doi:10.1111/j.0007-1013.2004.00400.x

Crawford, C. M. (2005). Towards digital equity within the learning environment: Overcoming the Digital Divide through openly accessible Internet-based tools. *The International Journal of Learning (v 12).*

Crawford, C. M. (2007). Developing multimedia architectural support within online learning environments: Reinventing modalities of meaning as society moves from the Information Age towards the Conceptual Age within the Knowledge Economy. *The International Journal of the Humanities*, *2*(5), 37–46.

Crawford, C. M., & Gannon Cook, R. (2008). Building autonomous and dynamic communities of learning within distance learning environments: Focusing upon making connections, knowledge creation and practice communities. *The International Journal of Technology . Knowledge and Society*, *4*(4), 47–58.

Design-Based Research Collective. (2003). Design-based research: An emerging paradigm for educational inquiry. *Educational Researcher, 32*(1), 5–8. doi:10.3102/0013189X032001005

Dewey, J. (1916). *Democracy and education*. New York: The Free Press.

Doyle, J., Radzicki, M., & Trees, W. (2008). Measuring change in mental models of complex systems. In H. Qudrat-Ullah, J. M. Spector, and P. I. Davidsen, (Eds.), *Complex Decision Making: Theory and Practice*. (269-294). New York: Springer-Verlag Press.

Eco, U. (1996). *From Internet to Gutenberg. A lecture presented at the Italian Academy for Advanced Studies in America.* Retrieved on January 2, 2008 from http://www.hf.ntnu.no/anv/Finnbo/tekster/Eco/Internet.htm

Frand, J. (2000 September/October). The information age mindset: Changes in students and implications for higher education. *Educause*, 15-24. Retrieved on November 28, 2008 from http://net.educause.edu/apps/er/erm00/articles005/erm0051.pdf

Gannon Cook, R., & Crawford, C. (2004). From silos to communities: Addressing electronic isolation through interactivities. *Society for Information Technology and Teacher Education International Conference 2004*(1), 445-452. [Online]. Available: http://dl.aace.org/14315

Gannon-Cook, R., & Crawford, C. (2006). What can cave walls teach us? In A. Edmondson, (Ed.), *Globalized E-Learning Cultural Challenges*. Minneapolis, MN: eWorld Learning.

Gannon-Cook, R., & Crawford, C. M. (2001). Metaphorical representation within a distributed learning environment. In J.D. Price, D.A. Willis, N. Davis & J. Willis (Eds.), *SITE 2001 Annual - Society for Information Technology and Teacher Education* (pp. 1086-1088). Charlottesville, VA: Association for the Advancement of Computing in Education (AACE).

Garcia, T. (1995). The role of motivational strategies in self-regulated learning. In P.R. Pintrich (Ed.), *New directions for college teaching and learning: Self-regulated learning in the college classroom* (pp. 29-42). San Francisco: Jossey-Bass.

Gardner, H. (1993). *Frames of mind: The theory of multiple intelligences*. New York: Basic Books

Gask, L. (2005). *Overt and covert barriers to the integration of primary and specialist mental health care*. University of Manchester, Manchester M13 9DL, UK: National Primary Care Research and Development Centre. Georgetown College Center for Advanced Study of Assessment (CASA). (n.d.). *Formative Assessments*. Retrieved on March 20, 2009, from http://www.georgetowncollege.edu/CASA/documents/FomativeAssessment.pdf

Gay, L. R. (1987). *Educational research: Competencies for analysis and application (3rd ed.)*. New York: Merrill Publishing Company.

Hamilton, E. (1969). *Mythology: Timeless tales of gods and heroes*. New York: Penguin Books.

Hillman, D., Willis, D. J., & Gunawardena, C. (1994). Learner-interface interaction in distance education: An extension of contemporary models and strategies for practitioners. *American Journal of Distance Education*, *8*(2), 30–42. doi:10.1080/08923649409526853

Holmberg, B. (1995). *Theory and practice of distance education*. London: Routledge.

Jung, C. G. (1964). *Man and his symbols*. New York: Dell Publishing.

Lewin, K. (1951). *Field theory in social science: Selected theoretical papers*. D. Cartright (Ed). New York: Harper and Row.

Linden Research, Inc. (2008). *Second life*. Retrieved on December 2, 2008, from http://secondlife.com/

MCH Training Trantee Network. (2007). *Leadership training*. Retrieved on January 5, 2007, from http://leadership.mchtraining.net/index.php?module=pagemaster&PAGE_user_op=view_page&PAGE_id=25

McKenney, S., & van den Akker, J. (2005). Computer-based support for curriculum designers: A case of developmental research. *Educational Technology Research and Development*, *53*(2), 41–66. doi:10.1007/BF02504865

McLuhan, M. (1964a). *The medium is the messenger*. New York: Signet Books.

McLuhan, M. (1964b). *Understanding media: The extension of man*. New York: Signet Books.

McLuhan, M., & Fiore, Q. (1967). *The medium is the massage*. New York: Touchstone.

McManus, T. F. (1995). *Testing learner self-regulation in a web based learning environment.* Retrieved from the World Wide Web on January 1, 2003: http://www.edb.utexas.edu/mmresearch/Students96/McManus/srltest.html

Miranda, S. M., & Saunders, C. S. (2003, March). The social construction of meaning: An alternative perspective on information sharing. *Information Systems Research, 14*(1), 87-106. Retrieved on April 14, 2008, from http://faculty-staff.ou.edu/M/Shaila.M.Miranda-1/M&S03.pdf

Moodle.org. (2008). *Moodle – A free open source course management system for online learning.* Retrieved on December 4, 2008, from http://moodle.org/

Moore, M. (1989). Three types of interaction. *American Journal of Distance Education, 3*(2), 1–7. doi:10.1080/08923648909526659

Moore, M., & Kearsley, G. (2005). *Distance education: A systems view,* (2nd ed.). New York: Thomson/Wadsworth.

MSNBC.com. (1 December 2008). *It's official: U.S. is in recession: Economy began shrinking in December 2007, panel declares.* Retrieved December 1, 2008, from http://www.msnbc.msn.com/id/27999557

National Aeronautics and Space Administration. (2008). *NASA – Johnson Space Center website.* Retrieved on December 1, 2008, from http://www.nasa.gov/centers/johnson/home/index.html

Nelson, B., Ketelhut, D. J., Clarke, J., Bowman, C., & Dede, C. (2005). Design-based research strategies for developing a scientific inquiry curriculum in a multi-user virtual environment. *Educational Technology, 45*(1), 21–28.

Nelson, G. (2004). Building capacity in community health action research: Towards a praxis framework for graduate education. *Action Research, 2,* 289–408. doi:10.1177/1476750304047982

Paas, F., Renkl, A., & Sweller, J. (2003). Cognitive load theory and instructional design: Recent developments. *Educational Psychologist, 38*(1), 1-4. Retrieved on October 7, 2008, from http://books.google.com/books?hl=en&id=DCcSZ7GC8lAC&dq=cognitive+load&printsec=frontcover&source=web&ots=3rCu__Aos4&sig=yzgoZmi4aTKHJD_bnd-EFUOlmrQ&sa=X&oi=book_result&resnum=3&ct=result#PPA1,M1

Pink, D. H. (2005). *A whole new mind: Moving from the information age to the conceptual age.* New York: Riverhead Books.

Pintrich, P. R. (1989). The dynamic interplay of student motivation and cognition in the college classroom. In C. Ames & M. Maehr (Eds.), *Advances in motivation and achievement: Motivation-enhancing environments* (Vol. 6, pp. 117-160).Greenwich, CT: JAI Press.

Pintrich, P. R., & De Groot, E. (1990). Motivational and self-regulated components of classroom academic performance. *Journal of Educational Psychology, 82,* 33–40. doi:10.1037/0022-0663.82.1.33

Reeves, T. C. (2005). Design-based research in educational technology: Progress made, challenges remain. *Educational Technology, 45*(1), 48–52.

Reis, R. M. (2001). To compete or collaborate? Career questions in science. *The Chronicle of Higher Education.* Retrieved on December 2, 2008, from http://chronicle.com/jobs/news/2001/05/2001052502c.htm

Richey, R. (2005). Validating instruction design and development models. In J. M. Spector & D. A. Wiley (Eds.), *Innovations in instructional technology: Essays in honor of M. David Merrill* (pp.171-185). Mahwah, NJ: Lawrence Erlbaum Associates, Publishers.

Richey, R., & Klein, J. (2007). *Design and development research.* Mahwah, NJ: Lawrence Erlbaum Associates, Publishers.

Rothstein, E. (1995). *Emblems of mind.* New York: Avon Books.

Ryan, M., Hodson-Carlton, K., & Ali, N. S. (2005). A model for faculty teaching online: Confirmation of a dimensional matrix. *The Journal of Nursing Education, 44*(8), 357–365.

Schlain, L. (1998). *The alphabet versus the goddess: The conflict between word and image.* New York: Penguin Books.

Schunk, D. H. (1994). Self-regulation of self-efficacy and attributions in academic settings. In D. H. Schunk & B. J. Zimmerman (Eds.), *Self-regulation of learning and performance: Issues and educational implications* (pp. 75-99). Hillsdale, NJ: Erlbaum.

Schunk, D. H., & Zimmerman, B. J. (1997). *Self-regulation.* Retrieved on January 10, 2009 from http://education.calumet.purdue.edu/vockell/EdPsyBook/Edpsy7/edpsy7_self.htm

Schutz, A. (1967/1997). *Phenomenology of the social world.* Evanston, IL: Northwestern University Press.

Seels, B., & Glasgow, Z. (1997). *Making instructional design decisions.* Mahwah, NJ: Lawrence Erlbaum Associates, Publishers.

Siemens, G. (12 December 2004). *Connectivism: A learning theory for the digital age.* Retrieved on October 7, 2008, from http://www.elearnspace.org/Articles/connectivism.htm

Singleton, R. A., & Straits, B. C. (2005). *Approaches to social research* (4th ed.). New York: Oxford University Press.

Sloodle. (2008). *Sloodle: Simulation Linked Object Oriented Dynamic Learning Environment.* Retrieved on December 2, 2008, from http://www.sloodle.org/moodle/

Texas Higher Education Coordinating Board. (2008). *Overview: Tuition deregulation.* Retrieved on December 1, 2008, from: http://www.thecb.state.tx.us/reports/PDF/1527.PDF

University of Houston – Clear Lake. (2008a). *About UHCL.* Retrieved on December 1, 2008, from http://prtl.uhcl.edu/portal/page/portal/HOMEPAGE/TAB_ABOUTUHCL

University of Houston – Clear Lake. (2008b). *Institutional Effectiveness: UHCL Strategic Plan FY05-FY11.* Retrieved on December 1, 2008, from http://prtl.uhcl.edu/portal/page/portal/OIE/PLANNING

University of Houston – Clear Lake. (2008c). *UHCL History.* Retrieved on December 1, 2008, from http://prtl.uhcl.edu/portal/page/portal/PRE/RTTC/UHCL_History

University of Victoria Counseling Services. (2003). *Bloom's taxonomy.* Retrieved on November 28, 2008, from http://www.coun.uvic.ca/learn/program/hndouts/bloom.html

U.S. Department of Education, National Center for Education Statistics. (2008). *The Condition of Education 2008* (NCES 2008-031). Retrieved on March 20, 2009, from http://nces.ed.gov/FastFacts/display.asp?id=16

Viklund, A. (2008). *Discussion of David Brooks' article on the cognitive age.* Retrieved on November 28, 2008 from http://neuroanthropology.net/2008/05/03/david-brooks-part-one-the-cognitive-age/

Vockell, E. (n.d.). *Educational Psychology: A Practical Approach.* Retrieved on March 20, 2009, from http://education.calumet.purdue.edu/vockell/EdPsyBook/Edpsy7/edpsy7_self.htm

Vockell, E. L., & Asher, J. W. (1995). *Educational research* (2nd ed.). Englewood Cliffs, NJ: Prentice-Hall, Inc.

Vygotsky, L. S. (1935). *Mental development of children during education.* Moscow: Uchpedzig.

Vygotsky, L. S. (1962). *Thought and language*. Cambridge, MA: MIT Press.

Vygotsky, L. S. (1978). *Mind in society*. Cambridge, MA: Harvard University Press.

Vygotsky, L. S. (1981). The genesis of higher mental functions. In J. V. Wertsch (Ed.), *The concept of activity in Soviet Psychology*. Armonk, NY: Sharpe.

Wenger, E., McDermott, R., & Snyder, W. (2002). *A guide to managing knowledge: Cultivating communities of practice*. Boston: Harvard Business School Press.

Wertsch, J. V. (1985). *Cultural, communication, and cognition: Vygotskian perspectives*. Cambridge, UK: Cambridge University Press.

Wikipedia. (2009). *Formative assessment*. Retrieved on February 20, 2009, from http://en.wikipedia.org/wiki/Formative_assessment

Wilson, B. (2008, 21 July). *Expert, novices, and cognitive-load.* Retrieved on February 1, 2009 from http://www.crucialthought.com/2008/07/21/experts-novices-and-cognitive-load/

Zimmerman, B. J. (1989). A social cognitive view of self-regulated academic learning. *Journal of Educational Psychology*, *80*, 329–339. doi:10.1037/0022-0663.81.3.329

Zimmerman, B. J. (1990). Self-regulated learning and academic achievement: An overview. *Educational Psychologist*, *25*, 3–17. doi:10.1207/s15326985ep2501_2

Zimmerman, B. J. (1994). Dimensions of academic self-regulation: A conceptual framework for education. In D. H. Schunk & B. J. Zimmerman (Eds.), *Self-regulation of learning and performance: Issues and educational implications* (pp. 3-21). Hillsdale, NJ: Erlbaum.

APPENDIX A

Student Postings in the Course (Arranged by Topic)

Cognitive Load

- Cognitive Load is an additional consideration to online learning that flows in line with self-regulation. As the student must feel that they are in control of their learning, so too must they feel that the information is not overwhelming. Designing a simple course that progressively builds a cognitive and conceptual framework whereby the student builds an understanding of the subject. By scaffolding the learning process, students build upon their knowledge piece by piece, moving from new information to new information that builds on the previous lesson. (Student C, 27 October 2008 at 9:52pm, paragraph 3)
- Cognitive load issues – Graphic organizers can assist students with cognize load issues by help them to organize their ideas, and thoughts, as they complete an assigned tasks. On a daily basis I can see students that have great ideas and before they get them completely developed, get lost and forget what their idea was. Graphic organizers can be used to minimize this common phenomenon for a lot of students. If students are taught to draw out their ideas using graphic organizers before they begin to develop them, students could effectively use their time to develop that idea. (Student D, 6 November 2008)

Communities of Learning

- Professionalism within the eLearning Environment is key to ensuring that instructors meet a set expectation of aptitude, ethics, procedure, and evaluation necessary for imparting knowledge to students and (meeting) set standards for the competencies, skills, or performance by which members are expected to adhere in the classroom and within the professional community.
- One very astute professor I once had provided the class with an opportunity to research the standards by which we as professionals would operate. Part of the lesson was to write a letter to our family whereby we discussed our roles as Instructional Designers and provided the professional standards by which we agreed to operate. This set the tone for the class as well as carries me today through my schoolwork and professional interactions. (Student C, 1 November 2008)

Higher Order Thinking Skills

- Authentic assessment: Authentic means real. If we really get real, we will find that we learned more about ourselves and the educational content when we did projects rather than tests. Authentic assessment is about making portfolios, unit projects, class discussions; not rote memorization activities. I can also say that as a parent, I love seeing and talking with my children about a fun activity they did at school. My daughter, who is in Kindergarten, takes reading assessment tests, but the testing is put into a portfolio and can be viewed by parents so that we see the growth in the amount of letter recognition and sight word reading abilities. This to me is authentic assessment. (Student E, 4 December 2008)

Impact of Audio Support within Online Learning Environments:

- The availability of the online course instructor is a factor in the online learner's success. I personally would not recommend instructors giving out home addresses, home or cell phone numbers or personal email addresses. Also the response time to the various medium is important. It is not always possible for the instructor to be at their office during the whole office hours period, but the availability should be a high percentage of the time. The instructor would also want to check phone messages and emails on a timely basis. I could put into my course syllabus how often I would check my phone messages and email messages. (Student A, 9 November 2008)
- Cognitive load issues – Having student post their responses to discussion questions in podcasts allows them to say what they mean. A lot of time when I am having conversations with students I can understand everything they are saying but when I ask them to write down those viewpoints it does not always come out at clear as when they were verbally expressed. A way that I can integrate this into my course is when I assign students a research topic have them verbally talk about their finding and post their responses as well as submitting the written version so that they can be compared. (Student D, 10 November 2008)
- Self regulation issues – Allowing students to post podcasts can make it a little easier for the teacher to capture the students' emotions on the topics being discussed. Since recording a podcast is easier than typing a response, for most students, this will also help students to stay on task and focus more on major projects while posting simple 1-2 minute responses to general discussion questions via podcasts. (Student D, 10 November 2008)

Impact of Graphic Support within Online Learning Environments

- Graphically representing knowledge is easier than ever with the use of modern software technology. Software has made use of pictures, colors, making graphs, and even the old standard of bulleted outlines much easier and faster to use; even for the novice user. The use of graphic organizers has exploded over the last 20 or so years. Software like Inspiration, and even the latest Microsoft Smart Art is useful and not too hard to learn. Graphically representing data is easier for the student to understand, comprehend, and make future reference to. Making a lesson, especially for use on the internet, if all text based is tedious (to say the least). The use of graphic organizers makes information easier to follow and understand. (Student E, 6 November 2008)

Instructor Interaction:

- Instructor Feedback: One of the most important things I have noticed in my experience with online courses is the impact of instructor feedback. It can provide students with the confidence to approach their instructor with questions and concerns about the course. It can also help students understand what they are required to do and what they can do to improve their work as the course progresses. Feedback does not need to be immediate (meaning that the second an instructor gets an e-mail it is answered), and I do not feel the instructor should be required to sit by a computer for 24 hours a day, 7 days a week. However, the instructor should provide students with what they can expect, as well as alternate ways to contact the instructor if a deadline is approaching and they have not received a response. Through my online course, I hope to model the type of behavior I have grown to expect from my online course instructors. (Student I, 2 November 2008)

Interactive Activity Issues:

- Interactive Activities in my opinion is a major contributor to the difference between an online course and a successful online course. Interactive activities in eLearning are necessary for supporting self-regulation, lessening student anxiety about content and success, as well as providing the ever-necessary encouragement that all students need to learn. Furthermore, interactive activities in the online classroom are essential to creating communities of learning that alleviate the student isolation that is so common in this type of learning environment. (Student C, 1 November 2008)
- Interactivity between learners and instructor: It is very important for there to be a certain level of interactivity between the learners, and between learners and the instructor. If the learner is not able to communicate with other members, through chat, discussions, or feedback, it is easy for the learner to feel disconnected and fall behind. I have been in courses before where I am blocked from communicating with the other students via chat or e-mail, and that always adds to my anxiety in the course. A student should be allowed to communicate with others to clarify or better understand the course's expectations in the eyes of their fellow students. Because of those experiences, I imagine any courses that I designed would make sure that communication with others is an easy and encouraged process. (Student I, 2 November 2008)

Learning Styles:

- Graphic organizers, as well as metaphoric representations and semiotics, when integrated into an online course, or any course, can assist educators in reaching that small percentage of students who usually don't do very well because they are not good at reading and regurgitate material. When these tools can be used to communicate important information to students in a course I thing that there will be a clarification of understanding for those students who be right there at the edge of understand but need a little something extra to bring them across. These tools can also help those students who do understand the concepts be able to see how the new knowledge can be used in other aspects of their lives. (Student D, 6 November 2008)

Learning Theories:

- In traditional learning the teacher is the center of instruction with a lot of structure. The teacher asks specific questions with expectation of specific answers. The students work individually in a quiet setting on structured tasks. In Constructivist environments, the students are the center of instruction with the teacher as a facilitator and coach. The students work on more open ended questions that do not have specific answers. They work collaboratively on projects and grade each other' work. I think in an online environment you would want a combination of the two styles. The teacher needs to have structure with specific expectations, but also allow the students to work individually to research about different topics and then discuss these topics collaboratively with the other students. (Student A, 26 October 2008)

Online Netiquette:

- Online etiquette (or netiquette) provides the opportunity to carry over good communication skills to the online environment. It is easy to forget basic rules for conduct when you are able to send immediate messages to others. It is important to provide review of the expected behavior for students with regards to others in the course. In order to make sure that students in a course understand the netiquette and expectations, a review session would be important. Younger students may require this more than adult learners, but even some adults who have only had informal experiences with e-mail and chatting may need these "gentle" reminders. (Student I, 2 November 2008)

Self-Reflection:

- It is very important for the learner to commit to learning. There needs to be a plan to schedule time and to designate a place to study, research, and complete requirements and deliverables. In my online course environment, a syllabus will be provided to make sure that the learner knows the expectation of the course, including a calendar, a list of assignments and deadlines. (Student G, 28 October 2008)
- Self-Regulation. I can integrate self regulation into my online course's learning environment by making sure to include plenty of checklists and evaluation rubrics that students can use as they complete their assigned tasks. (Student D, 28 October 2008)

Reflective Practice:

- I will integrate self-reflection into my online course by providing students with the evaluation rubrics for each assignment so that they can evaluate their assignments themselves before I evaluate it using the same rubric. I will also incorporate reflection essays to be written by students after major assignments are completed. (Student D, 28 October 2008)

Semiotic Tools:

- Semiotics is another useful tool for providing clarity or direction in a course by providing cultural symbols throughout the online environment to assist students in quickly understanding what may be required of them. One caution in using semiotics in any learning situation is to make sure that the signs or "language" you use to assist students truly relates to their cultural background. A student from China may not feel that a depiction of a tree is a sign for recyclable material. However a student from China and a student from Texas might recognize a keyboard icon as an indicator to type something. A method would be to provide icons for the appropriate programs such as email, or to use stoplights or road signs to help students navigate the course (4 November 2008)
- Metaphoric Representation is a great way to add clarity to any course, but especially an online course. Since the instructor is not always there to draw on a whiteboard or provide a summation of intertwined topics, metaphoric representation provides a great vehicle for accomplishing those types of tasks. One application of metaphoric representation could be providing a course map at the beginning of the year, and then showing which area or areas are being addressed by each lesson by displaying a cross-section of the entire map. (Student C, 4 November 2008)
- I hope to use metaphors to help find connections between difficult science concepts for children who have limited background knowledge. (Student I, 6 November 2008)

Chapter 20
Teaching Statistics and Operations Research Online:
Experiences at the Open University of Catalonia

A. Juan
Open University of Catalonia, Spain

J. Faulin
Public University of Navarre, Spain

P. Fonseca
Technical University of Catalonia, Spain

C. Steegmann
Open University of Catalonia, Spain

L. Pla
University of Lleida, Spain

S. Rodríguez
University of Lleida, Spain

S. Trenholm
Herkimer County Community College, USA

EXECUTIVE SUMMARY

This chapter presents a case study of online teaching in Statistics and Operations Research (OR) at the Open University of Catalonia (UOC). UOC is a purely online university with headquarters in Barcelona, Spain, with students from many countries. As common to most math-related knowledge areas, teaching and learning Statistics and OR present difficult challenges in traditional higher education. These issues are exacerbated in online environments where face-to-face interactions between students and instructors as well as among students themselves are limited or non-existent. Despite these difficulties, as evidenced in the global growth of online course offerings, Web-based instruction offers comparative benefits

DOI: 10.4018/978-1-60566-942-7.ch020

to traditional face-to-face instruction. While there exists a plethora of literature covering experiences and best practices in traditional face-to-face instruction in mathematics, there is a lack of research describing long-term successful experiences in Statistics and OR online courses. Based on the authors' experiences during the last decade, this chapter aims to share some insights on how to design and develop successful online courses in these knowledge areas.

INTRODUCTION

Information Technologies are changing the way in which higher education is delivered in developed countries. In the last decade, the use of instructional technologies has experienced steady growth in universities around the world (Ex: learning management systems for individual and collaborative learning, Internet-based academic resources, online repositories and databases, specific software for some knowledge areas, groupware and social networking software…). With the spread of these technologies, new pure-online universities have emerged and traditional face-to-face universities worldwide are witnessing transformations that affect the nature of the courses and degree programs they offer. These technological innovations have also driven the growth of distance learning and related teaching opportunities. On the one hand, students who are time constrained due to job or travel difficulties, or place constrained due to geographic location or physical disabilities are now able to access courses and degree programs at their convenience (Simonson et al., 2003). On the other hand, students and professors from one university can participate as learners or teachers, respectively, in online courses offered at other universities. This dynamic thus promotes virtual mobility and knowledge sharing among distant universities.

With the rapid growth of distance and global education, e-learning models are currently practiced widely all over the world (Nagy, 2005; Allen & Seaman, 2008). Current instructional technologies facilitate the shifting from a traditional educational paradigm centered on the "masterful instructor" to an emergent educational paradigm that considers students as active and central actors in their learning process. In this new paradigm, student learning outcomes are achieved with the help of instructors, technology and other students. The instructor's primary role shifts from one of knowledge transmission to learning facilitator and specialist responsible for course design, guidance and supervision. In Europe, for instance, this paradigm shift is officially promoted by the Bologna declaration and the subsequent development of a European Area of Higher Education which aims to increase the international competitiveness and employability of European citizens (Van der Wende, 2000).

Regarding the areas of Statistics and Operations Research (OR), educational reforms are widespread both in pure-online and face-to-face education. For example, many instructors are being encouraged to try new teaching strategies based on online support, inter-disciplinary collaborative learning, and integration of statistical and OR software in their courses (Hardin & Ellington, 2005; Leon et al., 2006; Faulin et al., 2009). University departments worldwide have also begun working on new, engaging curricula that promotes conceptual understanding versus simple procedural knowledge. The goal is to increase student's abilities to solve important real-life problems in different market sectors including solutions that yield improved efficiencies (Camm, 2007). Of course, this task is not easy and numerous challenges must be confronted. Some of these challenges are due to the intrinsic nature of the so called "Internet-generation" student while others are due to the intrinsic nature of Statistical and OR content (Leon et al., 2008).

BENEFITS AND CHALLENGES OF ONLINE EDUCATION

Most universities worldwide are currently integrating e-Learning Management Systems (LMS) –like Moodle (http://moodle.org/), Sakai (http://sakaiproject.org/portal) or Blackboard/WebCT (www.blackboard.com/), among others– in their higher education programs. These Web-based tools can be used to develop both alternative and complementary strategies to traditional face-to-face learning systems. These approaches permit delivery of instruction to students who are time- or place-constrained (Seufert, Lechner & Stanoevska, 2002). As Howell et al. (2003) also point out, in some developed countries the current higher-education infrastructure cannot easily accommodate the growing educational demand due to significant enrolment increases. Online education can be a useful and efficient means of mitigating this problem.

Today's e-learning platforms provide fresh possibilities for instruction. Among them, students may use current technology to conveniently access all or part of their course material, take tests, complete homework assignments, participate in various individual and/or collaborative learning activities, post questions for instructors or for collaborative group student problem solving...

Other significant advantages for students are the following:

- *More flexibility when selecting learning timetables and schedules:* In traditional face-to-face learning processes, students have to attend a class on campus at a scheduled time (Zirkle, 2003). In many cases, this system is only valid for full-time students. On the contrary, online learning processes delivered asynchronously, tend to offer more scheduling flexibility, which is an important factor for adult students who have work or family duties. Moreover, students enrolled in online programs usually have the possibility of self-pacing some of the course content and activities. Empirical studies of online courses confirm their ability to reach students with special needs. Robinson's 2005 study found that 43% of students across 18 disciplines at 13 universities took online courses because they were convenient for work schedules, while 22% chose them due to family duties.
- *Less geographical or time constraints to communication with other students or instructors:* More interaction among students and between students and instructors is facilitated which, in turn, encourages the development of collaborative and working-group activities (Daradoumis et al., 2006).
- *Promotion of continuous evaluation processes:* This, in turn, allows students to receive timely feedback about their academic progress during the course. As some authors suggest, interactive self-assessment might improve students' academic results as well as their perception of learning (Peat & Franklin, 2002; Lowry, 2005). While online platforms are very convenient for disseminating or publishing online, on the other hand, students are also able to work autonomously using any or all educational resources available on the platform.
- *Promotion of a multimedia representation of information:* By combining text, images, voice, and video, practical knowledge transmission is facilitated. Faulin et al. (2009) point out that these technologies can help reduce the gap between theory and practice.

At the same time, there are some important challenges typically associated with e-learning. Most of these challenges are especially relevant in the case of math-related online courses:

- *Significant differences in backgrounds and technical skills:* Generally speaking, students taking online courses are typically older than most undergraduate students. Consequently, according to Simonson, et al. (2003), it is somewhat typical to have to deal with students with a limited technological and mathematical background. Challenges also remain with regards to students with physical limitations that require alternative ways to access content (Schwartzman, 2007).

- *High dropout rates and isolation risk:* As Sweet (1986) and Truluck (2007) point out, distance-education programs tend to produce higher dropout rates than face-to-face education programs. The lack of a personal contact between the agents involved in the learning process increases the risk of a sense of isolation among students. Students may feel disconnected from the instructors as well as from other students. For that reason, modes of interactive communication need to be facilitated and continuously encouraged by instructors. Truluck (2007) proposes a few interesting measures for addressing dropout rates in online courses; among others he suggests the use of informal online meetings or "coffee shops" for conversation. Similarly instructors can also feel isolated, affecting their satisfaction, motivation, and potential long-term involvement in online learning (Childers & Berner, 2000).
- *Continuous feedback and accreditation requirements:* As previously alluded, online learning platforms tend to be associated with the use of continuous evaluation processes, individualized self-assessment instruments and the use of multimedia interactive and collaborative activities. Consequently, there is a need for instructors to provide just-in-time guidance and assistance to students as well as periodical and current feedback related to assessment of students' learning activities. This is not a trivial task. Finally, related to this problem there is the necessity for developing protocols that certify the authorship of students' academic activities (Trenholm, 2007; Juan et al., 2008).

THE OPEN UNIVERSITY OF CATALONIA

The Open University of Catalonia or UOC (http://www.uoc.edu/portal/english) is a fully online university with headquarters in Barcelona, Spain. It was founded in 1995 by the Catalan Government with the mission of "providing people with lifelong learning and education through intensive use of information and communication technologies". According to official data, the UOC offers educational services over the Internet to more than 50,000 students, distributed in several undergraduate and graduate programs (Figure 1).

UOC students belong to different parts of the world, but they are mainly located in Spain and South America. About 60% of UOC undergraduate students are adult students (over 30 years old) that typically combine their professional activity and/or family responsibilities with their academic duties. Educational services are delivered by a team composed of more than 2,200 instructors –including UOC faculty and UOC online collaborators, most of these professors from other Spanish universities– and 550 management staff. The UOC uses an asynchronous and student-centered educational model and has already received several international prizes, such as the 2001 ICDE Prize for the best virtual and distance university in the world or the 2004 OEA Prize for educational quality. Currently, up to 22 accredited degrees and official masters are offered via the UOC Virtual Campus, a learning management system entirely developed and maintained at the UOC (Figure 2). Some of the most popular degrees (in number of registered students) offered at the UOC are as follows: Computer Engineer-

Figure 1. Distribution of UOC students by type of studies

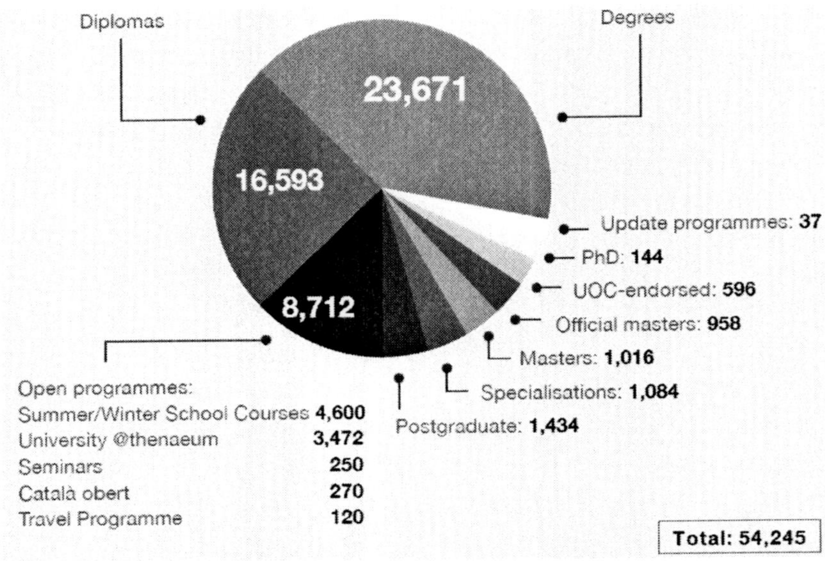

ing, Business Administration and Management, Psychology, Telecommunications, Information and Communication Sciences, Law, and Humanities. Due to the instrumental nature of the Statistics and Operations Research knowledge areas, related courses are offered in the first five out of the seven aforementioned degrees. For this reason, the Open University of Catalonia is officially supporting and funding several innovative projects intended to help develop high-quality

Figure 2. A screenshot of the UOC Virtual Campus with some of its functionalities

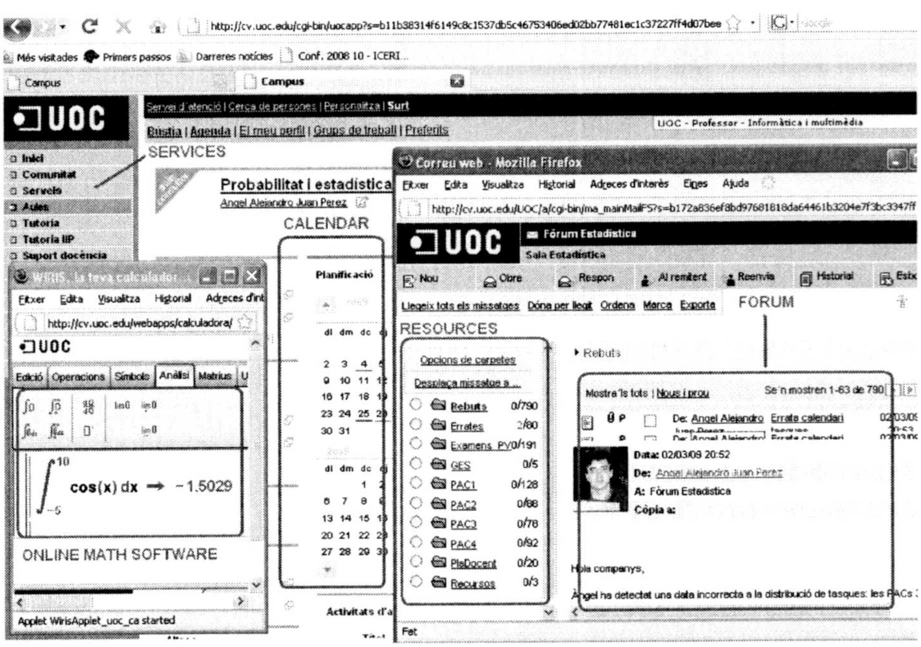

mathematics and statistics content to be shared among students and instructors in different degree programs utilizing online repositories and content management systems. The next sections of this chapter will describe the personal experiences of Statistics and OR instructors in the engineering degree programs.

GENERAL CHARACTERISTICS OF ALL UOC COURSES

At the beginning of the semester, students are enrolled in several classes inside the UOC Virtual Campus. There are no more than fifty students per class. Each of these classes is tutored by one instructor, who also gives guidance and support to students throughout the learning process. Each virtual class provides a private area inside the UOC Virtual Campus, This area contains course-related forums and academic resources. One vital resource is the course learning materials which can be accessed online or downloaded for printing. These materials have been carefully designed and written for students by the instructors and, therefore, include clear definitions, abundant examples and worked problems. A complete syllabus of the course, including academic objectives, methodology and evaluation system –with deadlines for deliverable homework and final exams– is also available for students at the beginning of the term.

All courses at the Open University of Catalonia make use of a continuous evaluation process. That is, throughout the semester, students are working on homework activities which they are responsible for sending to their respective instructors for evaluation and feedback. The number of activities varies depending on the specific course - usually between four and six,. Likewise, some of these activities are to be solved individually by each student, while others may be solved collaboratively in small groups. Students may also be required to participate in some form of discussion thread. Even when not strictly mandatory, students are encouraged to take part in the continuous evaluation process and to actively participate in course-related discussion forums. Years of experience teaching statistics indicate that these practices are among the most effective means of helping students achieve course learning goals and effectively prepare for, what is in most cases, a face-to-face final exam.

STATISTICS COURSES OFFERED TO UOC ENGINEERS

Presently, there are two online undergraduate courses on Statistics that UOC students must take to complete their engineering degree. Specifically, apart from other related courses such as Algebra, Mathematical Analysis, Discrete Mathematics, Data Mining, etc., engineering students at UOC must complete a course in Applied Statistics as well as a course in Probability and Stochastic Processes. Each semester the two courses together comprise up to approximately 450 students. They require the collaboration of 11 instructors, including one course coordinator, as well as several tutors for each virtual class.

Applied Statistics is a first course in Statistics which covers basic statistical concepts –descriptive statistics, correlation and simple regression, popular statistical distributions, sampling distribution results, confidence intervals and hypothesis testing– as well as more advanced topics such as multiple regression, ANOVA and non-parametric tests. The course follows a professionally-oriented approach, i.e., the focus is centered on professional applications of statistical concepts and techniques instead of on the mathematical theory that supports them. Thus, throughout the course students are confronted with several realistic situations where an engineer might need to apply statistics to solve a problem related to either Computer or Electrical Engineering. As it happens in most real-life situations, manual calculations are substituted by computer-aided calculations, i.e.: statistical software such as Minitab (www.minitab.com) or R (http://www.r-project.org) is used intensively during the course. Each student is free to choose between these two programs to complete the assigned homework. Apart from the regular virtual classes, students are also assigned to a virtual laboratory class where an instructor provides support and guidance on the use of these statistical packages. As in every other UOC courses, students must complete a set of homework activities during the course and send the corresponding work before a set deadline. For this course, they are also required to complete a two part final exam. The first part is a practical case-study to be solved with the help of a statistical package, while the second part is a two hour face-to-face summative evaluation. The final score is obtained by averaging the score obtained at the continuous evaluation and that from the final exam. Results from previous years consistently indicate that students who actively participate during the course and successfully complete the continuous evaluation process are likely to pass the course without any great difficulties. On the contrary, those who do not participate in the continuous evaluation process, rarely, if ever, pass the final summative exam.

Probability and Stochastic Processes is another introductory course that provides students with basic concepts about probability, random variables, probability distributions, stochastic processes, Markov chains and queueing theory. Engineering students, particularly telecommunications degree majors, require all of this content for upcoming courses. As previously, the course follows a professionally-oriented approach: realistic case studies are discussed and solved with the help of theories in probability and stochastic processes. Use of mathematical software, such as Wiris (http://www.wiris.com) or Matlab (http://www.mathworks.com) is also considered and promoted. Several homework assignments are posted through the course and students must also take a final face-to-face exam to prove that they have reached all required knowledge, skills and competencies.

With regards to graduate online courses in Statistics, a new Masters in Computer Engineering also includes a course on Advanced Data Analysis, mainly covering multivariate statistics. UOC is also currently offering an online Masters in Bioinformatics and Biostatistics. The methodology in all of these programs are very similar to those earlier explained for related undergraduate courses. The main difference is an increase in interdisciplinary collaborative activities and the realization of a final Master's Thesiswhich reinforces the professional orientation of the coursework.

COURSES ON OPERATIONS RESEARCH FOR UOC ENGINEERS

The UOC also offers two online undergraduate courses in Operations Research. The first one is an introduction to basic OR concepts such as linear programming, duality, sensibility analysis, dynamic programming, integer programming, flow optimization in networks, metaheuristics, queuing theory and simulation. This course follows a methodology similar to those previously described. With OR being an applied discipline by nature, the course is also professionally-oriented, case-study driven, and intensive in the use of OR software such as Lindo/Lingo (www.lindo.com) and Excel/VBA. The second course is a project-based class focused on discrete-event simulation where the student has to develop his/her own simulation project. Generally, the student can choose among different working lines and software, e.g.: simulation of computer networks with Opnet (www.opnet.com), development of simulation-based algorithms with general-purpose programming languages like Java or C/C++, or simulation of industrial and service processes with simulation packages such as GPSS (http://www.minutemansoftware.com) (Figure 3). Both

Figure 3. Simulating a service process in one of our OR courses

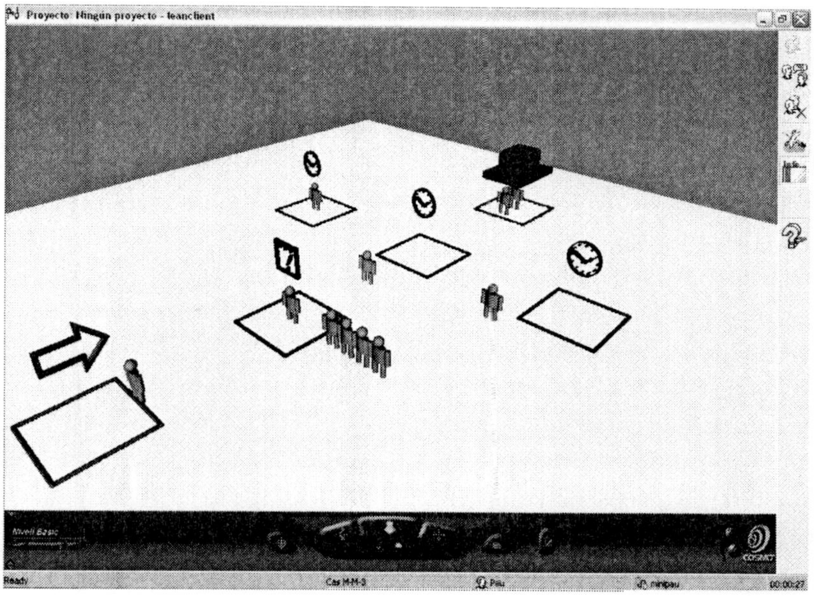

courses are optional inside the engineering degree curricula and generally, every academic year, more than 70 students enroll. Feedback has largely been positive. In some cases, advanced students have participated in research projects with instructors and, as a result, they have appeared as co-authors in several publications related to top-conferences such as the Winter Simulation Conference (www.wintersim.org) or the EURO (http://www.euro-online.org). This, in turn, acts as an additional motivation for students, who wish to enjoy the experience of publishing and presenting his/her work at international forums.

For graduate students, the UOC also offers an online course in Advanced Simulation, which covers topics such as modeling input data, random number generation, verification and validation, experimental design and output analysis. This course is research-oriented in the sense that it constitutes a formation complement for all those students who wish to start a PhD in the Operations Research area.

EVOLUTION OF STATISTICS AND OR COURSES AT THE UOC

In this section we will discuss some experiences regarding the pedagogical evolution of our Statistical and OR courses at the UOC. In particular, within the Computer Sciences Studies (CS) program, we will focus on how, over the last few years, these courses have evolving. These Studies have been offered at the UOC since 1996. At the onset, they had only a few dozen registered students. Today, they have more than 3,000 registered students. Each term about 400 students follow an introductory course in Probability and Statistics and about 70 students follow a course on Operations Research. During the past years at UOC, there has been a continuous improvement process involving the way these courses are taught and learned. We can distinguish three major stages in this process:

Stage 1 (1996 - 1999): Direct Implementation from Face-to-Face Courses

When the CS degree started in 1996, the common pedagogical model of traditional face-to-face universities was directly implemented in the new online environment. For instance, course materials were simply traditional books for distance education with some additional document in digital format –such as PDF or HTML– and, at best, some videos. Soon, it became obvious that the instructor's role in an online environment was quite different than a traditional one (Daradoumis et al., 2006). Consequently, a new methodological approach was necessary.

Stage 2 (2000 - 2004): Use of Technology and Innovative Methodologies

At the second stage, many innovative experiments were tested. The goal was to improve the overall quality of the learning-process at UOC. To pursue that goal, more information and communication technology resources were employed in the development of new math materials and courses. Complementary materials, which reinforced practical applications of theoretical concepts, were developed and published online. Also, some projects regarding the development of open online materials were developed. The e-Math project (www.uoc.edu/in3/e-math) was conceived to promote the efficient use and integration of instructional technologies –Internet and specialized software– as a fundamental part of most mathematical courses at the UOC, including: Algebra, Calculus, Probability and Statistics, Operations Research, Discrete Mathematics, etc. New learning materials, divided into individual modules named math-blocks, were developed and published online, usually in PDF or HTML format. These materials were designed as additional learning resources, and were particularly oriented to students with a poor background in mathematics or to students looking for complementary, practical and software-oriented learning resources. Usually, each math-block had an associated file containing computer data or computer laboratories with step-by-step guidance. Additionally, online homework and tests were included at different times during the semester. In the Statistics and OR courses, these tests and homework were particularly designed to promote the use of mathematics software among students.

Stage 3 (2005 - 2007): Curricula Redefinition Following a Top-Down Approach

During the third stage instructional technology resources were fully integrated into the Statistical and OR courses. Special attention was paid to the use of mathematical software and Java applets. They were used: (a) to perform real-life calculations that illustrate applications of mathematics to computer science problems, and (b) as interactive tools that help students to understand mathematical concepts by experimentation and visualization. One more major innovation developed at this stage was the complete redefinition of the Computer Science curriculum. The CS faculty staff was grouped in interdisciplinary work teams, each of them composed of lecturers from different knowledge areas. Each of these teams worked for months in order to identify the concepts, techniques and skills that our students required for graduation

Recommendations from the Association for Computing Machinery (ACM, www.acm.org/education/curricula.html) were considered and different curricula from universities worldwide were analyzed and discussed. Finally, members of each team met to share results and extract conclusions regarding the educational needs of our students. Employing these educational needs as a starting point, all CS curricula were redefined using a top-down approach, i.e.: starting with the "top subjects" –those located at the last semester in the CS curricula, and descending to the "bottom ones" –those located in the first semester. The major dependencies among subjects were identified and a dependencies map of contents linking different subjects was established. According to this map, all subjects in the CS degree were redefined, both in contents –giving priority to those contents that students will need in other subjects or in their future professional activity–, and in orientation –promoting a practical and updated approach to all subjects instead of a more theoretical and traditional one. Our goal was to redesign all subjects taking into consideration a global vision of the CS curricula instead of a more fractional vision restricted to each individual subject.

Stage 4 (2007 - 2009): Adaptation to the European Area of Higher Education

As stated at the beginning of this chapter, most European universities are currently involved in the creation of a European Area of Higher Education. In our particular context, this means a revision of all undergraduate and graduate courses so that they agree with the new EU directives. Among other things, there is a redefinition of the evaluation process where the focus remains on the evaluation of competences –both transversal and course-specific– attained by students during the semester. For example, more collaborative-learning practices are being introduced in our courses so that students can acquire transversal competencies related to the development of group projects. Likewise, some of our courses are being offered in English, instead of in Spanish, so that students can acquire foreign language competencies and also as a means to facilitate virtual mobility among students from different European universities.

LESSONS LEARNED FROM EXPERIENCE

From previous discussion, and based on long-term experience offering Statistical and OR courses online at the UOC, there are five fundamental factors or "golden rules" that, in our opinion, should be considered when designing and developing successful online courses in this particular knowledge area. In particular (Figure 4):

- *Course core learning materials:* They constitute the main source of information for

Figure 4. Key design factors of our Statistics and OR courses at the UOC

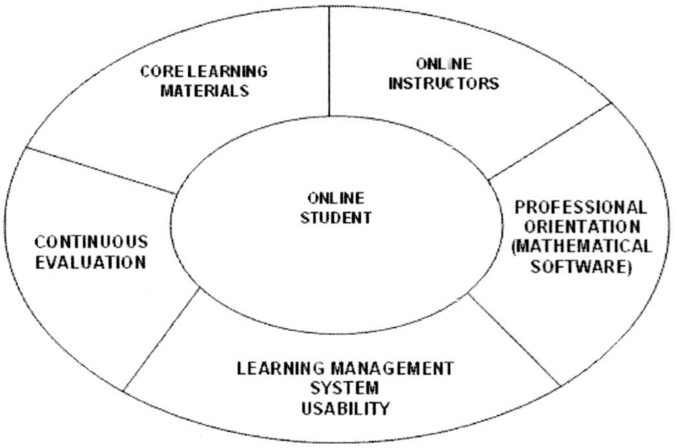

students during the learning process. They must be accurately designed for independent learning processes and must provide the student with an insight into all relevant aspects of the course. Ideally, these notes and learning materials should be designed and written by the instructors themselves in order to promote a strong correlation between the materials and the evaluation activities (homework and final exam). Due to the intrinsic nature of these knowledge areas, where students need to do a lot of thinking and annotations during the learning process, it is strongly recommended that these core materials should also be available in PDF format for easy printing. Of course, these core learning materials can and should be complemented with additional learning materials and resources, such as applets, related articles, simulations, etc.

- *Online instructors:* The role of online instructors is critical to the success of online instruction. Instructors should be responsible for designing and writing the core learning materials and, moreover, the homework activities that constitute the continuous evaluation process. While they should necessarily be responsible for designing the final exam, there is a pedagogical imperative for the provision of orientation, ongoing guidance and support as well as continuous feedback throughout the learning process. This guidance should be developed through posted messages (e.g. at the beginning of each week) with clear instructions about which contents and activities must be completed in the short-term. While they are working on the course material or with the math software, support should be provided with quick responses to student posts in shared forums and e-mail. This feedback should be provided no later than 48 hours from the posting of the question. Finally, coordination among different instructors of the same course is important in order to guarantee homogeneity.
- *Professional-oriented approach using mathematical software:* Distance-learning students, and particularly those with professional or family duties, need continuous motivation so that they feel that it is worthy to invest their time completing a degree program. This is especially true in the case of some knowledge areas, such as mathematics, which sometimes are presented to the student as a theoretical corpus without visible application to their professional careers. Theoretically-oriented math instruction might make sense for students in a pure Math degree program, but usually is not the best way to motivate students completing other degrees such as Computer Science or Telecommunications. Hence, it is important that students understand what Statistics and OR courses provide for them in terms of practical concepts and skills. Therefore, whenever possible, a professionally-oriented applied approach is likely to be more appreciated by students and will likely contribute to higher levels of motivation. However, it is completely understood that in order to be able to analyze and solve realistic problems and scenarios, use of mathematical software is mandatory. The available statistics and OR-software is of such high quality (Swain, 2009) that the issue is not as much which specific software to use but how to effectively integrate it into the course curriculum.
- *Continuous evaluation process:* At least in some European countries, most face-to-face courses were traditionally evaluated through a single exam at the end of the semester. At times, a mid-term exam was also included in the evaluation process. One of the aspects promoted by the European Area of Higher Education is the

generalization of continuous evaluation processes throughout degree programs. In the case of online students, the use of a continuous evaluation system is even more necessary since it is highly related to the issue of motivation and can significantly contribute to reducing dropout rates during the semester (Lera et al., 2009).

- *Learning Management System usability:* As stated before in this chapter, there a several LMS available today –both commercial and free. With regards to Statistics and OR courses, any of them may be an excellent alternative, however, one problem not yet solved is the challenge of effectively communicating with mathematical symbols and equations. Nevertheless, the most important aspect of any LMS is its usability, i.e., students and instructors should feel comfortable using the LMS and all main options should be intuitive and easy to find. For Statistics and OR courses, no specific requirements are needed, but the system should be able to facilitate an online space for posting instructor's notes –official messages from instructors to students– and another space for students to post notes and hold debates and discussions regarding the course contents. Other desirable LMS options would be the inclusion of a native equation editor and a monitoring feature that could provide regular feedback on students' activity and performance (Juan et al., 2009).

CONCLUSION

The current worldwide higher education endeavor is experiencing increased growth in the use of e-Learning Management Systems. This growth is challenging traditional pedagogy and re-defining the traditional roles of instructors and students in the current knowledge-based society. In this new context, with the lack of face-to-face interaction, teaching and learning online courses, particularly in mathematics, requires a unique approach both for faculty and students. This chapter has presented an analysis of some of the advantages and challenges associated with offering online courses in mathematics based upon long-term experience teaching online Statistics and Operations Research at the Open University of Catalonia. After discussing the evolution of design, developement and management of these courses over the last decade, the chapter highlights the main factors– according to the experience at UOC– that need to be carefully considered when offering online courses in mathematics or related subjects. This case study demonstrates not only the viability of teaching Statistics and OR courses online but also principles on how it can be taught successfully.

ACKNOWLEDGMENT

This work has been partially financed by the Spanish Ministry of Education and Science under grant EA2008-0151, and also by the Catalan Government under grant 2008MQD 00177.

REFERENCES

Allen, I., & Seaman, J. (2008). *Staying the course: online education in the United States*. Needham, MA: The Sloan Consortium.

Allen, T. H. (2006). Raising the question #1: Is the rush to provide online instruction setting our students up for failure? *Communication Education, 55*(1), 122–126. doi:10.1080/03634520500343418

Camm, J. (2007). O.R. in the Classroom – Get Real! *OR/MS Today, 34*(4), 34-37.

Childers, J. L., & Berner, R. T. (2000). General education issues, distance education practices: building community and classroom interaction through the integration of curriculum, instructional design, and technology. *The Journal of General Education*, *49*(1), 53–65. doi:10.1353/jge.2000.0001

Daradoumis, T., Xhafa, F., & Juan, A. (2006). A framework for assessing self, peer and group performance in e-learning. In *Self, peer, and group assessment in e-learning* (pp. 279-294). Hershey, PA: Idea Group Press.

Del Moral, M. E., & Villalustre, L. (2008). Wikis and collaborative education by means of Webquest in higher education. *Revista Latinoamericana de Tecnologia Educativa*, *7*(1), 73–83.

Faulin, J., Juan, A., Fonseca, P., Pla, L. M., & Rodriguez, S. V. (2009). Learning operations research online: benefits, challenges, and experiences. *International Journal of Simulation and Process Modelling. Special Issue on Modelling and Simulation Education*, *5*(1), 42–53.

Hardin, J., & Ellington, A. (2005). Using Multimedia to Facilitate Software Instruction in an Introductory Modeling Course. *INFORMS Transactions on Education, 5*(2). Retrieved Oct 19, 2009 from http://archive.ite.journal.informs.org/Vol5No2/HardinEllington/

Howell, S. L., Williams, P. B., & Lindsay, N. K. (2003). Thirty-two trends affecting distance education: An informed foundation for strategic planning. *Online Journal of Distance Learning Administration*, *6*(3), 1–18.

Juan, A., Daradoumis, T., Faulin, J., & Xhafa, F. (2009). SAMOS: A model for monitoring students' and groups' activity in collaborative e-learning. *International Journal of Learning Technology*, *4*(1/2), 53–72. doi:10.1504/IJLT.2009.024716

Juan, A., Huertas, M., Steegmann, C., Corcoles, C., & Serrat, C. (2008). Mathematical e-learning: state of the art and experiences at the Open University of Catalonia. *International Journal of Mathematical Education in Science and Technology*, *39*(4), 455–471. doi:10.1080/00207390701867497

Leon, L.; Przasnyski, Z. & Seal, K. (2008). Teaching OR/MS to Net-Gens: A Paradigm Gap? *OR/MS Today, 35*(5), 10-11.

Leon, L., Seal, K., & Przasnyski, Z. (2006). Captivate Your Students' Minds: Developing Interactive Tutorials to Support the Teaching of Spreadsheet Modeling Skills. *INFORMS Transactions on Education, 7*(1). Retrieved Oct 19, 2009 from http://archive.ite.journal.informs.org/Vol7No1/LeonSealPrzasnyski/

Lera, F., Juan, A., Faulin, J., & Cavaller, V. (2009): Monitoring Students' Activity and Performance in Online Higher Education: A European Perspective. In *Monitoring and Assessment in Online Collaborative Environments: Emergent Computational Technologies for E-Learning Support* (pp. 132-148). IGI Global, Hershey, Pennsylvania, USA.

Lowry, R. (2005). Computer-aided self assessment -an effective tool. *Chemistry Education Research and Practice*, *6*(4), 198–203.

Nagy, A. (2005). The Impact of E-Learning. In P. A. Bruck, A. Buchholz, Z. Karssen, A. Zerfass (Eds.), *E-Content: Technologies and Perspectives for the European Market,* (pp.79-96). Berlin: Springer-Verlag.

Peat, M., & Franklin, S. (2002). Supporting student learning. The use of computer-based formative assessment modules. *British Journal of Educational Technology*, *33*(5), 45–55. doi:10.1111/1467-8535.00288

Robinson, L. A. (2005). Consumers of online instruction. *Issues in Information Systems*, *6*, 170–175.

Schwartzman, R. (2007). Refining the question: how can online instruction maximize opportunities for all students? *Communication Education, 56*(1), 113–117. doi:10.1080/03634520601009728

Seufert, S., Lechner, U., & Stanoevska, K. (2002). A reference model for online learning communities. *International Journal on E-Learning, 1*(1), 43–54.

Simonson, M., Smaldino, S., Albright, M., & Zvacek, S. (2003). *Teaching and learning at a distance*. Upper Saddle River, NJ: Merrill Prentice Hall.

Swain, J. (2009). A long way from flip charts. *ORMS Today, 36*(1), 44–47.

Sweet, R. (1986). Student drop-out in distance education: An application of Tinto's model. *Distance Education, 7*(2), 201–213. doi:10.1080/0158791860070204

Trenholm, S. (2007). An investigation of assessment in fully asynchronous online math courses. *International Journal for Educational Integrity, 3*(2), 41–55.

Truluck, J. (2007). Establishing a mentoring plan for improving retention in online graduate degree programs. *Online Journal of Distance Learning Administration, 10*(1), 1–6.

Van der Wende, M. (2000). The Bologna declaration: enhancing the transparency and competitiveness of European higher education. *Higher Education in Europe, 25*(3), 305–310. doi:10.1080/713669277

Zirkle, C. (2003). Distance education in career and technical education: a review of the research literature. *Journal of Vocational Education Research, 28*(2), 151–171.

Compilation of References

Aboites, H. (1999). *Viento del norte: TLC y privatización de la educación superior en México*, (2nd ed.). México, DF: UAM / Plaza y Valdés.

Abrami, P. C., Cohen, P. A., & d'Appolonia, S. (1990). Validity of student ratings of instruction: what we know and what we do not. *Journal of Educational Psychology*, *82*(2), 219–231. doi:10.1037/0022-0663.82.2.219

Activeworlds Inc. (2008). *Active Worlds: Home of the 3D Internet, virtual worlds and community chat*. Retrieved on December 4, 2008, from http://www.activeworlds.com/

Advanogy. (2007). *Learning styles*. Retrieved on November 4, 2008 from http://www.learning-styles-online.com/overview/

Afaneh, M., Basile, V., & Bennett, J. (2007). *E-Learning Concepts and Techniques*. Retrieved 1st December 2008 from http://iit.bloomu.edu/Spring2006_eBook_files/index.htm

Ahrlberg, M. (2008). *Practical methods and techniques of knowledge representation in particular those related to concept mapping and mind mapping*. History, theoretical background, software, and comparison table. Paper presented October 6, 2008. Retrieved February 20, 2009 from http://itech1.coe.uga.edu/itforum/paper109/Ahlberg_manuscript.pdf

Akbaba-Altun, S. (2006). Complexity of integrating computer technologies in education in Turkey. *Educational Technology & Society*, *9*(1), 176–187.

Akindele, D., & Trennepohl, B. (2008). Breaking the culture of silence: Teaching writing and oral presentation skills to Botswana University students. *Language, Culture and Curriculum*, *21*(2), 154–166.

Akintunde, M. (2006). Diversity.Com: Teaching an online course on white racism and multiculturalism. *Multicultural Perspectives*, *8*(2), 35–45. doi:10.1207/s15327892mcp0802_7

Akkutay, Ü. (1996). *Foreign Expert Reports in National Education*. Avni Akyol Kültür ve Eğitim Ümit Vakfı.

Alberts, D. S., & Papp, D. S. (1997). *The information age: An anthology on its impact and consequences*. Washington, DC: CCRP Publication Series. Retrieved on November 10, 2008 from http://www.dodccrp.org/files/Alberts_Anthology_I.pdf

Albright, M. J. (1997). Web Course Authoring and Management Systems. *MC Journal: The Journal of Academic Media Librarianship*, *5*(1), 14–23.

Alcántara, C. H. (2001). *The Development Divide in a Digital Age An Issues Paper*. United Nations Research Institute for Social Development.

Ali, D. L. (1998) *Appendix 4: Literature Search to address some commonly expressed questions regarding student ratings of instruction*. Retrieved March 5, 2009 from http://www.ucalgary.ca/UofC/departments/VPA/usri/appendix4.html

Allen, I. E., & Seaman, J. (2006). *Making the grade: Online education in the United States, 2006*. Needham, MA: The Sloan Consortium

Allen, I. E., & Seaman, J. (2007). *Online nation: Five years of growth in online learning*. Needham, MA:

Sloan Consortium. Retrieved on December 1, 2008, from http://www.sloan-c.org/publications/survey/pdf/online_nation.pdf

Allen, I., & Seaman, J. (2008). *Staying the course: online education in the United States*. Needham, MA: The Sloan Consortium.

Allen, T. H. (2006). Raising the question #1: Is the rush to provide online instruction setting our students up for failure? *Communication Education, 55*(1), 122–126. doi:10.1080/03634520500343418

Ally, M. (2004). Foundations of educational theory for online learning. In Anderson, T. & Elloumi, F. (eds.). *Theory and practice of online learning*. Athabasca University (online book). Retrieved November 10, 2007, from http://cde.athabascau.ca/online book/index.html

Althaus, S. L. (1997). Computer-mediated communication in the university classroom: An experiment with online discussion. *Communication Education, 7*, 158–174. doi:10.1080/03634529709379088

Anderson, T. (2003). Modes of interaction in distance education: Recent developments and research questions. In M. Moore & G. Anderson (Eds.), *Handbook of distance education* (pp. 129-144). Mahwah, NJ: Erlbaum

Anderson, T. (2004). Toward a theory of online learning. *Theory and practice of online learning*. Retrieved August 21, 2007 from http://cde.athabascau.ca

Anderson, T., & Elloumi, F. (Eds.). (2004). *Theory and Practice of Online Learning*. Edmonton, Canada: Athabasca University Press. http://www.athabascau.ca (Accessed on May 15, 2005).

Anderson-Mejías, P. L. (2005). Online training for English as a second/foreign language teachers. In M.O. Thirunarayanan & A. Perez-Prado (Eds.), *Integrating technology in higher education* (pp. 143 - 154). Lanham, MD: University Press of America.

Anderson-Mejías, P. L. (2006). A case study in peer evaluation for second language teachers in training. In T. Roberts (Ed). *Self, peer, and group assessment in e-learning* (pp. 17-63). Hershey, PA: IGI Global.

Anonymous. (2007). *Women get more than men from e-learning*. Retrieved November 12, 2008, from http://findarticles.com/p/articles/mi_qa5402/is_200703/ai_n21284857

ANUIES. (1999). *La educación superior hacia el siglo XXI: Plan Maestro de Educación Superior Abierta y a distancia*. XIV Asamblea General Ordinaria de ANUIES. Universidad Autónoma del Estado de Hidalgo.

ANUIES. (2002). *Hacia la construcción de un sistema virtual para la educación superior*. Informe de la tercera reunión nacional de educación superior a distancia. Cozumel, Quintana Roo.

Apex Learning. (2009). Retrieved March 17, 2009, from http://www.apexlearning.com/courses.html

Aragon, S. R., Johnson, S. D., & Shaik, N. (2003). The influence of learning style preferences on student success in online versus face-to-face environments. *American Journal of Distance Education, 16*(4), 227–244. doi:10.1207/S15389286AJDE1604_3

Argyris, C., Putnam, R., & Smith, D. (1985). *Action science: Concepts, methods and skills for research and intervention*. San Francisco, CA: Jossey-Bass.

Asher, G. (2005). Inadequate Infrastructure and the Infusion of Technology into K-12 Education. In C. Howard, J.V. Boettcher, L. Justice, K. Schenk, P. L. Rogers, & G.A. Berg, (eds.), *Encyclopedia of Distance Learning*, (pp. 1061-1063). Hershey, PA: Idea Group Reference.

Attewell. (2001). The first and second digital divides. *Sociology of Education, 74*(3), 252-259.

August 4, 2006, from http://www.ssu.missouri.edu/ssu/AgEd/NAERM/s-e-4.htm

Avison, D., Lau, F., Meyers, M., & Nielsen, P. A. (1999). Action research: To make action research relevant, researchers should try out their theories with practitioners in situations and real organizations. *Communications of the ACM, 42*, 93–97. doi:10.1145/291469.291479

Azeta, A. A., Ikhu-Omoregbe, N. A., Ayo, C. K., & Atayero, A. A. (2008). Development and Deployment of VoiceXML-Based Banking Applications. *Journal of Nigeria Computer Society, 15*(1), 59–72.

Baldwin-Edwards, M. & Kiriakou, I. (2004). *Statistical Data for the Immigrants in Greece: Analytic Study on the Available Elements and Proposals for the Conformity with Standards of the European Union*. Mediterranean Observatory of Immigration, Panteios University, Revised publication.

Ball, S. J. (2003). The teacher's soul and the terrors of performativity. *Journal of Education Policy, 18*(2), 215–228. doi:10.1080/0268093022000043065

Bandura, A. (1991). Social cognitive theory of self-regulation. *Organizational Behavior and Human Decision Processes, 50*, 248–287. doi:10.1016/0749-5978(91)90022-L

Baptiste, I., & Heaney, T. (1996). *The political construction of adult education*. Paper presented at the Midwest Research-to-Practice Conference in Adult, Continuing, and Community Education, October 17-19, 1996, Lincoln, NE. Retrieved on November 23, 2006 from http://www.nl.edu/academics/cas/ace/facultypapers/TomHeaney_IanBaptiste_Construction.cfm

Barnett, R. (2003). *Beyond All Reason: Living with ideology in the university*. Buckingham, UK: Open University Press.

Barton, R., & Haydn, T. (2006). Trainee teachers and 'impact' learning. A study of trainees' views on what helps them to use ICT effectively in their subject teaching. *Journal of Computer Assisted Learning*, (22): 257–272. doi:10.1111/j.1365-2729.2006.00175.x

Bates, T. R. (1975). Gramsci and the Theory of Hegemony. *Journal of the History of Ideas, 36*(2), 351–366. doi:10.2307/2708933

Beard, C., Clegg, S., & Smith, K. (2007). Acknowledging the affective in higher education. *British Educational Research Journal, 33*(2), 235–252. doi:10.1080/01411920701208415

BECTA. (2006). *Getting the most from your interactive whiteboard: A guide for secondary schools*. Retrieved October 15, 2008, from http://foi.becta.org.uk/content_files/corporate/resources/foi/archived_publications/getting_most_whiteboard_secondary.pdf

Bejerano, A. R. (2008). The genesis and evolution of online degree programs: Who are they for and what have we lost along the way? *Communication Education, 1*(3), 408 – 414.

Bentley, R., Appelt, W., Busbach, U., Hinrichs, E., Kerr, D., & Sikkel, S. (1997). Basic Support for Cooperative Work. *International Journal of Human-Computer Studies, 46*, 827–846. doi:10.1006/ijhc.1996.0108

Berge, Z. (1998). Barriers to online teaching in post-secondary institutions: Can policy changes fix it? *Online Journal of Distance Learning Administration*, 1. Retrieved from http://www.westga.edu/~distance/ojdla/summer12/berge12.html

Berge, Z. L. (1998). Technology and Changing Roles in Education. In Z.L. Berge & M. Collins (Eds.), *Wired together: Computer-mediated communication in K-12, (1): Perspective and instructional design*, (pp. 1-13). Cresskill, NJ: Hampton Press.

Berge, Z., & Muilenburg, L. (2001). Obstacles faced at various stages of capability regarding distance education in institutions of higher learning. *TechTrends, 46*, 40–45. doi:10.1007/BF02784824

Berger, D., & Kam, R. (1996). *Training and instructional design*. Retrieved October 24, 2007, from http://www.umich.edu/~ed626/define.html

Berners- Lee. T. (2007, April). Interview with Tim Berners-Lee. *Business Week*. Retrieved August 29, 2007 from http://www.w3.org/2001/sw/

Berners-Lee, T. (2007). Time Berners-Lee on the Semantic Web. *Video on MIT Technology Review*, March 2007. Retrieved Aug. 29, 2007 from http://www.w3.org/2001/sw/

Berners-Lee, T. (n.d.). Retrieved September 8, 2003 from http://www.w3.org/People/Berners-Lee/ShortHistory.html

Bernstein, B. (1996). *Pedagogy, symbolic control and identity. Theory, research, critique*. Maryland: Rowman Littlefield Published.

Bertelsen, O. W., & Bodker, S. (2003) Activity theory. In J. M. Carroll (Ed.), *HCI models theories, and frameworks: Toward a multidisciplinary science*, (pp. 292-324). San Francisco: Morgan Kaufmann.

Biggs, J. B. (1987). *Student Approaches to Learning and Studying*. Melbourne, Australia: Australian Council for Educational Research.

Biggs, J. B. (1993, November). The quality of teaching and learning in an expanding tertiary system. *New Horizons (Baltimore, Md.)*, *34*, 102–104.

Biggs, J. B. (1994). What are effective schools? Lessons from East and West. *Australian Educational Researcher*, *21*, 19–39.

Biggs, J. B. (1996a). Approaches to Learning of Asian Students: A Multiple Paradox. In J. Pandey (Ed.), *Asian Contributions to Cross-Cultural Psychology*, (pp. 18 – 199). New Delhi: Sage Publications.

Biggs, J. B. (1999). *Teaching for quality learning at university*. Milton Keynes, UK: Open University Press.

Biggs, J. B., & Telfer, R. (1987). *The Process of Learning* (2nd. Edition). Sydney: Prentice Hall of Australia.

Biggs, J., & Tang, C. (2007). *Teaching for Quality Learning at University*. Maidenhead, UK: Open University Press.

Billings, D., & Connors, R. (2006). *NLN living book: best practices in online learning*. Retrieved November 20, 2008, from www.electronicvision.com/nln/chapter02/index.htm

Bitner, M. J. (1990). Evaluating service encounters: the effects of physical surroundings and employee responses. *Journal of Marketing*, *54*(2), 69–82. doi:10.2307/1251871

Blackboard (2009). *Blackboard web site*. Retrieved from http://www.blackboard.com

Blackboard Inc. (2008). *Blackboard: Educate. innovate. everywhere*. Retrieved on December 2, 2008, from http://www.blackboard.com/us/index.bbb

Blair, T. (1995). A week in politics. *Channel*, *4*(February), 18.

Blair, T. (1997). *Foreword in Connecting the learning society*. London: DfEE.

Bloom, B. S. (1956). *Taxonomy of Educational Objectives Handbook I: The Cognitive Domain*. New York: David McKay Co Inc.

Bloom, B. S. (1984). *Taxonomy of educational objectives*. Boston, MA: Allyn and Bacon.

Bloom, B. S., Engelhart, M. D., Furst, E. J., Hill, W. H., & Krathwohl, D. R. (1956). *Taxonomy of Educational Objectives – Handbook I: Cognitive Domain*. London: Longmans Green.

Bloom, B. S., Englhart, M. D., Furst, E. J., Hill, W. H., & Krathwohl, D. R. (Eds.). (1956). *Taxonomy of educational objectives, the classification of educational goals, Handbook I: Cognitive domain*. New York: Longmans.

Boer, N., Van Baalen, P., & Kumar, K. (2002). An activity theory approach for studying the dynamics of knowledge sharing. In *Proceedings of the Thirty-fifth Annual Hawaii Internatioinal Conference on System Sciences (HICSS 35)*, (pp. 90-93). Los Alamitos, CA: IEEE Computer Society Press.

Bohm, D., Factor, D., & Garrett, P. (1991). *Dialogue: A proposal*. Retrieved April 24, 2006, from http://www.infed.org/archives/e-texts/bohm_dialogue.htm

Bolton, R. N., & Drew, J. H. (1991). A longitudinal analysis of the impacts of service changes on customer attitudes. *Journal of Marketing*, *55*(1), 1–9. doi:10.2307/1252199

Boltuc, P. (2008, July22). Chronicle article focuses on distance-ed academic integrity. *Message posted to Sloan-C Electronic mailing list.*

Bond, M. H. (1991). *Beyond the Chinese Face: Insights from Psychology*. Hong Kong: Oxford University Press.

Bond, M. H. (Ed.). (1996). *The Handbook of Chinese Psychology*. Hong Kong: Oxford University Press.

Bonk, C. J., & Reynolds, T. H. (1997). Learner-centred Web instruction for higher-order thinking, teamwork and apprenticeship. In B. H. Khan, (Ed.). *Web-based instruction*. (pp. 167 – 178). Englewood Cliffs, NJ: Educational technology Publications.

Boud, D., & Feletti, G. (Eds.). (1997). *The Challenge of Problem-based Learning*, (2nd Ed. pp. 93-102). London: Kogan Page.

Bourhis, A., Dube, L., & Jacob, R. (2005). The success of virtual communities of practice: The leadership factor. [Available online at www.ejkm.com]. *Electronic Journal of Knowledge Management*, *3*(1), 23–24.

Boyd, D. (2005). *The bias of hyperlinks*. Retrieved December 3, 2008, from http://www.zephoria.org/thoughts/archives/2005/08/07/the_biases_of_links.html

Boyd, H. R., & Cowan, J. (1986). The case for self-assessment based on recent studies of student learning. *Assessment & Evaluation in Higher Education*, *10*(3), 225–236.

Boyer, E. L. (1990). *Scholarship Reconsidered*. San Francisco: Jossey-Bass.

Boyle, T., Bradley, C., Chalk, P., Jones, R., Haynes, R., & Pickard, P. (2003). *Can learning objects contribute to pedagogical improvement in higher education: lessons from a case study?* Paper based on presentation given at CAL '2003, April 2003. Retrieved February 23, 2009 from http://www.londonmet.ac.uk/ltri/learningobjects/papers_pres/CAL_Obje cts_paper.doc

Boyle, T., Bradley, C., Chalk, P., Jones, R., Haynes, R., & Pickard, P. (2003). *Can learning objects contribute to pedagogical improvement in higher education: lessons from a case study?* Paper based on presentation given at CAL '2003, April 2003. Retrieved February 23, 2009 from http://www.londonmet.ac.uk/ltri/learningobjects/papers_pres/CAL_Obje cts_paper.doc

Bransford, J., Brown, A., & Cocking, R. (1999). *How people learn: Brain, mind, experience and school*. The National Academy of Sciences. Retrieved Aug. 28, 2007 from http://www.nap.edu/html/howpeoplel

Bresnahan, T. F., Brynjolfsson, E., & Hitt, L. M. (2002). Information Technology, Workplace Organization and the Demand for Skilled Labor: Firm-level Evidence. *The Quarterly Journal of Economics*, *117*(1), 339–376. doi:10.1162/003355302753399526

Brett, J. (2006). 360 Degree feedback to leaders: Does it relate to changes in employee attitudes? *Group and Organization Management 36*(5). 578-600. Los Angeles: SAGE Publications. Retrieved on December 1, 2008, from http://gom.sagepub.com/cgi/content/abstract/31/5/578

Britton, J. (1993). *Language and Learning*. Portsmouth, NH: Boynton/Cook.

Brookfield, S. D. (1990). *The Skilful Teacher*. San Francisco: Jossey Bass.

Brooks, D. (2008). Demography is king. *The New York Times*, May 3. Retrieved on November 18, 2008 from http://neuroanthropology.net/2008/05/03/david-brooks-part-two-demography-is-king/

Brooks, D. (2008). The cognitive age. *The New York Times*, May 2, 2008. Retrieved on November 15, 2008, from http://www.nytimes.com/2008/05/02/opinion/02brooks.html?_r=1&em&ex=1209960000&en=1628bc39165590dc&ei=5087%0A

Brooks, F. P. (1999). *The mythical man-month: Essays on software engineering*. Reading, MA: Addison-Wesley.

Brown, J., Sheppard, B., & Stevens, K. (2001). *Final Report: Effective schooling in a telelearning environment*. St. John's, Canada: The Centre for TeleLearning and Rural Education.

Brown, M. (2004). *The Study of Wired Schools: A Study of Internet-Using Teachers*. Palmerston North, New Zealand: Massey University.

Bucy, E. P. (2000). Social Access to the Internet. *Press/Politics, 5*, 50-61.

Buffington, J. (2007). *Learning communities as an instructional model*. Retrieved on January 12, 2008, from http://projects.coe.uga.edu/epltt/index.php?title=Learning_Communities_as_an_Instructional_Model

Burbules, N. C. (2000). Does the Internet constitute a global educational community? In N. C. Burbules & C. Torres (Eds.), *Globalization and education: Critical perspectives* (pp. 323-355). New York: Routledge.

Burke, B. (1999, 2005). *Antonio Gramsci and informal education.* Retrieved December 3, 2008, from http://www.infed.org/thinkers/et-gram.htm

Burnette, G., & Buerkle, H. (2004). Information exchange in virtual communities: A comparative study. *Journal of Computer Mediated Communication, 9*(2). Retrieved March 5, 2009, from http://www.ascusc.org/jcmc/vol9/issue2/burnett.htm

Burnham, B., & Walden, B. (1997). Interactions in distance education: A report from the other side. In *Proceedings of the Adult Education Research Conference.* Oklahoma State University. Available: http://www.edst.educ.ubc.ca/aerc/1997/97burnham.html

Bushweller, K. (1999, April). Generation of cheaters. *The American School Board Journal, 186*(4), 24–32.

Butts, D., Cowan, J., & Bingham, E. G. (1976). *Learning systems, a case study.* Glasgow: Jordanhill College of Education.

Buzzetto-More, N., & Pinhey, K. (2006). Guidelines and Standards for the Development of Fully Online Learning Objects. *Interdisciplinary Journal of Knowledge and Learning Objects, 2*, 95–104.

California State University. Chico (2006). *Rubric for Online Instruction.* ROI Rubric for Online Instruction. Retrieved March 5, 2009, from http://www.csuchico.edu/celt/roi/

Callender, C., & Wilkinson, D. (2003). *2002/03 Student Income and Expenditure Survey, Research Report 487,* DfES, Nottingham, UK.

Camm, J. (2007). O.R. in the Classroom – Get Real! *OR/MS Today, 34*(4), 34-37.

Carliner, S. (2005).*CMSs versus LMSs.* Retrieved February 25, 2008 from ASTD Worldplace Learning & Performance http:www.astd.org

Carnes, L. W., Awang, F., & Marlow, J. (2003). Can instructors ensure the integrity and quality of online courses? *Delta Pi Epsilon Journal, 45*, 162–172.

Carr, S. (2000). As distance education comes of age, the challenge is keeping the students. *The Chronicle of Higher Education, 46*(A), 39-41.

Cazden, C. B. (1988). *Classroom discourse.* New York: Heinneman.

Chandler, D. (1994). *The Transmission Model of Communication.* Retrieved May 24, 2006 from http://www.aber.ac.uk/media/Documents/short/trans.html#A

Chang, H., & Holt, R. (1994). A Chinese perspective on face as inter-relational concern. In S. Ting-Toomey (ed.), *The Challenge of Facework* (pp. 95–132). Albany, NY: State University of New York Press.

Cheong, P. H. (2008). The young and techless? Internet use and problem solving behaviors among young adults in Singapore. *New Media & Society, 10*(5), 771–791. doi:10.1177/1461444808094356

Cherns, A. (1976). The principles of sociotechnical design. *Human Relations, 29*(8), 783–792. doi:10.1177/001872677602900806

Chickering, A. W., & Gamson, Z. F. (1987). Seven Principles for Good Practice in Undergraduate Education. *AAHE Bulletin, 39*(7), 3–7.

Childers, J. L., & Berner, R. T. (2000). General education issues. distance education practices: building community and classroom interaction through the integration of curriculum, instructional design, and technology. *The Journal of General Education, 49*(1), 53–65. doi:10.1353/jge.2000.0001

Chin, C. C., Hock, G. T., & Veerappan, C. M. (2006). *VoiceXML as Solution for Improving Web accessibility and Manipulation for e-Education.* School of Computing and IT, INTI College Malasia. Retrieved July 2008 from http://intisj.edu.my/INTISJ/InfoFor/StaffResearch/10.pdf

Chiu, Y. C. (2001, June). *Online support for journalistic news course.* Paper presented at Developing College

Specialization Seminar, Wenzao, Kaohsiung, Taiwan.

Chiu, Y. C. (2002, November). Exploring the learning effectiveness of internet news reading. In *Selected Proceedings of the 4th Pacific Asian Conference, and 11th English Teachers' Association Annual Conference,* Crane, Taipei.

Chiu, Y. C. (2006). *Exploring student and teacher interactions for critical thinking in face-to-face and online environments in an EFL course in Taiwan.* Unpublished doctoral dissertation, National Australian Catholic University, Melbourne, Australia.

Chiu, Y. C. (2007, July). Using the Shepherd Metaphor in Crossing Cultural Boundary Online for CHC Students. In *Proceedings of the 2007 E-Learning IADIS Conference, Lisbon, Portugal, International Association for the Development of the Information Society.*

Chiu, Y. C. (2008). Technical and Pedagogical Support for EFL Online Interactions. *Journal of Foreign Language Instruction, 2*(1), 83–102.

Chiu, Y. C. (2009). Facilitating Asian students' critical thinking in online discussions. *British Journal of Educational Technology, 40*(1), 42–57. doi:10.1111/j.1467-8535.2008.00898.x

Chomsky, N. (2004). *Language & thought.* Wakefield, RI: Moyer Bell Publishers.

CIEES. (2005). La evaluación interinstitucional de la educación superior en México. Comités Interinstitucionales para la Evaluación de la Educación Superior. Retrieved from http://www.ciees.edu.mx/ciees/publicaciones.php

Cizek, G. J. (1999). *Cheating on tests: how to do it, detect it, and prevent it.* Mahwah, NJ: Lawrence Erlbaum.

Clark, C. (2002). E-learning Strategy document. Version 2.1 (20.5.2002). *Information Technology Policy Committee.* Chair, e-learning steering Group of ITPC. Retrieved May 12, 2007 from http://www2.warwick.ac.uk/insite/forum/archive/elearning/stategydocument/

Clarke, C. (1999). Quoted in *The Guardian,* March 9.

Clerkin, M. J. (2004). *Practice: A comprehensive plan for preparing online students.* Needham, MA: The Sloan Consortium. Retrieved May 5th, 2008 from http://www.sloan.org/effective/details.asp?SS_ID=133

Cobley, P., & Jansz, L. (2003). *Introducing semiotics.* Royston, UK: Totem Books.

Cochrane, P. (1995). Quoted in *The Times Educational Supplement,* 23 June.

Cockburn, A., & Haydn, T. (2004). *Recruiting and retaining teachers: understanding why teachers teach.* London: RoutledgeFalmer.

Cohen, M. (1999). Turning teacher into a computer. *The Guardian,* October 12.

Cole, M. (1988). Cross-cultural research in the sociohistorical tradition. *Human Development, 31,* 137–151.

Cole, M., & Engeström, Y. (1993). A cultural-historical approach to distributed cognition. In G. Salomon (Ed.), *Distributed cognitions: Psychological and educational considerations,* (pp. 1-46). New York: Cambridge University Press.

Connaughton, S. L., & Shuffler, M. (2007). Multinational and multicultural distributed teams: A review and future agenda. *Small Group Research, 38,* 397–412. doi:10.1177/1046496407301970

Cook, A. (1985). *Changing the signs.* Lincoln, NE: University of Nebraska Press.

Cooper, J., & Mueck, R. (1992). Student involvement in learning: Cooperative learning and college instruction. In Goodsell, A. Maher, M. & Tinto, V. (Eds.)

COPAES. (2009). Programas educativos de licenciatura y técnico superior universitario reconocidos por su calidad. Enero 2009. Consejo para la acreditación de la educación superior, México, D.F.

Cordingley, P., Bell, M., Rundell, B., Evans, D., & Curtis, A. (2004). How does collaborative continuing professional development for teachers of the 5-16 age range affect teaching and learning? EPPI Centre, Institute of Education, University of London. Retrieved

December 6, 2008 from http://eppi.ioe.ac.uk/cms/Default.aspx?tabid=133&language=en-US

Corno, L., & Mandinach, E. B. (1983). The role of cognitive engagement in classroom learning and motivation. *Educational Psychologist, 18*, 88–108.

Cortazzi, M., & Jin, L. (1996). Cultures of learning: language classrooms in China. In H. Coleman (Ed.), *Society and the Language Classroom*. Cambridge, UK: Cambridge University Press.

Cowan, J. (1978). Freedom in selection of course content: a case study of a course without a syllabus. *Studies in Higher Education, 3*(2), 139–148. doi:10.1080/03075077812331376199

Cowan, J. (1984). *Learning contract design; a lecturer's perspective*. Royal Society of Arts - Occasional Paper No 7, London, Royal Society of Arts

Cowan, J. (1987). *Education for Capability*. Unpublished higher doctoral dissertation, Heriot-Watt University, Edinburgh.

Cowan, J. (1988). Struggling with self-assessment. In D. Boud (Ed.), *Student Autonomy in Learning*, (pp.192-210). London: Kogan Page.

Cowan, J. (2004). Beyond reflection: where next for curricula which concentrate on abilities? In C. Baillie & I. Moore (Eds.), *Effective Learning and Teaching in Engineering* (pp 202-19). Abingdon, UK: RoutledgeFalmer.

Cowan, J. (2005) Atrophy of the affect in academia. In S. Robinson & C. Katalushi. (Eds). *Values in Higher Education*. Leeds, UK: Aureus.

Cowan, J. (2006a). *On Becoming an Innovative University Teacher* (2nd Ed.). Maidenhead, UK: Open University Press.

Cowan, J. (2006b). Introduction. In J. O'Donoghue (Ed.), *Technology Supported Learning and Teaching*, (pp. 1-14). Hershey, PA: Information Science Publishing.

Cowan, J. (2007). Quoted at p. 62 in E. J. Burge, (Ed.), *Flexible Higher Education*. Maidenhead, UK: Open University Press.

Cowan, J., Morton, J., & Bolton, A. (1973). An experimental learning unit for structural engineering studies. *The Structural Engineer, 51*(9), 337–339.

Crawford, C. M. (2001). Developing webs of significance through communications: Appropriate interactive activities for distributed learning environments. [CWIS]. *Campus-Wide Information Systems Journal, 18*(2), 68–72. doi:10.1108/10650740110386675

Crawford, C. M. (2003). Emerging learning environments: Enhancing the online community. *Academic Exchange Quarterly, 7*(4), 131–135.

Crawford, C. M. (2004). Nonlinear instructional design model: Eternal, synergistic design and development. *British Journal of Educational Technology, 35*(4), 413–420. doi:10.1111/j.0007-1013.2004.00400.x

Crawford, C. M. (2005). Towards digital equity within the learning environment: Overcoming the Digital Divide through openly accessible Internet-based tools. *The International Journal of Learning (v 12).*

Crawford, C. M. (2007). Developing multimedia architectural support within online learning environments: Reinventing modalities of meaning as society moves from the Information Age towards the Conceptual Age within the Knowledge Economy. *The International Journal of the Humanities, 2*(5), 37–46.

Crawford, C. M., & Gannon Cook, R. (2008). Building autonomous and dynamic communities of learning within distance learning environments: Focusing upon making connections, knowledge creation and practice communities. *The International Journal of Technology . Knowledge and Society, 4*(4), 47–58.

Cuban, L. (2001). *Oversold and Underused: Computers in the Classroom*. Cambridge, MA: Harvard University Press

Dahlberg, L. (2001). Computer-mediated communication and the public sphere: A critical analysis, journal of computer-mediated communication.

Dalziel, J. (2002) Reflections on the COLLIS (Collaborative Online Learning and Information Systems)

demonstrator project and the "learning object lifecycle". In A. Williamson, C. Gunn & T. Clear (Eds.), *Winds of change in the sea of learning, Proceedings of ASCILITE 2002 Annual Conference*. Retrieved 23 February 2009 from http://www.unitec.ac.nz/ascilite/proceedings/papers/207.pdf

Daradoumis, T., Xhafa, F., & Juan, A. (2006). A framework for assessing self, peer and group performance in e-learning. In *Self, peer, and group assessment in e-learning* (pp. 279-294). Hershey, PA: Idea Group Press.

Davies, C. (2008). Teachers denied a say in school design. *The Observer*, September 21.

Day, R. J. F. (2004). From hegemony to affinity: The political logic of the newest social movements. *Cultural Studies*, *18*(5), 716–748. doi:10.1080/0950238042000260360

Del Moral, M. E., & Villalustre, L. (2008). Wikis and collaborative education by means of Webquest in higher education. *Revista Latinoamericana de Tecnologia Educativa*, *7*(1), 73–83.

Dell, L. A. B. (2005). Connecting K-12 Schools in Higher Education. In C. Howard, J.V. Boettcher, L. Justice, K. Schenk, P. L. Rogers, & G.A. Berg, (eds.), *Encyclopedia of Distance Learning* (pp. 374-378). Hershey, PA: Idea Group Reference

Delone, W. H., & Mclean, E. R. (1992). Information system success: The quest for the dependent variable. *Information Systems Research*, *3*(1), 61–95. doi:10.1287/isre.3.1.60

Delone, W. H., & Mclean, E. R. (2003). The Delone and Mclean information system success: A ten years update. *Journal of Management Information Systems*, *19*(4), 30–36.

Dennen, V. P. (2005). From message posting to learning dialogues: Factors affecting learner participation in asynchronous discussion. *Distance Education*, *26*(1), 127–148. doi:10.1080/01587910500081376

Department for Education and Skills. (2005a). *White Paper: Higher standards, better schools for all*. London: DfES.

Department for Education and Skills. (2005b). *Secondary National Strategy for school improvement: pedagogy and practice, teaching and learning in secondary schools*. London: DfES.

Department for Education and Skills. (2005c). *Secondary National Strategy for school improvement: leading and coordinating CPD in secondary schools*. London: DfES.

Design-Based Research Collective. (2003). Design-based research: An emerging paradigm for educational inquiry. *Educational Researcher*, *32*(1), 5–8. doi:10.3102/0013189X032001005

Dewey, J. (1916). *Democracy and education*. New York: The Free Press.

DfEE (Department for Education and Employment). (1998). *Teaching: high status, high standards. Requirements for courses of initial teacher training, Circular 4/98*. London: DfEE.

DfES (Department for Education and Skills). (2002). *Qualifying to teach: professional standards for Qualified Teacher Status and requirements for initial teacher training*. London: DfES.

Diaz, D. P. (2002). Online drop rates revisited. *The Technology Source*. Retrieved December 4, 2008, from http://technologysource.org/article/online_drop_rates_revisited/

Dickinson, A. Gordon, P. & Lee, P. (2001). *Raising standards in history education*, London: Woburn Press.

DiMaggio, P., & Hargittai, E. (2001). *From the 'Digital Divide' to 'Digital Inequality': Studying Internet Use as Penetration Increases*. Princeton, NJ: Woodrow Wilson School, Princeton University, Center for the Arts and Cultural Policy Studies.

Division of Instructional Innovation and Assessment. (2008). *Criteria for Evaluating Online Courses*. The University of Texas at Austin. Retrieved March 5, 2009, from http://www.utexas.edu/academic/cit/howto/resources/assessment/evallearning.html

Doll, W. J., & Torkzadeh, G. (1989). The measurement of end-user computing satisfaction. *MIS Quarterly, 12*(2), 259–274. doi:10.2307/248851

Donegan, M. (2000). *Voice Recognition Technology in Education: Factors for Success.* Oxford, UK: Ace Centre.

Doo, M. Y. (2006). A problem in online interpersonal skills training: do learners practice skills? *Open Learning, 21*(3), 263–272. doi:10.1080/02680510600953252

Dorniden, A. (2005). K-12 Schools and Online Learning. In C. Howard, J.V. Boettcher, L. Justice, K. Schenk, P. L. Rogers, & G.A. Berg, (eds.), *Encyclopedia of Distance Learning,* (pp. 1182-11880. Hershey, PA: Idea Group Reference.

Dougiamas, M., & Taylor, P. (2003). Moodle: Using Learning Communities to Create an Open Source Course Management System. In D. Lassner & C. McNaught (Eds.), *Proceedings of World Conference on Educational Multimedia, Hypermedia and Telecommunications 2003* (pp. 171-178). Chesapeake, VA: AACE.

Doyle, J., Radzicki, M., & Trees, W. (2008). Measuring change in mental models of complex systems. In H. Qudrat-Ullah, J. M. Spector, and P. I. Davidsen, (Eds.), *Complex Decision Making: Theory and Practice.* (269-294). New York: Springer-Verlag Press.

DPT. (2006). *Knowledge Society Strategy.* Retrieved October 11, 2008, from http://www.bilgitoplumu.gov.tr/btstrateji/Strateji_Belgesi.pdf

Drury, D. H. (1998). A hierarchical structural model of information systems success. *Infor,* (3). Retrieved January 15, 2008 http://findarticles.com/p/articles/mi_qa3661/is_199802

Dupin-Bryant, P. A. (2004, December). Pre-entry variables related to retention in online distance education. *American Journal of Distance Education, 18*(4), 199–206. doi:10.1207/s15389286ajde1804_2

Dutton, W. H., Cheong, P. H. & Park, N. (2004b). The social shaping of a virtual learning environment: The case of a university-wide course management system. *Electronic Journal of e-Learning, 2*(1), 69-80.

Dutton, W. H., Cheong, P. H., & Park, N. (2004a). An ecology of constraints on e-learning in higher education: The case of a virtual learning environment. *Prometheus, 22*(2), 131–149. doi:10.1080/0810902042000218337

Earwaker, J. (1992). *Helping and Supporting Students.* Buckingham, UK: SRHE and Open University Press.

Ecclestone, K. (2004). Learning or Therapy? The Demoralization of Higher Education. *British Journal of Educational Studies, 52*(2). doi:10.1111/j.1467-8527.2004.00258.x

Eco, U. (1996). *From Internet to Gutenberg. A lecture presented at the Italian Academy for Advanced Studies in America.* Retrieved on January 2, 2008 from http://www.hf.ntnu.no/anv/Finnbo/tekster/Eco/Internet.htm

Edelman, P., Holzer, H. J., & Offner, P. (2006). *Reconnecting disadvantaged young men.* Washington, DC: Urban Institute Press.

Education Review Office. (2005). *E-learning in Secondary Schools.* Wellington, New Zealand: Education Review Office.

Elliott, J. (1991). *Action research for educational change.* Buckingham, UK: Open University Press.

Elliott, J. (2001). Characteristics of performative cultures: their central paradoxes and limitations as resources for educational reform. In D. Gleeson & C. Husbands (Eds.) *The Performing School,* (pp. 192-209). RoutledgeFalmer: London and New York.

Engeström, Y. (1987). *Learning by expanding: An activity-theoretical approach to developmental research.* Helsinki, Finland: Orienta-Konsultit.

Engeström, Y. (2002). *The horizontal dimension of expansive learning: Weaving a texture of cognitive trails in the terrain of health care in Helsinki.* New Challenges to Research on Learning, Helsinki: University of Helsinki.

Ertl, H., & Plante, J. (2004). *Connectivity and Learning in Canada's Schools.* Ottawa, Canada: Statistics Canada, Government of Canada.

Ess, C., & Sudweeks, F. (2005). Culture and computer-mediated communication: Toward new understandings. *Journal of Computer-Mediated Communication, 11*(1), 179–191. doi:10.1111/j.1083-6101.2006.tb00309.x

Eurostat. (2009). Retrieved from http://epp.eurostat.ec.europa.eu/portal/page/portal/eurostat/home

Eysenck, M. W., & Piper, D. W. (1987). In J. T. E. Richardson, M. W. Eysenck, & D. W. Piper, (Eds), *Student Learning* (p. 215). Milton Keynes, UK: Society for Research into Higher Education and Open University Press.

Faulin, J., Juan, A., Fonseca, P., Pla, L. M., & Rodriguez, S. V. (2009). Learning operations research online: benefits, challenges, and experiences. *International Journal of Simulation and Process Modelling. Special Issue on Modelling and Simulation Education, 5*(1), 42–53.

Fayol, H. (1960). General principles of management. In H.F. Merrill (Ed.), *Classics in management* (pp. 216-224). New York: American Management Association.

Federman, M. (2005). *Why Johnny and Janey can't read, and why Mr. and Ms. Smith can't teach: The challenge of multiple media literacies in a tumultuous time.* Paper presented at the University of Toronto Senior Alumni Association Lecture Series, Toronto, Canada, November 2005.

Federman, M. (2006a). *Online1141/EntryClass.* Retrieved on December 3, 2008 from http://online1141.pbwiki.com/EntryClass

Federman, M. (2006b). *What is the (next) message: Reflections of an adult educator - part 5.* Retrieved December 3, 2008 from http://whatisthemessage.blogspot.com/2006/10/reflections-of-adult-educator-part-5.html

Federman, M. (2006c). *Online1141/ExitClass.* Retrieved on December 3, 2008 from http://online1141.pbwiki.com/EntryClass

Feenberg, A. (1989). The written world: On the theory and practice of computer conferencing. In R. Mason & A. Kaye (Eds.), *Mindweave: Communication, computers, and distance education.* Toronto: Pergamon Press.

Feldman, K. A. (1976). The superior college teacher from the student's view. *Research in Higher Education, 5*, 243–288. doi:10.1007/BF00991967

Finn, J. D. (1989). Withdrawing from school. *Review of Educational Research, 59*(2), 117–142.

Flowerdew, J., & Miller, L. (1996). Lectures in a second language: Notes towards a cultural grammar. *English for Specific Purposes, 15*(2), 121–140. doi:10.1016/0889-4906(96)00001-4

Flowerdew, L. (1998). A cultural perspective on group work. *ELT Journal, 52*(4), 323–329. doi:10.1093/elt/52.4.323

Flowerdew, L. (n.d.). *Methodological Issues in Teaching English in Hong Kong.* Retrieved May 27, 2004 from http://ettu18.edu.polyu.edu.hk/abstarcts/10.html

Foucault, M. (1980). *Power/Knowledge.* New York: Pantheon Books.

Francis, H., & Cowan, J. (2007). Fostering an action-research dynamic amongst student practitioners. *Journal of European Industrial Training, 32*(5), 336–346. doi:10.1108/03090590810877067

Frand, J. (2000 September/October). The information age mindset: Changes in students and implications for higher education. *Educause*, 15-24. Retrieved on November 28, 2008 from http://net.educause.edu/apps/er/erm00/articles005/erm0051.pdf

Fry, H., Ketteridge, S., & Marshall, S. (2000). *A Handbook for Teaching and Learning in Higher Education.* London: Kogan Page.

Fung, Y. H. (2004). Collaborative online learning: interaction patterns and limiting factors. *Open Learning: the Journal of Open and Distance Learning, 19*(2), 135 – 149. Retrieved Oct. 13, 2007 from http://dx.doi.org/10.1080/0268051042000224743

Furey, D. (2008). From radio broadcasts to virtual reality: A case study of distance education in Hermitage Bay Schools. Paper presented May 2, 2008 at Symposium 2008, *From the Rhetoric to the Reality, Fifty Years of Educational Change in Newfoundland and Labrador.*

St. John's, Canada: Memorial University of Newfoundland.

Furey, D., & Murphy, E. (2005). An introduction to K-12 virtual schools and schooling: Factors enabling growth and impacts on traditional systems, schools and roles. *The Morning Watch: Educational and Social Analysis, 32*, 3–4.

Gaceta, U. A. B. C. (2007). Reconocen a UABC por su calidad certificada. Gaceta UABC, num. 195, 6 de octubre de 2007. Available from http://www.uabc.mx/gaceta/gacetas/g195.pdf

Gallivan, P., Hong, Q., Jordan, L., Li, E., Mathew, G., Mulyani, Y., et al. (2002). VoiceXML Absentee System. In *Proceedings of MASPLAS'02*. The Mid-Atlantic Student Workshop on Programming Languages and Systems, Pace University, USA.

Galusha, J. M. (1997). Barriers to Learning in Distance Education. *Interpersonal Computing and Technology: An Electronic Journal for the 21st Century, 5*(3-4), 6-14. Retrieved April 28, 2007, from http://www.emoderators.com/ipct-j/1997/n4/galusha.html

Galway, G. (2004). E-Learning as a means of promoting educational equity in rural and remote communities: A Canadian case study. Paper presented at the *2003 OECD Workshop: Promoting Equity Through ICT in Education*, Budapest (pp. 70-77).

Ganesh, S., Zoller, H., & Cheney, G. (2005). Transforming resistance, broadening our boundaries: Critical organizational communication meets globalization from below. *Communication Monographs, 72*(2), 169–191. doi:10.1080/03637750500111872

Gannon Cook, R., & Crawford, C. (2004). From silos to communities: Addressing electronic isolation through interactivities. *Society for Information Technology and Teacher Education International Conference 2004*(1), 445-452. [Online]. Available: http://dl.aace.org/14315

Gannon-Cook, R., & Crawford, C. (2006). What can cave walls teach us? In A. Edmondson, (Ed.), *Globalized E-Learning Cultural Challenges*. Minneapolis, MN: eWorld Learning.

Gannon-Cook, R., & Crawford, C. M. (2001). Metaphorical representation within a distributed learning environment. In J.D. Price, D.A. Willis, N. Davis & J. Willis (Eds.), *SITE 2001 Annual - Society for Information Technology and Teacher Education* (pp. 1086-1088). Charlottesville, VA: Association for the Advancement of Computing in Education (AACE).

Garcia, T. (1995). The role of motivational strategies in self-regulated learning. In P.R. Pintrich (Ed.), *New directions for college teaching and learning: Self-regulated learning in the college classroom* (pp. 29-42). San Francisco: Jossey-Bass.

Gardner, H. (1993). *Frames of mind: The theory of multiple intelligences*. New York: Basic Books

Garity, E. J., Glassberg, B., Kim, Y. J., Sanders, L., & Shin, S. K. (2005). An experimental investigation of web-based information system success in the context of electronic commerce. *Decision Support Systems, 39*(3), 485–503. doi:10.1016/j.dss.2004.06.015

Garrison, D. R., & Shale, D. (1990). A new framework and perspective. In D. R. Garrison & D. Shale (Eds.), *Education at a distance: From issues to practice* (pp. 123-133). Malabar, FL: Robert E. Krieger.

Gask, L. (2005). *Overt and covert barriers to the integration of primary and specialist mental health care*. University of Manchester, Manchester M13 9DL, UK: National Primary Care Research and Development Centre. Georgetown College Center for Advanced Study of Assessment (CASA). (n.d.). *Formative Assessments*. Retrieved on March 20, 2009, from http://www.georgetowncollege.edu/CASA/documents/FomativeAssessment.pdf

Gatian, A. W. (1994). Is user satisfaction a valid measure of system effectiveness. *Information & Management, 26*(3), 119–131. doi:10.1016/0378-7206(94)90036-1

Gay, G., & Hembrooke, H. (2004). *Activity-centered design: An ecological approach to designing smart tools and usable systems*. Cambridge, MA: MIT Press.

Gay, L. R. (1987). *Educational research: Competencies for analysis and application (3rd ed.)*. New York: Merrill Publishing Company.

George, J. W. (1996). Action Research for Quality Development in Rural Education in Scotland. *Journal of Research in Rural Education, 12*(2), 76–82.

George, J. W., Cowan, J., Hewitt, L. A., & Cannell, P. (2004). Failure dances to the tune of insecurity: affective issues in the assessment and evaluation of access learning. *Journal of Access Policy and Practice, 1*(2), 119–133.

Gibbons, A., & Fearweatther, P. (2000).Computer-based instruction. In T. S. Fletcher, J.(eds.), *Training & retraining: A Handbook for Business, Industry, Government, and the Military* (pp. 410-442). New York: Macmillan Reference.

Gibbs, G. (1988). *Learning by Doing: A Guide to Teaching and Learning Methods.* Birmingham, UK: SCED.

Gilbert, S. W. (February 2, 2002). The Beauty of Low Threshold Applications. *Syllabus*. Retrieved November 25, 2008 from campustechnology.com/articles/38981_2/. See also www.tltgroup.org/ltas.htm.

Gill, S. (2008, October). The Comprehension Matrix: A Tool for Designing Comprehension Instruction. [from Professional Development Collection database.]. *The Reading Teacher, 62*(2), 106–113. Retrieved December 1, 2008. doi:10.1598/RT.62.2.2

Glass, J. C. Jr. (1996). Factors affecting learning in order adults. *Educational Gerontology, 22*(4), 359–372. doi:10.1080/0360127960220405

Gorard, S., Smith, E., May, H., Thomas, L., Adnett, N., & Slack, K. (2006). *Review of widening participation research: Addressing the barriers to participation in higher education.* Report to HEFCE by the University of Yorke, HEA and Institute for Access Studies, UK.

Government of Newfoundland and Labrador. (2000). *Supporting Learning: Report on the Ministerial Panel on Educational Delivery in the Classroom.* St John's. New Zealand: NL, Department of Education.

Gow, L. Balla, J. Kember, D. & Hau, K.T. (1996). The Learning Approaches of Chinese People: A Function of Socialization Processes and the Context of Learning. In M.H. Bond, (Ed.), *The Handbook ofChinese Psychology.* Hong Kong, Oxford University Press.

Grace, A. P. (1998). Parameters, pedagogy and possibilities in changing times. In S. M. Scott, B. Spencer, & A. M. Thomas, (Eds.), *Learning for life: Canadian readings in adult education* (pp. 114-24). Toronto, Canada: Thompson Educational Publishing.

Grace, A. P. (2004). *Mediating lifelong learning in Canada in inclusive contexts.* Paper presented at the 3rd International Lifelong Learning Conference, Central Queensland University, Yeppoon, Central Queensland, Australia.

Graves, E. G. (2006). Saving our young black men. *Black Enterprise, 36*(6), 10.

Green, H., & Hannon, C. (2007). *Their Space: Education for a Digital Generation.* London: Demos.

Green, J. (1998). Androgogy: Teaching adults. In B. Hoffman (Ed.), *Encyclopedia of Educational Technology.* Retrieved March 5, 2009, from http://coe.sdsu.edu/eet/Articles/andragogy/start.htm

Greene, J. P., & Winters, M. A. (2006, April). *Leaving boys behind: Public high school graduation rates* (Civic Report No. 48). New York: Manhattan Institute for Policy Research. Retrieved September 10, 2006, from http://www.manhattan-institute.org/pdf/cr_48.pdf

Griffin, P., & Cole, M. (1984). Current activity for the future: The zo-ped. In B. Rogoff & J. V. Wertsch (Eds.), *Children's learning in the zone of proximal development.* San Francisco: Jossey-Bass.

Gurstein, M. (2003). Effective Use: A Community Informatics Strategy beyond the Digital Divide. *First Monday, 8*(12). i2010 European Strategic Plan. (2005). *i2010 Communication.* Retrieved on October 11, 2007 from http://ec.europa.eu/information_society/eeurope/i2010/inclusion/index_en.htm

Hagedorn, L. S. (2005, January) Square pegs: adult students and their "fit" in postsecondary institutions. *Change.* Retrieved December 15, 2008, from http://www.coe.ufl.edu/Leadership/ihe/TRUCCS/Files/Square_pegs_for_Change. pdf

Haggis, T., & Pouget, M. (2002). Trying to be motivated:

Perspectives on learning from younger students accessing higher education. *Teaching in Higher Education, 7*(3), 323–336.

Hall, E. T. (1959). *The silent language*. Garden City, NY: Anchor Press.

Hall, E. T. (1966). *The hidden dimension*. Garden City, NY: Anchor Press.

Hamilton, E. (1969). *Mythology: Timeless tales of gods and heroes*. New York: Penguin Books.

Hannay, L. M., & Ross, J. A. (2001). Internalizing Change Capacity in Secondary Schools Through Organizational Change. *The Alberta Journal of Educational Research, 47*(4), 325–340.

Hara, N. and R. Kling (2002). *IT supports for communities of practice: An empirically-based framework*. Center for Social Informatics, 2003.

Harasim, L. (1989). Online education: A new domain. In R. Mason & A. Kaye (Eds.) *Mindweave: Communication, computers, and distance education*. Toronto: Pergamon Press.

Harasim, L., Calvert, T., & Groeneboer, C. (1996). Virtual-U™: a web-based environment customized to support collaborative learning and knowledge building in post secondary courses, (pp. 120-127). In *Proceedings of the 1996 International Conference on Learning Sciences*, Evanston, Illinois.

Harasim, L., Hiltz, S., Teles, L., & Turoff, M. (1995). *Learning networks: A field guide to teaching and learning online*. London: MIT Press.

Hardin, J., & Ellington, A. (2005). Using Multimedia to Facilitate Software Instruction in an Introductory Modeling Course. *INFORMS Transactions on Education, 5*(2). Retrieved Oct 19, 2009 from http://archive.ite.journal.informs.org/Vol5No2/HardinEllington/

Hargittai, E. (2002). Second-Level Digital Divide: Differences in People's Online Skills. *First Monday, 7*(4), Retrieved Sep.2002 from http://firstmonday.org/issues/issue7_4/hargittai/index.html

Harrison, C., Comber, C., Fisher, T., Haw, K., Lewin, C., Lunzer, E., et al. (2002). *ImpaCT2: The impact of ICT on pupil learning and attainment*. Coventry, UK: BECTa.

Harrison, S. (2003). The use of ICT for history teaching: slow growth, a few green shoots. Findings of HMI inspection 1999-2001. In T. Haydn and C. Counsell (Eds.) *History, ICT and learning in the secondary school* (pp. 38-51). London: RoutledgeFalmer.

Hauben, M., Hauben, R., & Hauben, J. (1994, 1996). *Netizens: On the history and impact of USENET and the internet*. Retrieved December 3, 2008, from http://www.columbia.edu/~rh120/

Hawkes, M., & Halverson, P. (2002). Technology Facilitation in the Rural School: An Analysis of Options. *Journal of Research in Rural Education, 17*(3), 162–170.

Haydn, T. (2004). *The use of ICT in history teaching in England and Wales, 1970-2003*, PhD thesis, University of London, London.

Haydn, T., & Barton, R. (2007). 'First do no harm': factors influencing teachers' ability and willingness to use ICT in their subject teaching. *Computers & Education, 51*(1), 439–447. doi:10.1016/j.compedu.2007.06.001

Haythornthwaite, C. (2002). Building social networks via computer networks: Creating and sustaining distributed learning communities. In K.A. Renninger & W. Shumar, (eds.). *Building virtual communities: Learning and change in cyberspace*, (pp. 159-190). Cambridge, UK: Cambridge University Press.

Haythornthwaite, C. (2006). Facilitating collaboration in online learning. *Journal of Asynchronous Learning Networks, 10*(1), 7–24.

Healey, D., & Stevens, K. (2002). Student Access to Information Technology and Perceptions of Future Opportunities in Two Small Labrador Communities. *Canadian Journal of Learning and Technology / La Revue Canadienne de l'Apprentissage et de la Technologie, 28*(1,Winter), 7-18.

Hersey, P., Blanchard, K. H., & Natemeyer, W. E. (1979). Situational leadership, perception, and the impact of power. *Group & Organization Studies*, *4*(4), 418–428. doi:10.1177/105960117900400404

Hewling, A. (2005). Culture in the online class: Using message analysis to look beyond nationality-based frames of reference. *Journal of Computer-Mediated Communication*, *11*(1), 337–356. doi:10.1111/j.1083-6101.2006.tb00316.x

Heywood, J. (2000). *Assessment in Higher Education.* London: Jessica Kingsley Publishers.

Higgison, C. A. (Ed.). (2002). *Online Tutoring e-book.* Aberdeen, Scotland: Heriot-Watt University and The Robert Gordon University. Retrieved 1st December 2008 from http://otis.scotcit.ac.uk/onlinebook/

Hillman, D. C., Willis, D. J., & Gunawardena, C. N. (1994). Learner-interface interaction in distance education: An extension of contemporary models and strategies for practitioners. *American Journal of Distance Education*, *8*(2), 30–42. doi:10.1080/08923649409526853

Hillman, D., Willis, D. J., & Gunawardena, C. (1994). Learner-interface interaction in distance education: An extension of contemporary models and strategies for practitioners. *American Journal of Distance Education*, *8*(2), 30–42. doi:10.1080/08923649409526853

Hiltz, S. R., & Turoff, M. (2005). Education goes digital: The evolution of online learning and the revolution in higher education. *Communications of the ACM*, *48*(10), 59–65. doi:10.1145/1089107.1089139

Hine, C. (200). *Virtual Ethnography.* London: Sage.

Ho, D. Y. F. (1991). *Cognitive socialization in Confucian heritage cultures.* Paper presented to Workshop on Continuities and Discontinuities in the Cognitive Socialization of Minority Children, US Department of Health and Human Services, Washington, DC, June 29 – July 2.

Ho, D. Y. F. (1996). Filial Piety and Its Psychological Consequences. In M.H. Bond, (Ed.) *The Handbook of Chinese Psychology.* Hong Kong: Oxford University Press.

Ho, I. T. (2001). Are Chinese teachers authoritative? In J. B. Biggs & D. A. Watkins (Eds.), *Teaching the Chinese learner: Psychological and pedagogical perspectives.* (pp. 99-114). Hong Kong: Comparative education research centre, University of Hong Kong.

Hockings, C., Cooke, S., & Bowl, M. (2007). Academic engagement' within a widening participation context – a 3D analysis. *Teaching in Higher Education*, *12*(5-6), 721–733. doi:10.1080/13562510701596323

Hofstede, G. (1980). *Culture's consequences: International differences in work-related values.* Beverly Hill, CA: Sage.

Hofstede, G. (1997). *Cultures and organizations: Software of the mind* (Rev. Ed.). New York: McGraw-Hill.

Hofstede, G., & Bond, M. (1984). Hofstede's culture dimensions: an independent validation using Rokeach's value survey. *Journal of Cross-Cultural Psychology*, *15*, 417–433. doi:10.1177/0022002184015004003

Holmberg, B. (1995). *Theory and practice of distance education.* London: Routledge.

Horrigan, J., & Jones, S. (2008). *When technology fails.* Retrieved Nov. 15, 2008 from: http://www.pewinternet.org/pdfs/PIP_Tech_Failure.pdf

Howell, S. L., Williams, P. B., & Lindsay, N. K. (2003). Thirty-two trends affecting distance education: An informed foundation for strategic planning. *Online Journal of Distance Learning Administration*, *6*(3), 1–18.

Huang, S. J., & Liu, H. F. (2000). Communicative Language Teaching in a Multimedia Language Lab. *The Internet TESL Journal*, *6*(2), February 2000. Retrieved February 19, 2005 from http://iteslj.org/Techniques/Huang-CompLab.htm

Hunt, D. (1995, March 27). Quoted in 'Stemming the decline in science.' *The Guardian.*

Hussein, R., Abdu-Karim, N. S., Mohamed, N., & Ahlan, A. R. (2007). The influence of organizational factors on information system success in e-government agencies in Malaysia. *Electronic Journal of Information Systems in Developing Countries EJISDC*, *29*(1), 1–17.

Huyton, J.L. (2009). Significant Personal Disclosure: exploring the support and development needs of HE tutors engaged in the emotion work associated with supporting students. *Journal of Learning Development in Higher Education,* (1).

Huzzard, T. (2004). Communities of domination? Reconceptualising organisational learning and power. *Journal of Workplace Learning: Employee Counselling Today, 16*(6), 350–361.

Hwang, K. (1987). Face and favour: the Chinese power game. *Journal of Sociology (Melbourne, Vic.), 94*(2), 944–974.

Ikhu-Omoregbe, N. A. (2007). *A Formal Framework for usable Operations Support Systems in E-Health.* Unpublished doctoral dissertation, Covenant University, Nigeria.

Information Highway Advisory Council. (1995). *The Challenge of the Information Highway.* Ottawa, Canada: Industry Canada.

Information Highway Advisory Council. (1997). *Preparing Canada for a Digital World.* Ottawa, Canada: Industry Canada.

Information Technology Faculty Roles & Rewards Action Team (ITFRR). (2000). *Information Technology Promotion, Tenure, and Merit Guidelines.* Retrieved November 25, 2008 from cstl.semo.edu/ITFRR/.

ISO 9241-11. (1998). Ergonomic requirements for office work with visual display terminals (VDTs) - Part 11: Guidance on usability.

Israel, G. D. (2003). *Determining sample size*. Retrieved from February 10, 2007, http://edis.ifas.edu

Ives, B., Olson, M. J., & Baroudi, J. J. (1983). The measurement of user information satisfaction. *Communications of the ACM, 26*(10), 785–793. doi:10.1145/358413.358430

Jakubowicz, P. (2006). *Can Online Learning be Fun?* [Unpublished paper]. Paper presented at the 4th Asia TEFL Conference in Fukuoka, Japan, Aug. 18 – 21.

James, J. (2005). The Global Digital Divide in the Internet: Developed Countries Constructs and Third World Realities. *Journal of Information Science, 31*(2), 114–123. doi:10.1177/0165551505050788

James, R., McInnis, C., & Devlin, M. (2002). *On-line assessment.* Retrieved April 25, 2005, from http://www.adelaide.edu.au/myuni/instruct/planning/assessment.html

Johnson, C. M. (2001). A survey of current research on online communities of practice. *The Internet and Higher Education, 4*(1), 45–60. doi:10.1016/S1096-7516(01)00047-1

Johnson, D. W. Johnson, R.T. & Smith, K.A. (1998). *Active Learning: Cooperation in the college classroom.* Interaction Book Co.

Jonassen, D. H. (1991). Evaluating constructivistic learning. *Educational Technology, 31*(10), 28–33.

Jonassen, D. H., Tessmer, M., & Hannum, W. H. (1999). *Task analysis methods for instructional design.* Mahwah, NJ: Erlbaum.

Joseph, R. (2001). Understanding the Digital Divide. *Prometheus, 19*(4), 333–336. doi:10.1080/08109020110091396

Juan, A., Daradoumis, T., Faulin, J., & Xhafa, F. (2009). SAMOS: A model for monitoring students' and groups' activity in collaborative e-learning. *International Journal of Learning Technology, 4*(1/2), 53–72. doi:10.1504/IJLT.2009.024716

Juan, A., Huertas, M., Steegmann, C., Corcoles, C., & Serrat, C. (2008). Mathematical e-learning: state of the art and experiences at the Open University of Catalonia. *International Journal of Mathematical Education in Science and Technology, 39*(4), 455–471. doi:10.1080/00207390701867497

Jung, C. G. (1964). *Man and his symbols.* New York: Dell Publishing.

Justice Policy Institute. (2002). *Cellblocks or classrooms: The funding of higher education and corrections and its impact on African American men.* Washington, DC: United States Justice Fund. Retrieved September 9, 2006, from http://www.soros.org/initiatives/justice/

articles_publications/publications/cellblocks_20020918/coc.pdf

Kanata, T., & Martin, J. N. (2007). Facilitating dialogues on race and ethnicity with technology: Challenging "Otherness" and promoting a dialogic way of knowing. *Journal of Literacy and Technology*, *8*(2), 1–40.

Kekkonen-Moneta, S. & Moneta, G.B. (2001). E-Learning in Hong Kong. *British Journal of Educational Technology.*

Kelly, B., Phipps, L., & Swift, E. (2004). Developing A Holistic Approach For E-Learning Accessibility. *Canadian Journal of Learning and Technology, 30*(3, Autumn). Retrieved December 1st, 2008 from www.ukoln.ac.uk/web-focus/papers/cjtl-2004/

Kelly, H. F., Ponton, M. K., & Rovai, A. P. (2007). A comparison on student evaluations of teaching between online and face-to-face courses. *The Internet and Higher Education*, *10*(2), 189–101. doi:10.1016/j.iheduc.2007.02.001

Kennedy, P. (2002). Learning cultures and learning styles: myth-understandings about adult (Hong Kong) Chinese learners. *International Journal of Lifelong Education*, *21*(5), 430–445. doi:10.1080/02601370210156745

Kerckhoff, A. C. (1995). Institutional Arrangements and Stratification Processes in Industrial Societies. *Annual Review of Sociology*, *21*, 323–347. doi:10.1146/annurev.so.21.080195.001543

Kılıç, E., & Özdemir, S. (2006). Evaluation of the Distribution and Keeping the Continuity of the IT Classrooms Established in the Scope of Ministry of National Education (MONE) Basic Education Project. *Journal of Turkish Educational Sciences*, *4*(2), 129–140.

Kim, K. J., Liu, S., & Bonk, C. J. (2005). Online MBA students' perceptions of online learning: Benefits, challenges, and suggestions. *Internet and Higher Education, 8*, 335, 344.

Kim, K., & Bonk, C. J. (2002). Cross-cultural comparisons of online collaboration. *Journal of Computer-Mediated Communication*, *8*(1).

King, J. C., & Doerfert, D. L. (1996). *Interaction in the distance education setting.* Retrieved

Klemm, W. R. (2002). *Extending the pedagogy of threaded-topic discussions.* Retrieved December 3, 2008, from http://technologysource.org/article/extending_the_pedagogy_of_threadedtopic_discussions/

Kneser, C., Pilkington, R., & Treasure-Jones, T. (2001). The tutor's role: An investigation of the power of exchange structure analysis to identify different roles in CMC seminars. *International Journal of Artificial Intelligence in Education*, *12*, 63–84.

Knowles, M. S. (1980). *The modern practice of adult education: From pedagogy to andragogy.* Englewood Cliffs: Prentice Hall/Cambridge.

Koss, F. A. (2001). Children falling into the digital divide. *Journal of International Affairs*, *55*(1), 75–90.

Kozma, R. B. (2001). Counterpoint theory of 'learning with media.' In R. E. Clark (Ed.). *Learning from media: Arguments, analysis and evidence*, (pp. 137 – 178). Greenwich, CT: Information Age Publishing Inc.

Kozma, R. B. (2003). Technology and Classroom Practices: An International Study. *Journal of Research on Technology in Education*, *36*(91), 1–14.

Krathwohl, D., Bloom, B. S., & Mersia, D. B. (1964). Taxonomy of Educational Objectives. In *Handbook 2: Affective Domain.* London: Longmans Green.

Krippendorff, K. (2004). *Content Analysis: An Introduction to Its Methodology*, (2nd. Ed.). Thousand Oaks, CA: Sage.

Kruse, K. (2004). *The benefits and drawbacks of e-learning.* Retrieved November 12, 2008, from http://www.e-learningguru.com/articles/art1_3.htm

Lai, E. (1993). Cooperative learning in a Geography class. In J. B. Biggs, & D.A. Watkins, (Eds.), *Learning and teaching in Hong Kong: What is and what might be.* Hong Kong: Educational Papers 17, Faculty of Education, The University of Hong Kong.

Lai, K.-W. (2005). *e-Learning Communities: Teaching and Learning With the Web*. Dunedin, FL: University of Otago Press

Laiken, M. (1998). *The Anatomy of High Performing Teams: A Leader's Handbook (3rd Edition)*. Toronto: University of Toronto Press.

Laiken, M. (2002). Models of organizational learning: paradoxes and best practices in the post-industrial workplace. *Organizational Development Journal, 21*(1, Fall), 8-19.

Laing, C., Chao, K., & Robinson, A. (2005). Managing the expectations of non-traditional students: a process of negotiation. *Journal of Further and Higher Education, 29*, 169–179. doi:10.1080/03098770500103572

Lammers, W., & Murphy, J. (2002). A profile of teaching techniques used in the university classroom. *Active Learning in Higher Education, 3*(1), 54–68. doi:10.1177/1469787402003001005

Lapadat, J. C. (2004). Online teaching: creating text-based environments for collaborative thinking. *The Alberta Journal of Educational Research, 50*(3), 236–251.

Lave, J., & Wenger, E. (1991). *Situated Learning: Legitimate Peripheral Participation*. Cambridge, UK: Cambridge University Press.

Law, K. K. M. (2004). *Research on finding effective ways to enhance the interest of using CU forum of Chinese speaking students in CUHK*. (Unpublished research report – The Chinese University of Hong Kong).

Lazarus, W., & Mora, F. (2000). *Online content for low-income and underserved Americans: The digital divide's new frontier*. CA: The Children's Partnership.

Leaderer, A. L., & Mendelow, A. L. (1988). Convincing top management strategic potential of information systems. *MIS Quarterly*, (Dec): 525–534. doi:10.2307/249127

Lears, T. J. J. (1985). The Concept of Cultural Hegemony: Problems and Possibilities. *The American Historical Review, 90*(3), 567–593. doi:10.2307/1860957

Leask, M. (2002). *Training for teachers and school librarians in the use of ICT*. London: TTA.

Lenhart, A., Hitlin, P., & Madden, M. (2005). *Teens and Technology: Youth Are Leading the Transition to a Fully Wired and Mobile Nation*. Washington, DC: Pew Internet and American Life Project.

Leon, L., Seal, K., & Przasnyski, Z. (2006). Captivate Your Students' Minds: Developing Interactive Tutorials to Support the Teaching of Spreadsheet Modeling Skills. *INFORMS Transactions on Education, 7*(1). Retrieved Oct 19, 2009 from http://archive.ite.journal.informs.org/Vol7No1/LeonSealPrzasnyski/

Leon, L.; Przasnyski, Z. & Seal, K. (2008). Teaching OR/MS to Net-Gens: A Paradigm Gap? *OR/MS Today, 35*(5), 10-11.

Lera, F., Juan, A., Faulin, J., & Cavaller, V. (2009): Monitoring Students' Activity and Performance in Online Higher Education: A European Perspective. In *Monitoring and Assessment in Online Collaborative Environments: Emergent Computational Technologies for E-Learning Support* (pp. 132-148). IGI Global, Hershey, Pennsylvania, USA.

Levin, T., & Wadmany, R. (2005). Changes in Educational Beliefs and Classroom Practices of Teachers and Students in Rich Technology-Based Classrooms. *Technology, Pedagogy and Education, 14*(93), 281–307. doi:10.1080/14759390500200208

Levy, J. (1998). *Some Observations on Distance Learning*. Ithaca, NY: Office of Distance Learning, Cornell University.

Lewin, K. (1951). *Field theory in social science: Selected theoretical papers*. D. Cartright (Ed). New York: Harper and Row.

Li, J. Z., & Bratt, S. E. (2004). *Activity as tool for analyzing asynchronous learning networks*. (. *LNCS, 3143*, 19–26.

Lincoln, Y., & Guba, E. (1985). *Naturalistic Inquiry*. Beverly Hills, CA: Sage.

Linden Research, Inc. (2008). *Second life*. Retrieved on December 2, 2008, from http://secondlife.com/

Lipman, M. (1991). *Thinking in education.* Cambridge, UK: Cambridge University Press.

Liu, N., & Littlewood, W. (1997). Why do many students appear reluctant to participate in classroom learning discourse? *System, 25*(3), 371–384. doi:10.1016/S0346-251X(97)00029-8

Livingston, J. (1997). *Metacognition: An overview.* Retrieved December 1st, 2008 from http://webu2.upmf-grenoble.fr/sciedu/grangeat/EnsI/Meta/MetaLivingstonOverview.pdf

Lowry, R. (2005). Computer-aided self assessment -an effective tool. *Chemistry Education Research and Practice, 6*(4), 198–203.

Luna Serrano, J., Pérez Fragoso, C., Morita Sandoval, O., & Ruiz Rodríguez, A. (2002). Evaluación de una herramienta computacional para la educación en línea. VIII Congreso Internacional de Informática en la Educación, La Habana, Cuba.

Luyt, B. (2004). Who Benefits from the Digital Divide? *First Monday, 9*(8). Retrieved from http://www.firstmonday.org/issues/issue9_8/

Lynch, M. (2003). Technology Education Through Online Mentoring: The Impact of an Online Course on Teachers' Beliefs and Attitudes about Being Mentors. *Dissertation Abstracts International, 63*(12), 4283A, (UMI No. 3075282).

MacDonald, P. J. (1997). In D. Boud & G. Feletti, (Eds.), *The Challenge of Problem- based Learning,* (2nd ed., pp. 93-102). London: Kogan Page.

Mandernach, B. J., Gonzales, R. M., & Garrett, A. L. (2006). *Journal of Online Learning and Teaching, 2*(4). Retrieved February 23, 2009 from http://jolt.merlot.org/vol4nol/Vol2_No4.htm

Many, J. (1991). The name assigned to the document by the author. This field may also contain sub-titles, series names, and report numbers.The Effects of Stance and Age Level on Children's Literary Responses. *Journal of Reading Behavior, 23*(1), 61–85.

Martin, J. N., & Davis, O. I. (2001). Conceptual foundations for teaching about Whiteness in Intercultural Communication courses. *Communication Quarterly, 50,* 298–313.

Martin, J. N., & Nakayama, T. K. (1999). Thinking dialectically about culture and communication. *Communication Theory, 9,* 1–25. doi:10.1111/j.1468-2885.1999.tb00160.x

Martyn, M. (2003). The hybrid online model: good practice. *Educause Quarterly, 1,* 18 – 23. McNeely, E. (n.d.). Computer-assisted instruction and the older-adult learner. *Educational Gerontology, 17*(3), 229–237.

MarylandOnline, Inc. (2005). *APPENDIX 1: References. Quality Matters Inter-Institutional Quality Assurance in Online Learning.* Retrieved March 5, 2009, from http://www.qualitymatters.org/Rubric.htm

MarylandOnline, Inc. (2006). The Quality Matters Rubric. *Quality Matters Inter-Institutional Quality Assurance in Online Learning.* Retrieved November 25, 2008, from http://qualitymatters.org/Rubric.htm

Mason, J., & Hart, G. (1997). *Effective use of asynchronous virtual learning communities.* Retrieved Aug. 1, 2007 from http://arch.usyd.edu.au/kcdc/conferences/VC97/papers/mason.html

McCluskey, F. (2008, July 22). *Chronicle article focuses on distance-ed academic integrity.* Message posted to Sloan-C Electronic mailing list.

McConnell, D. (2000). *Implementing Computer Supported Cooperative Learning,* (2nd Ed). London: Kogan Page.

McCormick, B., & Davenport, D. (2004). *Shepherd leadership; wisdom for leaders from Psalm 23.* (Y.C. Geh, trans.). Taipei: Apocalypse Press.

MCH Training Trantee Network. (2007). *Leadership training.* Retrieved on January 5, 2007, from http://leadership.mchtraining.net/index.php?module=pagemaster&PAGE_user_op=view_page&PAGE_id=25

McKenney, S., & van den Akker, J. (2005). Computer-based support for curriculum designers: A case of developmental research. *Educational Technology Research and Development, 53*(2), 41–66. doi:10.1007/BF02504865

McLuhan, M. (1964a). *The medium is the messenger.* New York: Signet Books.

McLuhan, M. (1964b). *Understanding media: The extension of man.* New York: Signet Books.

McLuhan, M., & Fiore, Q. (1967). *The medium is the massage.* New York: Touchstone.

McManus, T. F. (1995). *Testing learner self-regulation in a web based learning environment.* Retrieved from the World Wide Web on January 1, 2003: http://www.edb.utexas.edu/mmresearch/Students96/McManus/srltest.html

McPeck, J. (1990). *Teaching critical thinking.* New York: Routledge.

Mehan, H. (1979). *Learning lessons: The social organization of classroom instruction.* Cambridge, MA: Harvard University Press.

Mehan, H. (1998). The Study of Social Interaction in Educational Settings: Accomplishments and Unresolved Issues. *Human Development, 41,* 245–269. doi:10.1159/000022586

Mentkowski, M., et al. (2000). *Learning that Lasts.* San Francisco: Jossey Bass

Merryfield, M. (2003). Like a veil: Cross-cultural experiential learning online. *Contemporary Issues in Technology & Teacher Education, 3*(2), 146–171.

Merryfield, M. M. (2001). The paradoxes of teaching a multicultural education course online. *Journal of Teacher Education, 52*(4), 283–299. doi:10.1177/0022487101052004003

Mestre, L. (2006). Accommodating diverse learning styles in an online environment, (Guest Editorial). *Reference and User Services Quarterly, 46*(2), 27–32.

Miles, M. B. (1970). *Learning to work in groups: a program guide for educational leaders.* New York: Teachers College Press, Teachers College, Columbia University.

Miller, D., & Glover, D. (2002). The interactive whiteboard as a force for pedagogic change: The experience of five elementary schools in an English Education Authority. *Information Technology in Childhood Education Annual,* (1): 5–19.

Millett, A. (1998)... *Quoted in The Times Educational,* (May), 22.

Ministry of Education. (1998). *Interactive Education–An Information and Communication Technologies Strategy for Schools.* Wellington, New Zealand.

Miranda, S. M., & Saunders, C. S. (2003, March). The social construction of meaning: An alternative perspective on information sharing. *Information Systems Research, 14*(1), 87-106. Retrieved on April 14, 2008, from http://faculty-staff.ou.edu/M/Shaila.M.Miranda-1/M&S03.pdf

Mitchell, W. J. (1999). Equitable access to the online world. In D. A. Schon, B. Sanyal, & W. J. Mitchell (Eds.), *High technology and low-income communities: Prospects for the positive use of advanced information technology* (pp. 151-162). Cambridge, MA: MIT Press.

Mondada, L. (2006). Participants' online analysis and multimodal practices: Projecting the end of the turn and the closing of the sequence. *Discourse Studies, 8*(1), 117–129. doi:10.1177/1461445606059561

MONE. (2001). *National Education at the Beginning of 2002.* Republic of Turkey Ministry of National Education Research, Planning and Coordination Board (RPCB). Retrieved October 10, 2008, from http://www.meb.gov.tr/Stats/apk2002ing/apage00_0.htm

MONE. (2004a). *Basic Education Program ICT Politics Report.* Ankara, Turkey: Ministry of National Education.

MONE. (2004b). *Impact Research Report of ICT Classrooms.* Ankara, Turkey: Ministry of National Education.

Monge, P. R., & Contractor, N. S. (2003). *Theories of Communication Networks*. Oxford, UK: Oxford University Press. Retrieved Aug. 21, 2007 from http://hyperion.math.upatras.gr/commorg/nosh/HOCNets.html

Moodle.org. (2008). *Moodle – A free open source course management system for online learning.* Retrieved on December 4, 2008, from http://moodle.org/

Moore, M. (1989). Three types of interaction. *American Journal of Distance Education, 3*(2), 1–7. doi:10.1080/08923648909526659

Moore, M. (1994). Administrative barriers to adoption of distance education. *American Journal of Distance Education, 8*, 1–4.

Moore, M. G. (1991). Distance education theory. *American Journal of Distance Education, 5*(3), 1–6. doi:10.1080/08923649109526758

Moore, M., & Kearsley, G. (2005). *Distance education: A systems view,* (2nd ed.). New York: Thomson/Wadsworth.

Morrison, K. R. B. (2006). *Promoting the Conditions for Effective Learning in an e-Learning Environment.*

MSNBC.com. (1 December 2008). *It's official: U.S. is in recession: Economy began shrinking in December 2007, panel declares.* Retrieved December 1, 2008, from http://www.msnbc.msn.com/id/27999557

Mumby, D. K. (1997). The problem of hegemony: Rereading Gramsci for organizational communication studies. *Western Journal of Communication, 61*(4), 343–375.

Murphy, D. (1987). Offshore education: a Hong Kong perspective. *Australian Universities Review, 30*(2), 43–44.

Nagy, A. (2005). The Impact of E-Learning. In P. A. Bruck, A. Buchholz, Z. Karssen, A. Zerfass (Eds.), *E-Content: Technologies and Perspectives for the European Market,* (pp.79-96). Berlin: Springer-Verlag.

Napierkowski, H. (2001). *Collaborative learning and sense of audience in two computer-mediated discourse communities.* Paper presented at the 52nd Annual Meeting of the Conference on College Composition and Communication, Denver, CO, March 14-17.

Nardi, B. (1996). Studying context: A comparison of activity theory, situated action models, and distributed cognition. *Context and Consciousness.*

Nardi, B. (Ed.). (1996). *Context and Consciousness: Activity Theory and Human-Computer Interaction.* Cambridge, MA: MIT Press.

National Aeronautics and Space Administration. (2008). *NASA – Johnson Space Center website.* Retrieved on December 1, 2008, from http://www.nasa.gov/centers/johnson/home/index.html

National College of School Leadership. (2004). *Networked Learning Communities: learning about learning networks.* Cranfield, UK: NCSL.

National Education Association. (2005). *Focus on Blacks.* Retrieved September 18, 2006, from http://www.nea.org/assets/docs/mf_blacksfocus05.pdf

National Foundation for Educational Research (NFER). (2000). *Who would be a teacher? A review of the factors motivating and demotivating prospective and practising teachers.* Slough, UK: NFER.

Natriello, G. (2001). Bridging the second digital divide: What can sociologists of education contribute? *Sociology of Education, 74*(3), 260–265. doi:10.2307/2673278

Nelson, B., Ketelhut, D. J., Clarke, J., Bowman, C., & Dede, C. (2005). Design-based research strategies for developing a scientific inquiry curriculum in a multi-user virtual environment. *Educational Technology, 45*(1), 21–28.

Nelson, G. (1995). Cultural differences in learning styles. In J. Reid (Ed.), *Learning Styles in the ESL/EFL Classroom.* Boston: Heinle and Heinle.

Nelson, G. (2004). Building capacity in community health action research: Towards a praxis framework for graduate education. *Action Research, 2,* 289–408. doi:10.1177/1476750304047982

New Zealand Ministry of Education. (2002). *ICT Draft Strategy for Schools - 2002-2004.*

Newkirk, T. (2006, September). Media and literacy: What's good? *Educational Leadership, 64*(1), 62–66.

Nichol, J., & Watson, K. (2003). Editorial: Rhetoric and reality - the present and future of ICT in education. *British Journal of Educational Technology, 34*(1), 1–6.

Nisbet, P. D., & Wilson, A. (2002). *Introducing Speech Recognition in Schools: using IBM ViaVoice.* Pub. CALL Centre, University of Edinburgh, Scotland.

Noss, R., & Pachler, N. (1999). The challenge of new technologies: doing old things in a new way or doing new things? In P. Mortimore (Ed.) *Understanding pedagogy and its impact on learning,* (pp. 195-211). London: Paul Chapman.

Notess, G. (2002). *The Blogrealm: News sources, searching with Daypop, and content management.* Retrieved Aug. 30, 2007 from http://www.onlinemag.net/sep02/OnTheNet.htm

Nunan, D. (1991). Communicative tasks and the language curriculum. *TESOL Quarterly, 25*(2), 279–295. doi:10.2307/3587464

Nunan, D. (2004). *Task-based Language Teaching.* Cambridge, UK: Cambridge University Press.

Nunan, D. (2005) Important Tasks of English Education: Asia-wide and Beyond. *The Asian EFL Journal, 7*(3). Retrieved June 14, 2006 from http://www.asian-efl-journal.com/September_05_dn.php

Nunnally, J. C. (1978). *Psychometric Theory,* (2nd ed.), New York: McGraw- Hill.

Oates, J. (2007). *Researching information system and computing.* Los Angeles: Sage Publications.

Oblinger, D. G. (October 20, 2004). *Educating the Net Generation.* Featured Speaker Address at the Educause 2004 Conference, Denver, CO.

Odabaşı, F. (1998). Computer Assisted Instruction. In Y. Hoşcan (Ed.), *Computer.* Eskişehir. Anadolu University, Open Education Faculty Press.

OECD. (2001). *E-learning The Partnership Challenge.* Centre for Educational Research and Innovation. OECD Publishing.

OECD-Organisation for Economic Co-operation and Development. (2001). *Understanding the Digital Divide.* Retrieved January 17, 2005 from http://www.oecd.org/dataoecd/38/57/1888451.pdf

Ofsted (Office for Standards in Education). (2002). ICT *in schools: the impact of government initiatives.* London: Ofsted.

Ogborn, J. (2000). Shaping the future: revitalising physics Education. In A. Morris (Ed.), *Post-16 physics for tomorrow.* Bristol: Institute of Physics.

Ohler, J. (2006). The World of Digital Storytelling. *Educational Leadership, 63*(4), 44–47.

Ohler, J. (2008). *Digital storytelling in the classroom: New media pathways to literacy, learning, and creativity.* Thousand Oaks, CA: Corwin Press.

Olaniran, B. A. (2001) The effects of computer-mediated communication on transculturalism. In V. H. Milhouse, M. K. Asante, & P. O. Nwosu (Eds.) *Transcultural realities: Interdisciplinary perspectives on cross-cultural relations* (pp.83-105). Thousand Oaks, CA: Sage.

Olaniran, B. A., & Agnello, M. F. (2008). Globalization, educational hegemony, and higher education. *Multicultural Education & Technology Journal,* 2(2), 68–86. doi:10.1108/17504970810883351

Oliver, R. L. (1993). Cognitive, affective, and attribute bases of the satisfaction response. *The Journal of Consumer Research, 20*(3), 418–430. doi:10.1086/209358

Özdemir, S., & Kılıç, E. (2007). Integrating information and communication technologies in the Turkish primary school system. [BJET]. *British Journal of Educational Technology, 38*(5), 907–916. doi:10.1111/j.1467-8535.2006.00678.x

Paas, F., Renkl, A., & Sweller, J. (2003). Cognitive load theory and instructional design: Recent developments. *Educational Psychologist, 38*(1), 1-4. Retrieved on October 7, 2008, from http://books.

google.com/books?hl=en&id=DCcSZ7GC8lAC&dq=cognitive+load&printsec=frontcover&source=web&ots=3rCu__Aos4&sig=yzgoZmi4aTKHJD_bnd-EFUOlmrQ&sa=X&oi=book_result&resnum=3&ct=result#PPA1,M1

Pallof, R., & Pratt, K. (1999). *Building Learning Communities in Cyberspace: Effective Strategies for the Online Classroom*. San Francisco, CA: Jossey-Bass Publishers.

Pallof, R., & Pratt, K. (2001). *Lessons from the Cyberspace Classroom: The Realities of Online Teaching*. San Francisco, CA: Jossey-Bass Publishers.

Palm, J. Colombet, Sicotte, C. & Degoulet, P. (2006). Determinants of user satisfaction with clinical information system. *Amia Annu Symposium Proceedings,* (pp. 614- 618).

Pang, P. M. N., & Hung, D. W. L. (2001). Activity theory as a framework for analyzing CBT and E-Learning environments. *Educational Technology, 41*(4), 36–42.

Pask, G. (1975). *Conversation, Cognition, and Learning*. New York: Elsevier.

Paul, D. (2003). Speech recognition for students with disabilities. *Nisbet, CALL Centre, University of Edinburgh,* (pp. 1 – 6). Retrieved January 2009 from http://callcentre.education.ed.ac.uk/downloads/speech_recognition/hitmesse2003.pdf

Peat, M., & Franklin, S. (2002). Supporting student learning. The use of computer-based formative assessment modules. *British Journal of Educational Technology, 33*(5), 45–55. doi:10.1111/1467-8535.00288

Pelgrum, W. J. (2001). Obstacles to the integration of ICT in education: results from a worldwide educational assessment. *Computers & Education, 37*(2), 163–178. doi:10.1016/S0360-1315(01)00045-8

Pelz, B. (2004). (My) three principles of effective online pedagogy. *Journal of Asynchronous Learning Networks, 8*(3), 33–46.

Pennington, M. C. Balla, J. Detaramani, C. Poon, A. & Tam, F. (1992). *Towards a model of language choice among Hong Kong tertiary students (Research Report No. 18)*. Hong Kong: Department of English, City Polytechnic of Hong Kong.

Performance and Innovation Unit (UK). (2001). *Better policy delivery and design*. London: DfEE.

Petrides, L. A. (2002). Web-based technologies for distributed (or distance) learning: Creating learning-centered educational experiences in the higher education classroom. *International Journal of Instructional Media, 29*(1), 69–77.

Pettit, B., & Western, B. (2004). Mass imprisonment and the life course: Race and class inequality in U.S. incarceration. *American Sociological Review, 69,* 151-169. Retrieved September 16, 2006, from http://www.princeton.edu/~western/ASRv69n2p.pdf

Phelps, R., & Ellis, A. (2003). *From page turning to deep learning: A case history of four years of continual development of an ICT course*. Paper presented to E-Learn 2003: World Conference on E-Learning in Corporate, Government, Healthcare and Higher Education, Phoenix, Arizona. Chesapeake, VA: Association for the Advancement of Computing in Education (AACE). Retrieved Dec. 16, 2004 from http://www.aace.org/conf/elearn/

Phillips, J. (2002). The dinosaurs didn't see it coming, but historians had better: computer aided activities in the history classroom. *History Computer Review, 18*(1), 27–36.

Pilkington, R. (2001). Analysing educational dialogue interaction: Towards models that support learning. *International Journal of Artificial Intelligence in Education, 12,* 1–7.

Piñera Ramírez, D. (2006). *Los primeros cincuenta años de educación superior en Baja California*. México, DF: Miguel Ángel Porrua-UABC.

Pink, D. H. (2005). *A whole new mind: Moving from the information age to the conceptual age*. New York: Riverhead Books.

Pintrich, P. R. (1989). The dynamic interplay of student motivation and cognition in the college classroom. In C.

Ames & M. Maehr (Eds.), *Advances in motivation and achievement: Motivation-enhancing environments* (Vol. 6, pp. 117-160).Greenwich, CT: JAI Press.

Pintrich, P. R., & De Groot, E. (1990). Motivational and self-regulated components of classroom academic performance. *Journal of Educational Psychology*, *82*, 33–40. doi:10.1037/0022-0663.82.1.33

Poftak, A. (2006). Blackboard-WebCT merger draws concern. *Technology & Learning*, *26*(6), 6.

Polsani, P. R. (2003). Use and abuse of reusable learning objects. *Journal of Digital Information*, *3*(4). Retrieved 23 February 2009 from http://jodi.ecs.soton.ac.uk/Articles/v03/i04/Polsani/

Prensky, M. (2001). Digital natives, digital immigrants. *On the Horizon*, *9*(5). Retrieved November 2, 2008, from http://www.marcprensky.com/writing/Prensky %20-%20Digital%Natives,%20Digital%20Immigrants%20-%20Part1.pdf

Preston, C. (2004). *Learning to use ICT in classrooms; teachers' and trainees' perspectives: an evaluation of the English NOF training programme 1999-2003*. London: MirandaNet and the TTA.

Proulx, S. (2001). Usages de l'Internet: la « pensée-réseaux » et l'appropriation d'une culture numérique In E. Guichard (Ed) *Comprendre les usages de l'Internet*, (pp. 139-145). Paris: Éditions Rue d'Ulm.

Quilter, S. M., & Chester, C. (2001). *The relationship between web-based conferencing and instructional outcomes*. Retrieved July 31, 2003, from Wilson SelectPlus database.

Quinn, C. N., Alem, L., & Eklund, J. (2005). *A pragmatic evaluation methodology for an assessment of learning effectiveness in instructional system*. Retrieved March 15, 2008 from http://www.testingcentre.com/jeklund/Interact.htm

Rahm, D., & Reed, B. J. (1998). Tangled Webs in Public Administration: Organizational Issues in Distance Learning. *Public Administration and Management Interactive Journal*, *3*(1). Retrieved May 13, 2005 from www.pamij.com/rahm/html

Rao, S. S. (2005). Bridging Digital Divide: Efforts in India . *Telematics and Informatics*, *22*(4), 361–375. doi:10.1016/j.tele.2005.01.007

Reeves, T. C. (2005). Design-based research in educational technology: Progress made, challenges remain. *Educational Technology*, *45*(1), 48–52.

Reis, R. M. (2001). To compete or collaborate? Career questions in science. *The Chronicle of Higher Education*. Retrieved on December 2, 2008, from http://chronicle.com/jobs/news/2001/05/2001052502c.htm

Renwick, W. (1993). The Future of Face-to-Face and Distance Teaching in Post-Secondary Education. *The Impact of Information and Communication Technologies on Post-Secondary Education*. Paris: CERI/OECD.

Reynolds, D., Treharne, D., & Tripp, H. (2003). ICT: the hopes and the reality. *British Journal of Educational Technology*, *34*(2), 151–167. doi:10.1111/1467-8535.00317

Rice, M. F. (2001). The Digital Divide and Race in the United States. *Politics . Administrative Change*, *36*, 20–31.

Richardson, J. T. E., Eysenck, M. W., & Piper, D. W. (1987) *Student Learning*. Milton Keynes, UK: Society for Research into Higher Education and Open University Press.

Richey, R. (2005). Validating instruction design and development models. In J. M. Spector & D. A. Wiley (Eds.). *Innovations in instructional technology: Essays in honor of M. David Merrill* (pp.171-185). Mahwah, NJ: Lawrence Erlbaum Associates, Publishers.

Richey, R., & Klein, J. (2007). *Design and development research*. Mahwah, NJ: Lawrence Erlbaum Associates, Publishers.

Ring, G., & Mathieux, G. (2002). *The key components of quality learning*. Paper presented at the ASTD Techknowledge 2002 Conference, Las Vegas.

Ritzhaupt, A., Hohlfeld, T., Barron, A., & Kemker, K. (2008). Trends in technology planning and funding in Florida K-12 public schools. *International Journal of Education Policy and Leadership*, *3*(8), 1–17.

Robertson, I. (2008). Sustainable e-learning, activity theory and professional development. In *Hello! Where are you in the landscape of educational technology? Proceedings ascilite Melbourne 2008,* (pp. 819-826). Retrieved from http://www.ascilite.org.au/conferences/melbourne08/procs/robertson.pdf

Robinson, L. A. (2005). Consumers of online instruction. *Issues in Information Systems, 6,* 170–175.

Robinson, S., & Katalushi, C. (Eds.). (2005). *Values in Higher Education.* Leeds, UK: Aureus.

Rodgers, M. (May 28, 2008). The Online Class Rubric and You. *TSL 2008 Summer Institute.* Retrieved November 25, 2008 from http://cstl.semo.edu/institute/2008Summer/Rubric/Rubric.htm

Rodgers, M., & Lockhart, F. (2008). The Dog Ate my e-Homework: Excuses, Academic Honesty, and the 21st Century Learner? *TSL 2008 Summer Institute.* Retrieved November 25, 2008 from http://cstl.semo.edu/institute/2008Summer/Edog/edogschedule.htm

Rodgers, M., & Starrett, D. (March 2000). *Technology Serving Learning Summer Institutes.* Paper presented at the Syllabus Conference, Orlando, FL.

Rodríguez Gómez, R. (2002). Continuidad y cambio de las políticas de educación superior. *Revista Mexicana de Investigación Educativa, 8*(14), 133–154.

Rogers, C. R. (1961). *On Becoming a Person.* London: Constable.

Rogers, C. R. (1969). *Freedom to learn; a view of what education might become.* Columbus, OH: Merrill.

Rogers, C. R. (1980). *A Way of Being.* Boston: Houghton Mifflin

Rosenblatt, L. (1982). The Literary Transaction: Evocation and Response. [from Professional Development Collection database.]. *Theory into Practice, 21*(4), 268. Retrieved December 1, 2008. doi:10.1080/00405848209543018

Rosenblatt, L. (1996). *Literature as Exploration.* New York: Modern Language Association of America.

Rosenblatt, L. (2005, March 1). From What Facts Does This Poem Teach You? *Voices from the Middle, 12*(3), 43–46.

Roskams, T. (1998). Collaborative interactions in networked writing classrooms: The student experience. In *The Japan Conference on English for Specific Purposes Proceedings,* Aizuwakamatsu City, Fukushima, Japan, November 8, 1997.

Rothstein, E. (1995). *Emblems of mind.* New York: Avon Books.

Rouillard, J. (2007). Web services and speech-based applications around VoiceXML. *Journal of Networks, 2*(1), 27–35.

Rovai, A. P., & Jordan, H. M. (2004). Blended Learning and sense of community: A comparative analysis with traditional and fully online graduate courses. *The international review of research in open and distance learning, 5*(2), 1-13. Retrieved November 5, 2008, from http://www.irrodl.org/index.php/irrodl/article/view/192/795

Rowe, N. C. (2004, Summer). Cheating in online student assessment: beyond plagiarism. *Online Journal of Distance Learning Administration, 7.* Retrieved April, 15, 2005, from http://www.westga.edu/~distance/ojdla/summer04/olt72.html

Ruan, J., & Beach, S. (2005). Using online peer dialogue journaling to promote reflection in elementary preservice teachers. *Action in Teacher Education, 27*(3), 64–75.

Ruiz Durán, C. (1997). *El reto de la educación superior en la sociedad del conocimiento.* México, DF: ANUIES.

Ryan, M., Hodson-Carlton, K., & Ali, N. S. (2005). A model for faculty teaching online: Confirmation of a dimensional matrix. *The Journal of Nursing Education, 44*(8), 357–365.

Salmon, G. (2005). Flying not flapping: A strategic framework for pedagogic innovation in higher education institutions. *ALT-J Research in learning technology, 13*(3), 201-218.

Salmon, G. (2006). Above the parapet for e-learning: Recognition, reward and celebration. *International Journal of Learning Technology, 2*(4), 342–356. doi:10.1504/IJLT.2006.011339

Samuel, A. (2004). *Hacktivism and the future of political participation.* Unpublished Ph.D. dissertation, Harvard University, Cambridge, MA.

Sandholtz, J. H., & Reilly, B. (2004). Teachers, Not Technicians: Rethinking Technical Expectations for Teachers. *Teachers' . College Record, 106*(3), 487–512. doi:10.1111/j.1467-9620.2004.00348.x

Santovec, M. L. (2003). A model for evaluating online courses. *Distance Education Report, 7*(8), 7–8.

Sauro, J., & Kindlund, E. (2005). A Method to Standardize Usability Metrics into a Single Score. *ACM, CHI,* April 2-7, Portland, OR.

SC Department of Education. (2009). *The State of South Carolina 2008 Annual School Report Card.* Retrieved February 26, 2009, from http://ed.sc.gov/topics/researchandstats/schoolreportcard/2008/

Scardamalia, M., & Bereiter, C. (2006). Knowledge-building: Theory, Pedagogy and Technology. In R. K. Sawyer (Ed.), *The Cambridge Handbook of the Learning Sciences.* New York: Cambridge University Press.

Schlain, L. (1998). *The alphabet versus the goddess: The conflict between word and image.* New York: Penguin Books.

Schön, D. (1987). *Educating the reflective practitioner.* San Francisco: Jossey-Bass.

Schuetze, H. G., & Slowey, M. (2000). Traditions and new directions in higher education. In H.G. Schuetze & M. Slowey (Eds.), *Higher Education and Lifelong Learners, International perspectives on change,* (pp.309-327). London: Routledge/Falmer.

Schunk, D. H. (1994). Self-regulation of self-efficacy and attributions in academic settings. In D. H. Schunk & B. J. Zimmerman (Eds.), *Self-regulation of learning and performance: Issues and educational implications* (pp. 75-99). Hillsdale, NJ: Erlbaum.

Schunk, D. H., & Zimmerman, B. J. (1997). *Self-regulation.* Retrieved on January 10, 2009 from http://education.calumet.purdue.edu/vockell/EdPsyBook/Edpsy7/edpsy7_self.htm

Schutz, A. (1967/1997). *Phenomenology of the social world.* Evanston, IL: Northwestern University Press.

Schwartzman, R. (2007). Refining the question: how can online instruction maximize opportunities for all students? *Communication Education, 56*(1), 113–117. doi:10.1080/03634520601009728

Scollon, R., & Scollon, S. (1995). *Intercultural Communication.* Oxford, UK: Blackwell.

Seddon, P. B. (1997). A Respecification and Extension of the DeLone and McLean Model of IS Success. *Information Systems Research, 8*(3), 240–253. doi:10.1287/isre.8.3.240

Seddon, P. B., & Kiew, M.-Y. (1994). A Partial Test and Development of DeLone and McLean's Model of IS Success. In *Proceedings of the International Conference on Information Systems,* Vancouver, Canada, (pp. 99-110).

Seels, B., & Glasgow, Z. (1997). *Making instructional design decisions.* Mahwah, NJ: Lawrence Erlbaum Associates, Publishers.

Selwyn, N. (2003). Exploring patterns of computer use in schools. *Welsh Journal of Education, 12*(1), 75–97.

Selwyn, N. (2004). Reconsidering Political and Popular Understandings of the *Digital Divide. New Media & Society, 6*(3), 341–362. doi:10.1177/1461444804042519

Semler, R. (1994). Why my former employees still work for me. *Harvard Business Review, 72*(1), 64–74.

Senge, P. (1990). *The Fifth Discipline: The Art & Practice of The Learning Organization.* New York: Doubleday.

Serce, F. C., & Yildirim, S. (2006). A web-based synchronous collaborative review tool: A case study of an on-line graduate course. *Educational Technology & Society, 9*(6), 166–177.

Servon, L. J. (2002). *Bridging the Digital Divide: Technology, Community, and Public Policy.* Oxford, UK: Blackwell Publishers.

Seufert, S., Lechner, U., & Stanoevska, K. (2002). A reference model for online learning communities. *International Journal on E-Learning, 1*(1), 43–54.

Sharpe, R., Benfield, G., & Francis, R. (2006). Implementing a university e-learning strategy: Levers for change within academic schools. *ALT-J Research in learning technology, 14*(2), 135-151.

Sheard, J., & Markham, S. (2005). Web-based learning environments: developing a framework for evaluation. *Assessment & Evaluation in Higher Education, 30*(4), 353–368. doi:10.1080/02602930500099169

Sieber, J. E. (2005). Misconceptions of and realities about teaching online. *Science and Engineering Ethics, 11*, 329–340. doi:10.1007/s11948-005-0002-7

Siemens, G. (12 December 2004). *Connectivism: A learning theory for the digital age*. Retrieved on October 7, 2008, from http://www.elearnspace.org/Articles/connectivism.htm

Siemens, G. Free and Open movement (2003). Part 1 – History and Philosophies. *Elearnspace, everything elearning* (pp. 1-4). Retrieved February 2009, from http://www.elearnspace.org/Articles/open_source_part_1.htm

Simeon, B., Skelton, J., & Lunn, K. (2005). *Schaum's Outlines UML, (*2nd Ed.). London: McGraw-Hill International, UK

Simonson, M., Smaldino, S., Albright, M., & Zvacek, S. (2003). *Teaching and learning at a distance*. Upper Saddle River, NJ: Merrill Prentice Hall.

Singleton, R. A., & Straits, B. C. (2005). *Approaches to social research* (4th ed.). New York: Oxford University Press.

Sirithumgul, P., Suchato, A., & Punyabukkana, P. (2007). *E-Learning Transformation For the Visually Impaired*. Assistive Technology Laboratory. Retrieved 3rd March 2009 from http://cu-ocw.eng.chula.ac.th/cu/eng/cp/e23e30e40e1ae35e22e1ae27e34e18e35e27e34e08e31e22e17e32e07e27e34e28e27e01e23e23e21e04e2de21e1ee34e27e40e15e2de23e4c-research-methods-in-computer-engineering/Assistive%20Paper%20I.pdf

Slavin, R. E. (1989-1990). Research on cooperative learning: Consensus and controversy. *Educational Leadership, 47*(4).

Sloodle. (2008). *Sloodle: Simulation Linked Object Oriented Dynamic Learning Environment*. Retrieved on December 2, 2008, from http://www.sloodle.org/moodle/

Smith, P., & Dillon, C. (1999). Comparing distance learning and classroom learning: conceptual considerations. *American Journal of Distance Education, 13*(2), 6–23. http://www.ajde.com/contents/vol13_2.htm Retrieved February 23, 2003. doi:10.1080/08923649909527020

Smith, R. (2005). Working with difference in online collaborative groups. *Adult Education Quarterly, 55*(3), 182–199. doi:10.1177/0741713605274627

Smith, R. S. (2004). *Guidelines for authors of learning objects* (p.8). The New Media Consortium. Retrieved December 1, 2008 from http://archive.nmc.org/guidelines/NMC%20LO%20Guidelines.pdf

Smithers, A., & Robinson, P. (2000). *Attracting teachers: past patterns, present policies, future prospects*. Liverpool, UK: Carmichael.

Somyürek, S., Atasoy, B., & Özdemir, S. (in press). Board's IQ. What makes a board smart? *Computers & Education*.

Spolsky, J. (2003, 2003/03/03). *Building communities with software*. Retrieved November 4, 2006, from http://www.joelonsoftware.com/articles/BuildingCommunitieswithSo.html

Spradley, J. P. (1979). *The Ethnographic Interview*. New York: Holt, Rinehart, and Winston.

Squires, D., & Preece, J. (1999). Predicting quality in educational software: Evaluating for learning, usability and the synergy between them. *Interacting with Computers, 5*(11), 467–483. doi:10.1016/S0953-5438(98)00063-0

Stanford-Bowers, E. (2008). Persistence in online classes: As study of perceptions among community college stakeholders. *Journal of Online Learning and Teaching, 4*(1). Retrieved February 23, 2009 from http://jolt.merlot.org/vol4no1/stanford-bowers0308.htm

Stanley, L. D. (2003). Beyond Access: Psychosocial Barriers to Computer Literacy. *The Information Society, 19*, 407–416. doi:10.1080/715720560

Stefansdottir, L. (1993). The Icelandic Educational Network – Ismennt. In G. Davies, & B. Samways, (Eds.), *Teleteaching – Proceedings of the IFIP TC3 Third Teleteaching Conference* (pp. 829-835). Amsterdam: Elsevier Science Publishers.

Stein, D. S., Wanstreet, C. E., Calvin, J., Overtoom, C., & Wheaton, J. E. (2005). Bridging the transactional distance gap in online learning environments. *American Journal of Distance Education, 19*(2), 105. doi:10.1207/s15389286ajde1902_4

Stenhouse, L. (1975). *An introduction to curriculum research and development.* London: Heinemann.

Stephenson, J. (2001). *Teaching and Learning Online – Pedagogies for New Technologies.* London: Kogan Page Limited.

Stevens, K. (1999). Telecommunications Technologies, Telelearning and the Development of Virtual Classes for Rural New Zealanders. *Open Praxis, 1,* 12–14.

Stevens, K. J. (1994). Some Applications of Distance Education Technologies and Pedagogies in Rural Schools in New Zealand. *Distance Education, 15*(4), 318–326. doi:10.1080/0158791940150210

Stevens, K. J. (1995a). Geographic Isolation and Technological Change: A New Vision of Teaching and Learning in Rural Schools in New Zealand. *The Journal of Distance Learning, 1*(1), 3–38.

Stevens, K. J. (1995b). The Technological Challenge to the Notion of Rurality in New Zealand Education Repositioning the Small School. In I. Livingstone (ed.) *New Zealand Annual Review of Education, 5,* 93 –102.

Stevens, K. J. (2003a). E-Learning and the Development of Open Classes for Rural Students in Atlantic Canada. In J. Bradley (ed.), *The Open Classroom - Distance Learning In and Out of Schools* (pp. 149-157). London: Kogan Page.

Stevens, K. J. (2003b). From Distance Education to TeleLearning – The Organization of Open Classes at Local, Regional and National Levels. In J.Bradley (Ed.) *The Open Classroom – Distance Learning In and Out of Schools* (pp. 171-180). London: Kogan Page.

Stevens, K., & Moffatt, C. *(1996). TeleLearning and New Zealand Secondary Schools* –Some Pedagogical and Management Issues in the Development of Courses for Flexible Delivery. *Technology for Flexible Delivery Conference.* Palmerston North, New Zealand: Massey University.

Stevens, K.J. & Starkey, L. (2006). Three Stages in the Digital Integration of New Zealand Schools. *New Zealand Annual Review of Education - Te Arotake a Tau o te Ao o te Matauranga i Aotearoa, 16,* 105-117.

Strover, S. (1999). *Rural Internet Connectivity.* Columbia, MO: Rural Policy Research Institute.

Swain, J. (2009). A long way from flip charts. *ORMS Today, 36*(1), 44–47.

Sweet, R. (1986). Student drop-out in distance education: An application of Tinto's model. *Distance Education, 7*(2), 201–213. doi:10.1080/0158791860070204

Tang, C. (1996). Collaborative learning. In D. Watkins & J. Biggs (eds.), *The Chinese Learner: Cultural, Psychological and Contextual Influences,* (pp. 183–204). Hong Kong: The Comparative Education Research Centre, Faculty of Education, University of Hong Kong.

Tate, O. (1993). *A Vision of Distance Education in New Zealand.* Inaugural conference of Distance Education and Information and Communications Technology Providers, National Library of New Zealand, Wellington, New Zealand.

Tatum, A. W. (2006, February). Engaging African American Males in Reading. *Educational Leadership, 57*(8), 44–49.

Taylor, F. W. (1911). *The principles of scientific management.* New York: Harper.

Technology Associates. (2008). *Online Course Best Practices Checklist.* Center for Scholarship in Teaching and Learning, Southeast Missouri State University.

Technorati. (2008). *Technorati: Popular blogs.* Retrieved December 3, 2008 from http://www.technorati.com/pop/blogs/

Tella, S. (1995). *Virtual School in a Networking Learning Environment.* Helsinki, Finland: University of Helsinki, Department of Teacher Education.

Texas Higher Education Coordinating Board. (2008). *Overview: Tuition deregulation.* Retrieved on December 1, 2008, from: http://www.thecb.state.tx.us/reports/PDF/1527.PDF

The IDEA Center. (2008). *IDEA Papers.* Retrieved March 5, 2009, from http://www.theideacenter.org/IDEAPapers

Thomas, M. J. (2002). Learning within incoherent structures: the space of online discussion forums. *Journal of Computer Assisted Learning, 18*(3), 351–366. doi:10.1046/j.0266-4909.2002.03800.x

Thompson, L., & Ku, H.-Y. (2005). Chinese graduate students' experiences and attitudes toward online learning. *Educational Media International, 42*(1), 33–47. doi:10.1080/09523980500116878

Toprakci, E. (2006). Obstacles at Integration of Schools into Information and Communication Technologies by Taking into Consideration the Opinions of the Teachers and Principles of Primary and Secondary Schools in Turkey. *e-Journal of Instructional Science and Technology (e-JIST), 9*(1).

Trenholm, S. (2007). An investigation of assessment in fully asynchronous online math courses. *International Journal for Educational Integrity, 3*(2), 41–55.

Trochim, W. M. (2001). *The research methods knowledge base* (2nd ed.). Cincinnati, OH: Atomic Dog Publishing.

Trompanaars, F. (1993). *Riding the Waves of Culture.* London: Nicholas Brealy Publishing.

Truluck, J. (2007). Establishing a mentoring plan for improving retention in online graduate degree programs. *Online Journal of Distance Learning Administration, 10*(1), 1–6.

Tsobanopoulou, F. (2008). Acrobats on Borderlines: The Konitsa Border Guards. *Balkan Border Crossings Network.* Lit-Verlag.

Tsui, A. (1996). Reticence and anxiety in second language teaching. In K. Bailey & D. Nunan (Eds.), *Voices from the language classroom,* (pp. 145–167). Cambridge, UK: Cambridge University Press.

Tu, C.H. & McIsaac, M. (2002). The Relationship of Social Presence and Interaction in Online Classes. *American Journal of Distance Education.*

Tuckman, B. (1965). Developmental Sequence in Small Groups. *Psychological Bulletin, 63*(6). doi:10.1037/h0022100

Tukiainen, J. (2004). *Access to Computer, Internet and Mobile Phone at Home in Finland, Ireland, Netherlands and Sweden.* Vatt Discussion Papers 324, Government Institute for Economic Research.

U.S. Department of Education, National Center for Education Statistics. (2008). *The Condition of Education 2008* (NCES 2008-031). Retrieved on March 20, 2009, from http://nces.ed.gov/FastFacts/display.asp?id=16

UABC. (2000). Plan de Desarrollo Institucional 1999-2002. Mexicali, BC: Universidad Autónoma de Baja California.

UABC. (2003). Informe de Rectoría 2003. Comisión de Planeación, Universidad Autónoma de Baja California.

UABC. (2008). Informe de labores 2008. Mexicali, B.C.: Universidad Autónoma de Baja California.

UAEU Laptop Project (2004). *Laptop Project: Students' Survey Analysis report.* Al-Ain, UAE: UAEU.

UBel. (2002). *E-learning.* An Unpublished paper, University of Botswana, Gaborone.

UNESCO. (1997). *Recommendation concerning the Status of Higher-Education Teaching Personnel.* Date of adoption November 11, 1997. Retrieved from http://unesdoc.unesco.org/images/0011/001102/110220e.pdf#page=32

UNESCO. (1998). *World Declaration on Higher Education: Higher Education in the Twenty-First Century: Vision and Action*, Paris, Fr., 5 - 9 October 1998. Retrieved from http://www.unesco.org/education/educprog/wche/declaration_eng.htm

UNESCO. (2002). *Open and Distance Learning. Trends, Policy and Strategy Considerations*. Paris: Division of Higher Education. Retrieved from http://unesdoc.unesco.org/images/0012/001284/128463e.pdf

University of Botswana. (2007). Facts and Figures. *University Fact Book*.

University of Botswana. (2007). *WebCT Report*. Botswana: Center for Academic Development,University of Botswana.

University of Fribourg. (2006). *Platform use - 2005*. Retrieved December 3, 2008 from http://www.edutech.ch/lms/inst-platforms.php

University of Houston – Clear Lake. (2008a). *About UHCL*. Retrieved on December 1, 2008, from http://prtl.uhcl.edu/portal/page/portal/HOMEPAGE/TAB_ABOUTUHCL

University of Houston – Clear Lake. (2008b). *Institutional Effectiveness: UHCL Strategic Plan FY05-FY11*. Retrieved on December 1, 2008, from http://prtl.uhcl.edu/portal/page/portal/OIE/PLANNING

University of Houston – Clear Lake. (2008c). *UHCL History*. Retrieved on December 1, 2008, from http://prtl.uhcl.edu/portal/page/portal/PRE/RTTC/UHCL_History

University of Victoria Counseling Services. (2003). *Bloom's taxonomy*. Retrieved on November 28, 2008, from http://www.coun.uvic.ca/learn/program/hndouts/bloom.html

University Strategic Planning Committee. (1996). *The Southeast Missouri State University Strategic Plan*. Office of the Provost, Southeast Missouri State University.

Usluel, Y. K., Mumcu, F. K., & Demiraslan, Y. (2007). ICT in the Learning Teaching Process: Teachers' Views on the Integration and Obstacles. *Journal of Hacettepe University Education Faculty, 32*, 164–178.

UTPA. (2005). *Fact book*. Retrieved December 16, 2008, from http://oire.utpa.edu/publications/minifactbook2005.pdf

UTPA. (2008). *Stats at a Glance*. Retrieved February 12, 2009, from http://oire.utpa.edu/publications/minifactbook2008.pdf

Uys, P. M. (2003a). E-learning at the University of Botswana: A Bi-Annual Reflection on Progress. *Lonaka Bulletin*, (pp.5-12). Centre for Academic Development, University of Botswana, Special Edition on educational . *Technology (Elmsford, N.Y.)*, (March): 2003.

Valentine, D. (2002). Distance Learning: Promises, Problems, and Possibilities. *Online Journal of Distance Learning Administration, 6*(3).

Van der Wende, M. (2000). The Bologna declaration: enhancing the transparency and competitiveness of European higher education. *Higher Education in Europe, 25*(3), 305–310. doi:10.1080/713669277

van Dijk, J. (2005).*The Deepening Divide*. Thousand Oaks, CA: Sage.

Van Dijk, J. (2006). Digital Divide Research, Achievements and Shortcomings. *Poetics, 34*(4-5), 221–235. doi:10.1016/j.poetic.2006.05.004

Van Manen, M. (2002). The Pedagogical Task of Teaching. *Teaching and Teacher Education, 18*(2), 135–138. doi:10.1016/S0742-051X(01)00058-0

VanSlyke, T. (2003). Digital natives, digital immigrants: some thoughts from the generation gap. *The Technology Source*. Retrieved December 16, 2008, from http://technologysource.org/article/digital_natives_digital_immigrants/

Vélez, W., & Saenz, R. (2001, December). Toward a comprehensive model of the school leaving process among Latinos. *School Psychology Quarterly, 16*(4), 445–467. doi:10.1521/scpq.16.4.445.19897

Viklund, A. (2008). *Discussion of David Brooks' article on the cognitive age*. Retrieved on November 28, 2008 from http://neuroanthropology.net/2008/05/03/david-brooks-part-one-the-cognitive-age/

Vockell, E. (n.d.). *Educational Psychology: A Practical Approach*. Retrieved on March 20, 2009, from http://education.calumet.purdue.edu/vockell/EdPsyBook/Edpsy7/edpsy7_self.htm

Vockell, E. L., & Asher, J. W. (1995). *Educational research* (2nd ed.). Englewood Cliffs, NJ: Prentice-Hall, Inc.

Voice, X. M. L. (2007), *VoiceXML application development life cycle*. Palo Alto, CA: Hewlett Packard Development Company.

Vonderwell, S. (2003). An examination of asynchronous communication experiences and perspectives of students in an online course: A case study. *The Internet and Higher Education, 6*, 77–90. doi:10.1016/S1096-7516(02)00164-1

Voxeo (2003). *Voice server*. Retrieved October, 2008 from http://community.voxeo.com

Vygotsky, L. (1986). *Thought and Language* (original Russian edition 1934). Cambridge, MA: MIT Press.

Vygotsky, L. S. (1935). *Mental development of children during education*. Moscow: Uchpedzig.

Vygotsky, L. S. (1962). *Thought and language*. Cambridge, MA: MIT Press.

Vygotsky, L. S. (1978). *Mind in Society: The development of higher psychological processes*. Cambridge, MA: MIT Press.

Vygotsky, L. S. (1981). The genesis of higher mental functions. In J. V. Wertsch (Ed.), *The concept of activity in Soviet Psychology*. Armonk, NY: Sharpe.

Wade, R. W. (2004). Bridging the Digital Divide: New Route to Development or New Form of Dependency? In Avgerou et al. (Eds.), *The social study of Information and Communication Technology – Innovation, Actors, Contexts*. Oxford, UK: Oxford University Press.

Wagner, E. D. (1994). In support of a functional definition of interaction. *American Journal of Distance Education, 8*(2), 6–26. doi:10.1080/08923649409526852

Walker, S. A. (2004). Socratic strategies and devil's advocacy in synchronous CMC debate. *Journal of Computer Assisted Learning, 20*, 172–182. doi:10.1111/j.1365-2729.2004.00082.x

Wall, K., Higgins, S., & Smith, H. (2005). The visual helps me understand the complicated things': pupil views of teaching and learning with interactive whiteboards. *British Journal of Educational Technology, 36*(5), 851–867. doi:10.1111/j.1467-8535.2005.00508.x

Wallace, J. (2000). Supporting and guiding students. In H. Fry, S. Ketteridge, & S. Marshall. (Eds), *A Handbook for Teaching and Learning in Higher Education*. London: Kogan Page.

Walther, J. B. (1996). Computer-mediated communication: Impersonal, interpersonal, and hyperpersonal interaction. *Communication Research, 23*(1), 3–43. doi:10.1177/009365096023001001

Waltonen-Moore, S., Stuart, D., Newton, E., Oswald, R., & Varonis, E. (2006). From virtual strangers to a cohesive learning community: The evolution of online group development in a professional development course. *Journal of Technology and Teacher Education, 14*(14), 287–311.

Wang, Y. S. (2003). Assessment of learner satisfaction with asynchronous electronic learning system. *Information & Management, 41*, 75–86. doi:10.1016/S0378-7206(03)00028-4

Wang, Y.S; Wang, H.Y & Shee, D.Y.(2007). *Measuring e-learning systems success in an organizational context: Scale development and validation.*

Warschauer, M. (2003). *Technology and Social Inclusion: Rethinking the Digital Divide*. Cambridge, MA: MIT Press

Watkins, D. A., & Biggs, J. B. (1996). *The Chinese Learner: Cultural, Psychological and Contextual Influences*. Comparative Education Research Centre (CERC) Faculty of Education, University of Hong Kong and The Australian Council for Educational Research Ltd. (ACER).

Watkins, D. A., & Biggs, J. B. (2001). Insights into teaching the Chinese learner. In D. A. Watkins & J. B. Biggs

(Eds.), *Teaching the Chinese learner: Psychological and pedagogical perspectives* (pp. 277-300). Hong Kong: Comparative Education Research Centre, University of Hong Kong.

Watkins, D. A., & Biggs, J. B. (2001). *Teaching the Chinese Learner: Psychological and Pedagogical Perspectives.* Comparative Education Research Centre (CERC) and The Australian Council for Educational Research Ltd. (ACER).

Weber, M. (1947). *The theory of social and economic organizations*, (T. Parsons (Ed.), A.M. Henderson & T. Parsons, Trans.). New York: Free Press.

Weber, R. (1990). *Basic Content Analysis*, (2nd. Ed.). Newbury Park, CA: Sage.

Weedon, E. M., & Cowan, J. (2001). Commenting Electronically on Students' Reflective Learning Journals. In *Proceedings, ISL conference*, Edinburgh, (pp. 134-42). Oxford, UK: Oxford Centre for Staff and Learning Development.

Weedon, E. M., & Cowan, J. (2002). The Kolb Cycle, reflection and all that what is new? In *Proceedings, ISL conference*, Brussels, (pp 97-108). Oxford, UK: Oxford Centre for Staff and Learning Development.

Wegmann, S. & McCauley, J. (2008). *Shouting through the fingertips: Computer-mediated discourse in an asynchronous environment.* Presentation at Society for Information Technology, March 3, 2008.

Wegmann, S., & McCauley, J. (2007). Can you hear us now? Stances toward interaction and rapport. In Y. Inoue (Ed.), *Online Education for Lifelong Learning.* Hershey, PA: Information Science Publishing.

Wellington, N. Z. Ministry of Education (2006). *Enabling the 21st Century Learner; An e-Learning Action Planfor Schools 2006-2010.* Wellington, New Zealand: Learning Media.

Wellman, G. S., & Marcinkiewicz, H. (2004, December). Online learning and time-on-task: impact of proctored vs. un-proctored testing. *Journal of Asynchronous Learning networks, 8.* Retrieved April 26, 2005, from http://www.sloan-c.org/publications/jalnv8n

Wenger, E. (1998). *Communities of Practice: learning, meaning and identity.* New York: Cambridge University Press.

Wenger, E. (n.d.). *Communities of Practice – a brief introduction.* Retrieved March 20, 2006, from http://www.ewenger.com/theory/index.htm

Wenger, E., McDermott, R., & Snyder, W. (2002). *A guide to managing knowledge: Cultivating communities of practice.* Boston: Harvard Business School Press.

Wertsch, J. V. (1985). *Cultural, communication, and cognition: Vygotskian perspectives.* Cambridge, UK: Cambridge University Press.

Westbrook, R. A., & Oliver, P. R. (1991). The dimensionality of consumption emotion patterns and consumer satisfaction. *The Journal of Consumer Research, 18*(1), 84–91. doi:10.1086/209243

Wighton, D. J. (1993). *Telementoring: An Examination of the Potential for an Educational Network.* Education Technology Centre of British Columbia.

Wikipedia. (2008). *Wiki.* Retrieved on December 3, 2008 from http://en.wikipedia.org/wiki/Wiki

Wikipedia. (2009). *Formative assessment.* Retrieved on February 20, 2009, from http://en.wikipedia.org/wiki/Formative_assessment

Wiley, D. (2003). *Learning objects: difficulties and opportunities.* Retrieved February 23, 2009 from http://wiley.ed.usu.edu/docs/lo do.pdf

Williams, R. S., & Humphrey, R. (2007). Understanding and fostering interaction in threaded discussion. *Journal of Asynchronous Learning Networks, 11*(2). Retrieved December 2, 2008 from http://www.aln.org/publications/jaln/v11n2/v11n2_williams_member.asp

Wilson, B. (2008, 21 July). *Expert, novices, and cognitive-load.* Retrieved on February 1, 2009 from http://www.crucialthought.com/2008/07/21/experts-novices-and-cognitive-load/

Wimba, Inc. (2009). Retrieved from http://www.wimba.com/products/wimba_collaboration_suite/

Witkin, H. A., Moore, C. A., Goodenough, D. R., & Cox, P. W. (1977). Field-dependent and field-independent cognitive styles and their educational implications. *Review of Educational Research, 47*(1), 1–64.

WIUS (World Internet Usage Statistics—News and Population Stats). (n.d.). *Miniwatts Marketing Group. 2001-2009.* Retrieved from www.internetworldstats.com/stats.htm

Wynn, M. (2005). *Empowering African American males: Teaching, parenting, and mentoring successful black males.* Marietta, GA: Rising Sun Publishing.

Xiao, Li., & Dasgupta, S. (2002). Measurement of user satisfaction with web-based information systems: An empirical study. *Eight Americas Conference on Information Systems* (pp. 1149-1155).

Young, J. R. (2005, August 12). Professors give mixed reviews of Internet's educational impact. *Chronicle of Higher Education.* Retrieved August 21, 2005 from http://chronicle.com/weekly/v51/i49/49a03201.htm

Youngman, F. (2000). *The political economy of adult education and development.* New York: Zed Books Ltd.

Yu, Y. L., & Chou, S. G. (2004). Integrating writing with reading in EFL writing classroom with metacognitive approach. *Selected papers from Thirteenth International Symposium and Book Fair on English Teaching, 1*, 708-719.

Zacharias, P., Vasslopoulou, K., & Poulymenakou, A. (2002). *Designing online learning courses: Implications for usability.* Retrieved March 15, 2008: http://wwwjapit.org/zacharias_etal02.pdf

Zhang, W., Perris, K., & Yeung, L. (2005). Online tutorial support in open and distance learning: students' perceptions. *British Journal of Educational Technology, 36*(5), 789–804. doi:10.1111/j.1467-8535.2004.00492.x

Zhao, Y., & Frank, K. (2003). Factors affecting technology use in schools: an ecological perspective. *American Educational Research Journal, 40*(4), 807–840. doi:10.3102/00028312040004807

Zhao, Y., Pugh, K., Sheldon, S., & Byers, J. (2002). Conditions for classroom technology innovations. *Teachers College Record, 104*(3), 482–515. doi:10.1111/1467-9620.00170

Zimmerman, B. J. (1989). A social cognitive view of self-regulated academic learning. *Journal of Educational Psychology, 80*, 329–339. doi:10.1037/0022-0663.81.3.329

Zimmerman, B. J. (1990). Self-regulated learning and academic achievement: An overview. *Educational Psychologist, 25*, 3–17. doi:10.1207/s15326985ep2501_2

Zimmerman, B. J. (1994). Dimensions of academic self-regulation: A conceptual framework for education. In D. H. Schunk & B. J. Zimmerman (Eds.), *Self-regulation of learning and performance: Issues and educational implications* (pp. 3-21). Hillsdale, NJ: Erlbaum.

Zirkle, C. (2003). Distance education in career and technical education: a review of the research literature. *Journal of Vocational Education Research, 28*(2), 151–171.

About the Contributors

Bolanle A. Olaniran is a professor in the Department of Communication Studies at Texas Tech University. His research includes: Communication technologies and Computer-Mediated Communication, Organization communication, Cross-cultural communication, and Crisis Management and Communication. He has authored several articles in discipline focus and interdisciplinary focus Journals (i.e., Regional, National, and International) and edited book chapters in each of these areas. He is an accomplished international scholar. He serves as consultant to organizations at local, national and government level. His works have gained recognition such as the American Communication Association's "Outstanding Scholarship in Communication field," Who is who among College Teachers. He has won many teaching, research and services award including TTU 2006 Office of the President's Diversity, 2007 President's Excellence in Teaching Awards and nominated for 2007 TTU Chancellor's Distinguished Teaching Award. He is well sought after internationally to give invited lectures and keynote address for technology, culture, and related areas.

* * *

Dr. Pamela L. Anderson-Mejías is Professor of Linguistics and English as a Second Language and chair of the English Department at the University of Texas-Pan American in Edinburg, Texas. Dr. Anderson-Mejías works closely with the Center for Online Learning Teaching & Technology to deliver courses to local and remote students. Via teleconferencing, Wimba, and now starting in SecondLife she has delivered master's level courses in English as a second language to multiple sites in China, Mexico and elsewhere. Since beginning her work in the virtual classroom, she has consistently contributed to scholarly literature interfacing language teacher training with online best practices.

Aderemi Aaron Anthony Atayero holds a B.Sc. degree (summa cum laude) in Radio Engineering and M.Sc. in Satellite Communication Systems (1992, 1994 respectively) both from the Moscow Institute of Technology, and a Ph.D. in Technical Sciences (Speech Processing / Satcom, 2000) from the Moscow State Technical University of Civil Aviation, Moscow, Russia. His current areas of research interest are: Speech Processing, Information theoretic analysis of written languages, FPGA Implementation of Digital Speech Processors. He is a member of a number of academic and professional organisations including the Institute of Electrical and Electronic Engineers (IEEE). Atayero is a Senior Lecturer in the department of Electrical and Information Engineering at Covenant University.

Charles K. Ayo holds a B.Sc., M.Sc. and Ph.D in Computer Science. His research interests include: mobile computing, Internet programming, e-business and government, and object oriented design and development. He is a member of the Nigerian Computer Society (NCS), and Computer Professional Registration Council of Nigeria (CPN). He is currently the Head of Computer and Information Sciences Department of Covenant University, Ota, Ogun state, Nigeria, Africa. Dr. Ayo is a member of a number of international research bodies such as the Centre for Business Information, Organization and Process Management (BIOPoM), University of Westminister. http://www.wmin.ac.uk/wbs/page-744; the Review Committee of the European Conference on E-Government, http://www.academic-conferences.org/eceg/; and the Editorial Board, Journal of Information and communication Technology for Human Development.

A. A. Azeta is a Ph.D student in the Department of Computer and Information Sciences, Covenant University, Ota, Nigeria. He holds B.Sc. and M.Sc. in Computer Science from University of Benin and Lagos respectively. His current research interests are in the following areas: Software Engineering, Algorithm Design and Mobile Computing. He currently lectures at Covenant University. He is a member of the Nigerian Computer Society (NCS) and Computer Professional Registration Council of Nigeria (CPN).

Dr Roy Barton is a Senior Lecturer at the School of Education and Professional Development, University of East Anglia. Prior to joining the university in 1990 he had spent 18 years as a secondary school science teacher. Since joining the university he has been mainly involved in the initial training of science teachers. However, he also has extensive experience in supervising and examining students studying for higher degrees. His research interests are centred on the use of ICT for teaching and learning, particularly in science education, initial teacher education and teachers' continuing professional development. He has produced a number of publications in these areas including editing a book on Teaching Secondary Science with ICT.

Pauline Hope Cheong is associate professor of New Media and Intercultural Communication in the Hugh Downs School of Human Communication at Arizona State University, USA. She is also an affiliate faculty member in the School of Justice and Social Inquiry, and the Department of Film & Media Studies. Her research interests focus on the social implications of communication technologies, including issues of access and power for minority and marginalized communities. She has published multiple articles on E-learning and the digital divide in national and international journals, including *New Media and Society, Journal of Computer-Mediated Communication, and Information, Communication and Society.*

Jean Chiu is a teacher of English as a Foreign Language in Taiwan. She has a bachelor's degree of English in Taiwan, a master's degree of communication studies in the United States, and a doctoral degree of education in Australasia. Currently her primary concern is in nurturing the cognitive and affective development Western-style critical thinking by her students of English as a Foreign Language. In this area she combines funded post-doctoral research with her innovative teaching commitments. She has published papers in the East and in Europe, describing several of her activities.

John Cowan practiced as a structural engineer before moving into academia to teach design. Since then his several higher degrees have been educational. He has published on educational topics for 40 years, and has international awards for his educational development activities, which he describes as pragmatic, useful, and evaluated developments. His reputation centres on student-centred learning, wherein he was a pioneer in the 1970's. He only deems it acceptable to publish educationally while he is himself engaging in validated innovations in his current undergraduate and postgraduate teaching. He is still managing to satisfy both of these personal criteria.

Jenny Craven, is a Research Associate at the Centre for Research in Library and Information Management at Manchester Metropolitan University. She has worked on a variety of library and information management related research projects, primarily concerned with web accessibility and usability, with a particular focus on access to information by blind or visually impaired people. She has facilitated workshops on this topic in Chile, Mexico, Vietnam, Brazil, Sabah, and Kenya. Jenny is currently working on an Erasmus Lifelong Learning funded project: web_access, which is developing a Joint Programme on Accessible Web Design with partners from 6 European countries. Jenny contributes to the teaching programme in the Department of Information and Communications.

Caroline M. Crawford, Ed.D., is an Associate Professor of Instructional Technology at the University of Houston-Clear Lake in Houston, Texas, USA. At this point in Dr. Crawford's professional career, her main areas of interest focus upon communities of learning and the appropriate and successful integration of technologies into the learning environment. Dr. Crawford's scholarly pursuits are varied, including a minimum of 40 journal articles, a minimum of 150 professional conference-focused presentations and thirteen poster sessions, with a history of honored positions as a Review Editor for six international refereed journals, and reviewer for innumerable regional, national and international refereed journals.

Javier Faulin is an Associate Professor of Statistics and Operations Research at the Public University of Navarre (Pamplona, Spain) and an E-Tutor at the UNED Centre in Pamplona, Spain. He holds a PhD in Economics, a MS in Operations Management, Logistics and Transportation and a MS in Applied Mathematics. His research interests include logistics, vehicle routing problems and simulation modeling and analysis. He is also a member of INFORMS and EURO.

Mark Federman has more than twenty-five years' experience in the high-technology industry, and has played the role of Chief Strategist at the McLuhan Program in Culture and Technology at the University of Toronto. He has a regular guest lecturer at Högskölan in Jönköping Sweden, and a visiting professor at Fachhochschule in Kiel Germany, and elsewhere. Mark is the co-author, with Derrick de Kerckhove, of *McLuhan for Managers*. Some of his recent explorations have examined "No Educator Left Behind: The present future of educator reform," "How Do We Know: The changing culture of knowledge," "Why Johnny and Janey Can't Read, and Why Mr. and Ms. Smith Can't Teach," and "A Brief, 3,000-Year History of Organization." Internationally, Mark has spoken about the future emergence of culture for UNESCO, and at universities throughout North America and Europe. Mark is currently completing a Ph.D. at the Ontario Institute for Studies in Education at the University of Toronto.

Pau Fonseca is an Associate Professor of Statistics and Computer Simulation at the Technical University of Catalonia (Barcelona, Spain). He is also a Lecturer of computer systems simulation at the Open University of Catalonia. He holds a MS and a PhD on Computer Science. His research interests include computer simulation and applied statistics. He is member of the LogiSim research group, which develops simulation tools.

Ruth Gannon Cook, Ed.D., is an Assistant Professor in the DePaul University School for New Learning. At this point in Dr. Gannon Cook's professional career, her main areas of interest focus upon deep online learning through semiotics, as well as adult learners and competency-based online learning. Dr. Gannon Cook's scholarly pursuits are varied, with innumerable peer-reviewed journal articles and book chapters, innumerable professional presentations and poster sessions, and a history of honors associated with her strong beliefs related to service to the profession and the community.

Dr. Stuart S. Gold is an experienced professor with extensive experience teaching both online and face-to-face classes. He has taught Management and MIS courses in both the United States and the Caribbean during the past several years. He also has over 25 years of corporate experience including senior level positions including CIO and Vice President / Director of Information Systems. Dr. Gold holds a B.A. in Physics and Mathematics from Northeastern Illinois University, M.B.A. from Loyola University of Chicago, and PhD from Northcentral University.

Tammy Graham, Ed.D., is an Assistant Professor of Education at The Citadel. She teaches undergraduate and graduate courses related to special education, educational psychology, and child development. She is also co-advisor of The Education Leaders Club. Dr. Graham has 17 years of experience in the special education field of public education, most recently as a special education coordinator. She has taught students with exceptionalities at every grade level. Additionally, as an educational diagnostician, she assessed students in preschool, elementary school, middle school, and high school. Dr. Graham's research interests include e-learning, culturally relevant pedagogy and disproportionately in special education.

Jillian R. Griffiths is a Research Associate at the Centre for Research in Library and Information Management at Manchester Metropolitan University. Since joining the CERLIM team in 1999 she has worked on a number of projects focussed on the distributed delivery of library and information services, evaluation of projects and programmes, and user testing and evaluation. Currently Jill is working on the Research Information Network (RIN) project to investigate the publication and dissemination behaviour of UK scholars; the JISC (Joint Information Systems Committee)/RIN funded project DiSCmap identifying the prioritisation of library special collections for digitisation; and is undertaking research on students' information literacy with her colleague Bob Glass as part of the HEFCE funded LearnHigher CETL at MMU.

Dr Terry Haydn is a Reader in Education and Course Tutor for Secondary History at the School of Education, University of East Anglia. His research interests are in the use of information technology in the classroom, student teachers' use of ICT and the use of new technology in history teaching. Before working at UEA, he taught history at an inner-city school in Manchester, and spent four years in the Department of History, Humanities and Philosophy at the Institute of Education, University of London.

Stephenie Hewett is in her twenty-first year of service at The Citadel. As an Associate Professor in The Citadel School of Education, she is responsible for teaching classes related to literacy, child development, and adolescent development. She is the Academic Coordinator for the GEAR UP Program, a United States Department of Education initiative. She has been involved in the development and implementation of school and higher education partnerships throughout her career at The Citadel. She has published and presented nationally and internationally on topics related to the academic achievement of males, school/university partnerships, and instructional technology.

Nicholas Ikhu-Omoregbe has a B.Sc degree in Computer science from the University of Benin, Benin city, an M.Sc degree in Computer Sciences from the University of Lagos, and a PhD degree in Computer Science from Covenant University, Ota, Nigeria. His research interests include: Software Engineering, Mobile Computing, Mobile Healthcare and Telemedicine Systems, and Soft Computing. He currently lectures at Covenant University. He is a member of the Institution of Electrical and Electronics Engineers (IEEE).

Peter Jakubowicz Ph.D.: Born and educated in Canada, Peter Jakubowicz attended Sir George Williams University and Concordia University in Montreal. He received his B.A. in English Literature, Certificate in TESL (Teaching English as a Second Language) and M.A. in Educational Studies. Recently he has completed a Ph.D. in Education, with a particular focus on online learning. His main areas of research include English as a Second Language, online learning/teaching and Education. He has been teaching at The Chinese University of Hong Kong since 1998.

Angel A. Juan is an Associate Professor of Simulation and Data Analysis in the Computer Science Department at the Open University of Catalonia (Barcelona, Spain). He also collaborates, as a Lecturer of Applied Statistics, with the Department of Applied Mathematics I at the Technical University of Catalonia. He holds a Ph.D. in Applied Computational Mathematics (UNED), an M.S. in Information Technology (Open University of Catalonia), and an M.S. in Applied Mathematics (University of Valencia). His research interests include computer simulation, educational data analysis and mathematical e-learning.

Marilyn Laiken is Chair of the OISE, University of Toronto department of Adult Education & Counselling Psychology, Professor of Adult Education in Workplace Learning and Change, and Director of the OISE/UT Certificate Programs in Adult Training and Development and Leadership Coaching. She is also Principal of Laiken Associates, a Toronto consulting firm that, since 1975, has served hundreds of clients in the public, private and not-for-profit sectors. Marilyn combines an interest in adult education and organizational change through her research, teaching and field development, and speaks internationally and has published widely in these areas. Her book, "The Anatomy of High Performing Teams: A Leader's Handbook" has been used by teams and team leaders since 1991. Marilyn has been honored with the prestigious OCUFA award for "Excellence in Teaching", and the Canadian Society for Training and Development President's Award for her contributions to the field of Adult Education, as well as their first research award.

Judith N. Martin is professor of Intercultural Communication in the Hugh Downs School of Human Communication at Arizona State University, USA. Her research interests focus on the role of culture in online communication; interethnic and interracial communication, as well as sojourner adaptation and reentry. She has published numerous research articles in Communication journals as well as other disciplinary journals and has co-authored three textbooks in intercultural communication with Thomas K. Nakayama: *Intercultural Communication in Contexts*, *Experiencing Intercultural Communication*, and *Readings in Intercultural Communication*. She has developed and taught various communication courses (including intercultural communication) online for the past 10 years.

Selçuk Özdemir (Ph.D.) joined Gazi University faculty in 2002 after nearly 4 years of assistantship at Middle East Technical University in Ankara. Since his arrival at Gazi University, he has served on the Department of Computer Education and Instructional Technologies for over seven years, and has written various articles for national and international journals on ICT integration in education, educational software design, and development and tacit knowledge problem on e-learning environments. He currently teaches students how to design and develop software on .NET platforms using ASP.NET and Visual Basic.NET. In addition, he teaches on the subjects of usability and design of SCORM compatible learning objects at graduate level. He has served in several projects as a national consultant to the Ministry of Education in Turkey.

Carmen Pérez-Fragoso has a doctorate in Communications from Stendhal University, Grenoble 3, and is a Professor of Distance Education at the Instituto de Investigación y Desarrollo Educativo, Universidad Autónoma de Baja California, Mexico. Her research interests are Distance education, particularly Online education, and Youth's use of ICTs (Information and Communication Technologies) inside and outside the classroom. She is a fellow of the research group Discourse, Identity and Educational Practice. She has published in several national and international journals.

Lluis M. Pla is an Associate Professor in the Department of Mathematics at the University of Lleida (Lleida, Spain) and a Senior Researcher in the Area of Animal Production at the UdL-IRTA Center. His research interests include operational research methods applied in agriculture and forest management, with special reference to simulation, dynamic programming, planning, Markov decision processes and production planning. He coordinates the EURO working group called Operational Research in Agriculture and Forest management. He is also a member of INFORMS and EURO.

Michael L. Rodgers, PhD: Mike is Professor of Chemistry and Director of the Advanced Placement Teacher Development Program at Southeast Missouri State University. He is also a Technology Associate of the University's Center for Scholarship in Teaching & Learning; in that capacity, he has contributed significantly to the planning and teaching of the Center's "Technology Serving Learning" Institutes each year since they began in 1997. Mike developed and taught one of the first online introductory chemistry courses with a lab component, and as an instructor for UCLA, he co-developed and taught the first online teacher training course for Advanced Placement in Chemistry. He is active nationally in faculty instructional technology development efforts, having given workshops, papers, and panel sessions at many national and international conferences, such as Syllabus, Educause, MERLOT, SITE, AAHE, Council of Colleges of Arts and Sciences, POD, ED-MEDIA, and the Midsouth Instructional Technology Conference. Mike coauthored the chapter, "Can Faculty Meet Institutional Goals Through Technology?" which appeared in "Enhancing Productivity through Faculty Development", Anker Publishing Co., Spring 2004. He became a Quality Matters Reviewer in 2007. He also contributes regularly to the TECHPED column in the National Teaching and Learning Forum. He holds the BS in Chemistry from the University of Virginia, and both the BA in English and PhD in Chemistry from Iowa State University. He regularly teaches online CH 180, Chemistry in Our World, CH 186, General Chemistry II, and CH 181, Dual-Credit AP Chemistry (for high school students).

Sara V. Rodriguez is a Lecturer of Operations Research at the University of Lleida (Lleida, Spain), where she is also a PhD student of Engineering. Her research interests include simulation and optimization methodologies. She is a member of INFORMS.

Neal Shambaugh is an associate professor and program coordinator of Instructional Design and Technology at West Virginia University. He received his Ph.D. in instructional systems design from Virginia Tech. His master's degree is in curriculum and instruction, while his undergraduate degree is in Management Science/Decision Support Systems. Prior to entering academia, his business interests included radio broadcasting management and customized training program development for energy companies. His academic research interests include instructional design, visual literacy, and mental models. He has published two textbooks on instructional design.

Cristina Steegmann is a Lecturer of Algebra at the Open University of Catalonia (UOC). She also collaborates, as a Lecturer of Geometry, with the Department of Applied Mathematics I at the Technical University of Catalonia (UPC). She holds a Master on the Information and Knowledge Society (UOC), and a Master in Mathematics (Universitat Autonoma de Barcelona, Spain). She is currently working on her PhD thesis in the field of mathematical e-learning at university level.

Ken Stevens is a New Zealander and a Professor of Education at Memorial University of Newfoundland where he was appointed to a chair in e-learning, funded by Industry Canada. His previous appointments were at James Cook University in Queensland, Australia, and at Victoria University of Wellington in New Zealand. In Australia, New Zealand and Canada he has specialized in the provision of education in rural communities and the application of information technologies for teaching and learning in and between schools in sparsely-populated areas. He recently co-authored with David Stewart *Cybercells – Learning in Actual and Virtual Groups*, Thompson-Dunmore Press, Melbourne, 2005. Currently he is cross-appointed from Memorial University of Newfoundland as Adjunct Professor of Education at Victoria University of Wellington. He lives in Canada and New Zealand.

Vassilis Syrris has studied Mathematics at the Aristotle University of Thessaloniki, Greece (B.A.), Business Management at the Hellenic Management Association (Postgraduate studies) and Artificial Intelligence: Knowledge-based Systems at the University of Edinburgh, U.K. (M.Sc.). He is currently completing his Ph.D. in the Automation and Robotics Lab, Faculty of Electrical and Computers Engineering, Aristotle University of Thessaloniki, Greece. He is also working as assistant tutor in the Aristotle University of Thessaloniki, as laboratorial instructor in the Technological Educational Institute of Thessaloniki and as a high-school teacher with a private college in Thessaloniki. He has participated in and run European Community Programs related to education, informatics and professional occupation. His research interests are: machine learning, robotics, automation, evolutionary algorithms, fuzzy logic, e-learning, intelligent tutoring systems, symbolic representation, data manipulation and statistical analysis.

Mr. Adeyinka Tella (Bed, Med, MLS, PhD *ongoing*) is commonwealth doctoral students and Teaching Assistant at the department of Library and Information Studies, University of Botswana. He has published several articles in many international journals around the world. He has also presented papers in learned international conferences around the world. His research interest includes: E-learning and life long learning, Information Communication Technology Impact and Development, Information Management and Psychology, Digital divide, and information system.

Sven Trenholm teaches mathematics fulltime at the community college level within the State University of New York (SUNY). He holds a BSc in pure mathematics and Diploma in Education from McGill University, an MSc in Curriculum Design and Instructional Technology from State University of New York at Albany and is currently a PhD student in Internet-Based Education at the University of Nebraska-Lincoln. His research interests focus on issues of learning effectiveness with computer-mediated mathematics instruction, specifically hybrid and fully asynchronous mathematics e-Learning. He has presented several times on different aspects of web-based mathematics instruction and published related articles.

Fenia Tsobanopoulou has studied English Language and Literature at the Aristotle University of Thessaloniki, Greece (B.A.), Women Studies at the University of Bologna, Italy (intensive postgraduate studies), Social Anthropology at the University of Edinburgh, U.K. (postgraduate studies) and Philosophy at the Aristotle University of Thessaloniki, Greece (M.A). Currently she is a Ph.D. candidate in philosophy at the School of Philosophy and Pedagogy, Faculty of Philosophy, Aristotle University of Thessaloniki, Greece. She is also employed as a high-school teacher with a private college in Thessaloniki. She has participated in and run European Community Programs related to inter-cultural affairs, art, education and professional occupation. Her research interests involve: systematic philosophy, philosophy of language, linguistics, politics and economics, gender issues, human rights, inter-cultural dialogue, literature and art.

Dr. Susan Wegmann is an Assistant Professor at the University of Central Florida in Orlando and has been teaching online for the past 8 years. She currently teaches Content Area Reading in the Reading program area. Her research interests are online interaction and the structures that instructors provide to enhance students' online course experiences. She is the co-Principal investigator for the Florida Literacy and Reading Excellence grant (FLaRE), working with Reading Coach Coordinators throughout the state of Florida. In addition, she serves as the Director of Programs for the upcoming Morgridge International Reading Center, a global consortium focusing on reading strategies and research. She was graduated from the University of Florida (Ph. D.), the University of Mississippi (MA), and the University of South Florida (B.A.).

Index

P

Q

R

S

T

U

V

W